Brain Signals

Brain Signals

Physics and Mathematics of MEG and EEG

Risto J. Ilmoniemi and Jukka Sarvas

The MIT Press
Cambridge, Massachusetts
London, England

This book was set in Stone Serif and Stone Sans by Westchester Publishing Services. Printed and bound in the United States of America.

Library of Congress Cataloging-in-Publication Data

Names: Ilmoniemi, Risto (Risto J.), author. | Sarvas, Jukka, author.
Title: Brain signals : physics and mathematics of MEG and EEG / Risto J. Ilmoniemi and Jukka Sarvas.
Description: Cambridge, MA : MIT Press, [2019] | Includes bibliographical references and index.
Identifiers: LCCN 2018036973 | ISBN 9780262039826 (hardcover : alk. paper)
Subjects: LCSH: Magnetoencephalography. | Electroencephalography. | Brain—Magnetic fields—Measurement.
Classification: LCC RC386.6.M36 I46 2019 | DDC 616.8/047547—dc23
LC record available at https://lccn.loc.gov/2018036973

10 9 8 7 6 5 4 3 2 1

Contents

Preface

The brain defines our existence, character, actions, and well-being; its understanding represents a fundamental scientific challenge. In addition to the intellectual appeal, the study of the brain is motivated by the urgent need to alleviate neurological and psychiatric disorders that cause immense suffering and a great economic burden to society. Numerous techniques to study the brain have been developed; here we concentrate on two techniques that are based on the measurement of electromagnetic fields: electroencephalography (EEG) and magnetoencephalography (MEG).

Signals measured by EEG and MEG provide real time, millisecond-scale information about brain activity. EEG, a routine diagnostic and monitoring technique in virtually all hospitals, is widely used as a research method. On the other hand, the more sophisticated MEG, with its unique advantages, is promising to becoming a major clinical tool as well.

MEG was first measured in 1968; after arduous and ingenious work by physicists, mathematicians, and engineers, the first whole-head devices saw light in the 1990s. Since then, sophisticated data-analysis methods have been developed and published. Numerous publications describe different aspects of the underlying physics, neurophysiology, and the mathematical analysis of brain signals. However, these publications are scattered in a multitude of journals, with wide variability of goals, approach, quality, terminology, and notations. It may thus be hard to obtain a good grasp of the essential fundamentals of MEG and EEG signal generation and the basis for advanced data analysis.

Our aim is to provide a unified approach and firm starting point for understanding the physical and mathematical principles of MEG and EEG.

Recent developments indicate that, when combined with precise anatomical images, EEG and MEG have the potential to provide reliable and accurate spatiotemporal information about millisecond-scale cerebral activity. However, the current clinical practice and also scientific use of MEG and EEG are far from optimal. We intend to present an up-to-date understanding of the intricacies of these techniques and try to pave the way toward the future.

This book is intended for students, researchers, and developers in biomedical engineering, neurophysiology, and cognitive neuroscience.

We have been fortunate to have worked in the area of electromagnetic brain imaging and stimulation since the 1970s and 1980s at Aalto University (which includes the former Helsinki University of Technology), participating in the pioneering work of biomagnetism carried out at the Low Temperature Laboratory and the Laboratory of Biomedical Engineering, as well as the BioMag Laboratory at the Helsinki University Hospital. The material for this book has grown out of the vibrant community of neuroscientific and biomedical research and education at Aalto University, with a wide network of national and international cooperation. We express our gratitude to the large number of colleagues and our wonderful students who have contributed to the development of the field. We thank our colleagues and students for much of the work discussed in the book and for many fruitful discussions. Drs. Johanna Metsomaa, Niko Mäkelä, Tuomas Mutanen, Matti Stenroos, and Narayan Subramaniyam made many valuable comments on parts of the text they reviewed. Most of the illustrations were prepared by Ida Ilmoniemi. We are grateful to Robert Prior at MIT Press, Gillian Dickens and JodieAnne Sclafani at Westchester Publishing Services, and their colleagues for their amazingly professional work and their kind attitude in helping us improve the book.

Risto Ilmoniemi and Jukka Sarvas
Espoo, March 12, 2019

1 Introduction

In this book, we will present the basic physical and mathematical principles of electroencephalography (EEG) and magnetoencephalography (MEG), describing what kind of information is available in the neuroelectromagnetic field, how the measurements are performed, and how the signals can be analyzed. We wish to provide a firm basis for dealing with the data obtained by MEG and EEG. We believe that with ingenuity, a new generation of instruments, and thorough understanding of the basic principles, these methods can be developed to a totally new level of sophistication for the advancement of neuroscience and for the benefit of patients. In this chapter, after an introduction to the human brain and its electromagnetic fields, a historical background and future prospects of EEG and MEG technology are given. In chapter 2, the physical basis of the generation of neuronal electromagnetic fields is described. Chapter 3, discussing the forward problem, goes into considerable detail in how EEG and MEG signals can be calculated from assumed source currents in different volume conductor models.

EEG and MEG data analysis is presented in chapters 4 to 10. We will start by introducing linear and matrix algebra and basic statistics. We then present several analysis methods: dipole fitting, the minimum-norm estimate (MNE), beamforming, the multiple signal classification (MUSIC) algorithm, independent component analysis (ICA), and blind-source separation (BSS) with joint diagonalization. Special attention is paid to clarifying the mathematical and statistical structure of the methods. Rather than delving in or requiring the reader to delve in the large amount of widely varying literature on these topics, we aim at helping the reader to understand and use a limited number of methods. Some analysis algorithms are also summarized in a form suitable for coding them. We present the analysis methods in the time domain, but they can also be applied in the frequency or, for example, wavelet domains, either with the same or appropriately modified structure.

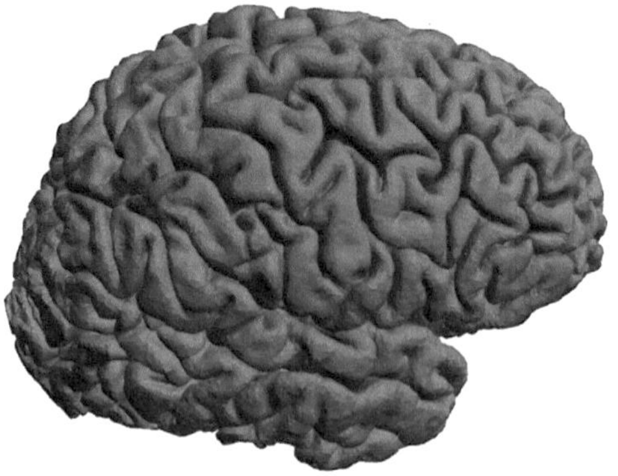

Figure 1.1
The human brain, here seen from the right side in a magnetic resonance image, or MRI, is an extremely complicated information processing system. The gray matter seen in the image is the cortex, which contains a highly interconnected system of neurons. These neurons are the source of electric currents in the tissue and the electromagnetic fields that we measure from the outside. This book is about these electromagnetic signals rather than about the brain itself.

1.1 The Brain

As far as we know, the brain (figure 1.1), which hides some of the greatest mysteries of nature, always behaves according to the known laws of physics: no exception has been found. The complexity of this powerful signal-processing system emerges from its genetically guided molecular and atomic-level operation. Some of the atoms of the brain are bound to relatively stable structures such as cell membranes or organelles, forming the slowly changing structure of the organ and providing the substrate for the analysis and storage of information. Other atoms, including those in transmitter molecules and those that are ions in the intra- or extracellular fluid such as potassium, sodium, and calcium, give rise to the real-time action: their movements produce our sensations, thoughts, and behavior. The signaling between neurons takes place at the time scale of milliseconds, pattern recognition and signal-pattern transformations may take tens of milliseconds, conscious thinking occurs on the order of hundreds of milliseconds, and many other processes such as emotions or learning are typically slower, particularly when they involve structural changes.

Cerebral activity in the time scales from milliseconds to seconds can be investigated by EEG and MEG (i.e., the measurement and analysis of the electromagnetic fields provide information about the movement of ions in the brain). They are the only methods that allow the direct measurement of brain electrical activity noninvasively.

These measurements offer a very clear window for us to observe the functioning of both healthy brains and those of neurological or psychiatric patients. However, these techniques are not trivial to use, and the interpretation of the resulting signals is not always straightforward. First, with the limited number of recording channels (at most several hundred), one can obtain only a low-dimensional projection of the multibillion-dimensional neuronal activity pattern. Second, the relationship between cerebral currents and the extracranial electromagnetic field is currently in practice inaccurately known because of insufficiently defined measurement geometry and lack of precise knowledge of the conductivity distribution of the head. Third, investigators have to deal with instrumental noise, electromagnetic interference, many kinds of physiological and nonphysiological artifacts, and the brain-state-dependent variability of responses to sensory stimuli. Fourth, it is a formidable task to transform correctly the information in the measured signals to a useful description of the underlying neuronal activity. Even if the paradigm is well designed, the measurements are carefully conducted, and the signal quality is high, the physiological variability of the data from one person to another or from trial to trial may prevent one from making useful statistical conclusions.

EEG and MEG are valuable clinical tools. EEG (figure 1.2) is used in virtually all hospitals to diagnose and characterize brain development disorders, trauma, epilepsy, and many other conditions. EEG can also inform us about states of consciousness, for example, for the purpose of anesthesia monitoring or to estimate the chances of recovering from coma. MEG (figure 1.3) can be used to determine loci of epileptic brain activity or to estimate nodes of the language network prior to surgery; it also has potential in characterizing abnormal oscillatory or connectivity patterns in degenerative or psychiatric disorders such as Parkinson's disease or schizophrenia [220, 262, 299].

1.2 Brain States

We experience our life through our brain states; apparently, there is nothing else to define the content of our mind. If we touch a hot stove, we would not feel the pain if the neuronal signals via nerve fibers would not bring the message to the brain. Our joys and sorrows, our intelligence, creativity, and passion, as well as our appreciation of art or science, are all experienced via the brain. To understand the brain and how it works, we must learn about brain states and the dynamics that determine the trajectory from one state to the next.

The remarkable capabilities of the brain include not only the elaborate transformation of ongoing input to perceptions to output but also its ability to be an active player

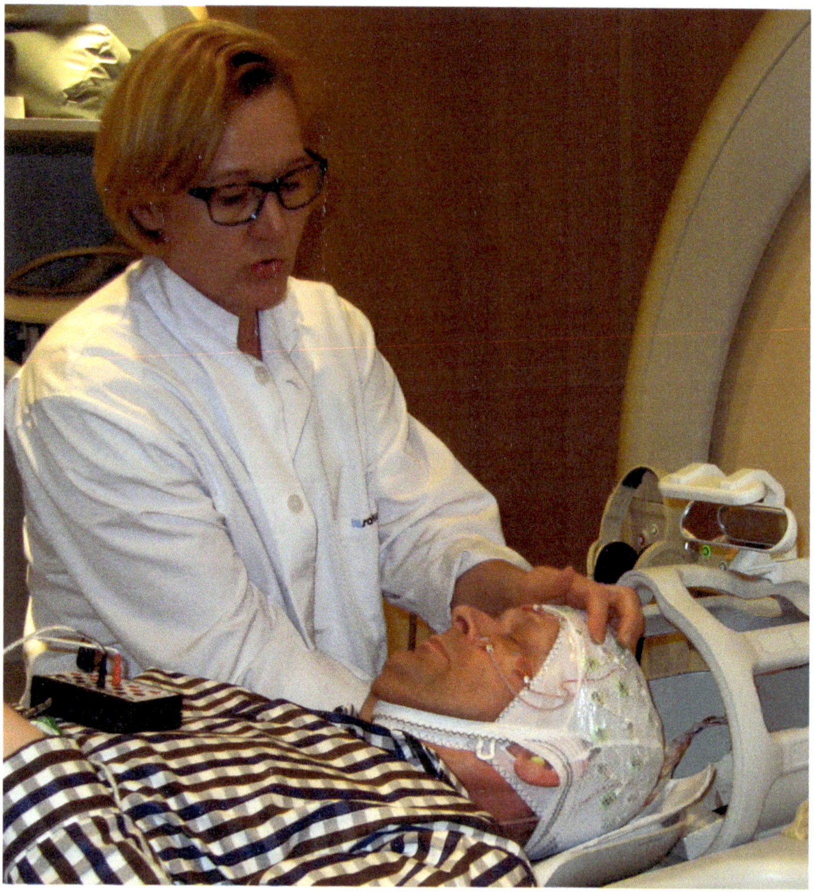

Figure 1.2
MRI-compatible EEG system "NeurOne Tesla MRI" (Bittium Plc, Kuopio, Finland). The flexible cap (Easycap; Brain Products GmbH, Munich, Germany) has sixty-four electrodes for simultaneous EEG and functional magnetic resonance imaging (fMRI) measurement. The 10-MHz clock signal from the MRI system (Philips Achieva 3T, Kuopio University Hospital) allows the NeurOne amplifiers to be synchronized with the MRI. In NeurOne, sixteen additional signals can be measured such as respiration and galvanic skin response (the black box over the subject's chest). Photo: Courtesy of Bittium.

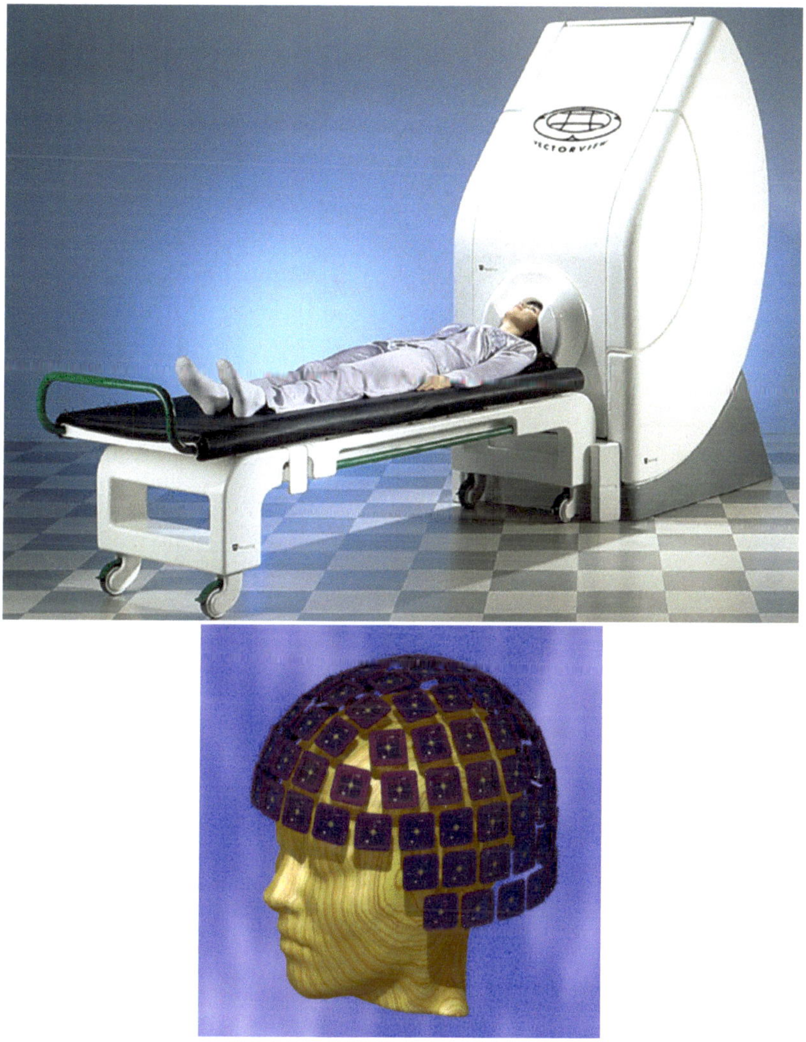

Figure 1.3

Whole-head MEG system "VectorView" (MEGIN, Helsinki, Finland). The magnetic field is recorded with a helmet-shaped array of superconducting SQUID sensors that are at liquid-helium temperature within the dewar. Photo: Courtesy of MEGIN Ltd.

itself. The brain has its intentions; it is an independent agent acting upon its environment, based on sensed data, memories, and instincts. Our studies of the nervous system reveal how information is processed and mental states created, but there is more to the brain than thinking, acting, or willing; there is the mystery of the mind and the puzzle of consciousness.

As far as anybody can tell, the brain is a physical system in the ordinary meaning of the word and nothing more. It obeys the ways of nature, and these ways are described at the most fundamental level by the laws of physics. Although we have reasonably good understanding of the physics of the cell and its constituents, we need higher-level concepts in describing the state of the brain when we discuss neuronal networks and their dynamics, learning and memory, perception, emotions, language, or motor activity. Depending on the level of study, we need to resort to different tools and methods. MEG and EEG are tools for the study of macrosopic electrical activity of the brain. These tools provide a direct measurement of neuronal activity, providing projections (in the mathematical or geometrical sense) of the electrical state of the brain as a function of time.

The activity of the brain as well as its maturation and decline can be described by the evolution of brain states. Conceptually, each brain state is simply the instantaneous distribution of matter and its motion. In classical physics, a material body is fully described by listing the locations and momenta of all its particles. In quantum formalism, the state of a system is described by a wave function. In practice, such detailed information is not available except for the simplest systems, and certainly never for the brain. More useful and practical descriptions can be made at the cellular level and at higher length scales.

At the cellular scale, cerebral information processing is based on both electrical and chemical signaling between and within neurons. Since neurons act on each other mainly via synapses, a key to monitoring and understanding brain function is to measure synaptic activity. Activity in individual cells can be measured only if brain tissue is exposed or if probes are inserted in the brain. On the macroscopic level, however, one can measure aggregate synaptic activity and thereby be informed about electrical states of the brain by MEG and EEG, which, as mentioned, measure projections of the electrical brain states. In this book, we will describe how these projections can be used and interpreted.

1.3 Electrical States of the Brain

Practically all signals from the senses to the brain and from neuron to neuron in the brain are transmitted electrically via axons. The arrival of an action potential to a synapse triggers the transmission of the signal to the postsynaptic cell; the consequent excitatory or inhibitory postsynaptic potentials (PSPs) are thought to give rise to the main generators of the primary current, the sources of EEG and MEG. The primary current, being roughly proportional to the postsynaptic currents, thus also reflects presynaptic activity (i.e., action potentials).

Each action potential originating at or near the soma of a neuron branches in the axon to multiple target synapses in a highly repeatable manner. Since MEG and EEG measure projections of the primary current distribution, one may say that they approximately measure a projection of the firing vector, with the components of this vector reflecting the firing frequencies of single neurons. It is quite remarkable to realize that MEG and EEG thus measure directly the information transmission state of the brain: we can noninvasively obtain a real-time measure of the signaling level of the neurons and thereby learn what the neurons are doing. This is very different from other techniques such as functional magnetic resonance imaging (fMRI), which inform us only indirectly about neuronal activity.

1.4 MEG and EEG Signals as Measures of Brain States

Signals measured by EEG and MEG *outside the head* inform us about electrical states of the brain. To interpret these signals, it is useful to transform them to a form that explicitly refers to cerebral activity instead of only providing field values outside of the head. The starting point in our analysis, the fundamental equation of MEG and EEG relating the measured electric or magnetic signals to neuronal events, is very simple:

$$\mathbf{Y} = \mathbf{A}\mathbf{X} + \eta\,. \tag{1.1}$$

Here, the *data matrix* **Y** denotes the set of signal values obtained from MEG or EEG measurements; **X** denotes the *amplitude, or time-course, matrix* of the unknown neuronal source currents; and η is the *noise matrix*. **A** is the *transfer, or lead field or mixing, matrix* that describes the linear (thus simple) relationship between the source currents and the measured electromagnetic fields; see figure 1.4. In more detail, **Y**, **X**, and **A** are the following matrices. The element y_{it}, $i = 1, \ldots, N_{\text{ch}}$, $t = 1, \ldots, N_{\text{t}}$ of the $N_{\text{ch}} \times N_{\text{t}}$ matrix **Y** is the sensor signal recorded at time t by the ith sensor (magnetic or voltage sensor).

The magnetic or electric fields, measured by the sensors, are due to the (joint) *primary source-current density* $\mathbf{J}^{\text{p}}(\mathbf{r}, t)$, which can be considered a weighted sum of several *component source-current density patterns* $\mathbf{J}_s^{\text{p}}$, $s = 1, \ldots, N_{\text{s}}$, which are spatially fixed and *time independent*; the weights are the *time-dependent amplitudes* x_{st}, which are the elements of the $N_{\text{s}} \times N_{\text{t}}$ source-amplitude, or time-course, matrix **X**. Accordingly, the joint primary

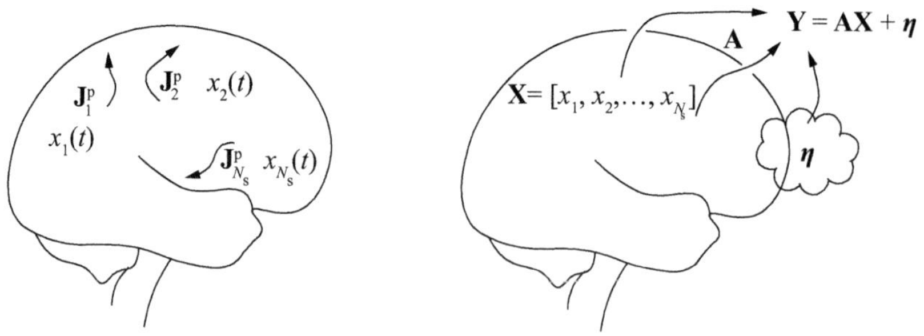

Figure 1.4
The meaning of equation $\mathbf{Y} = \mathbf{AX} + \boldsymbol{\eta}$. Brain activity is described here as a vector of source-current amplitudes $\mathbf{X} = [x_1(t), \ldots, x_{N_s}(t)]^T$, with amplitude $x_s(t)$ indicating the strength of current pattern $\mathbf{J}_s^p$ (right panel). The measured values $\mathbf{Y}$ are weighted sums or mixtures of x_s, the weighting being described by the mixing (or lead-field) matrix $\mathbf{A}$. In addition, the measured signal contains noise $\boldsymbol{\eta}(t)$, which comes from external disturbances, the sensors, and also the brain.

current is the sum

$$\mathbf{J}^p(\mathbf{r}, t) = \sum_{s=1}^{N_s} x_{st} \mathbf{J}_s^p(\mathbf{r}) . \tag{1.2}$$

As will be clarified later, $\mathbf{J}^p(\mathbf{r}, t)$ gives rise to variations of charge density $\rho(\mathbf{r}, t)$, which, in turn, gives rise to an electric field $\mathbf{E}(\mathbf{r}, t)$ according to Maxwell's first equation; see equation (2.1). The electric field drives another type of current called the *volume current* or *ohmic current* $\mathbf{J}^v(\mathbf{r}, t) = \sigma(\mathbf{r})\mathbf{E}(\mathbf{r}, t)$, where $\sigma(\mathbf{r})$ is the conductivity of tissue at $\mathbf{r}$. Currents $\mathbf{J}^p(\mathbf{r}, t)$ and $\mathbf{J}^v(\mathbf{r}, t)$ together produce the magnetic field $\mathbf{B}(\mathbf{r}, t)$ measured by MEG. EEG, on the other hand, records voltages (i.e., line integrals of the electric field).

The element A_{is}, $i = 1, \ldots, N_{ch}$, $s = 1, \ldots, N_s$, of the mixing matrix $\mathbf{A}$ is the signal that would be recorded by sensor i if the source-current pattern $\mathbf{J}_s^p(\mathbf{r})$ *were acting alone* with unit source amplitude $x_{st} = 1$. Element A_{is} depends on $\mathbf{J}_s^p(\mathbf{r})$, on the conductivity structure of the head, and on what spatial "component" of the electric or magnetic field pattern channel i is sensitive to: a typical magnetometer measures the flux of a magnetic field through a loop of wire while a voltage sensor measures the potential difference between two electrodes. Specifically,

$$\Phi_i = \int_{S_i} \mathbf{B}(\mathbf{r}) \cdot \mathbf{dS} ; \tag{1.3}$$

$$V_j = \int_{\mathbf{r}_{ref}}^{\mathbf{r}_j} \mathbf{E}(\mathbf{r}) \cdot \mathbf{dl} , \tag{1.4}$$

where Φ_i is the magnetic flux through the wire loop S_i of magnetic sensor i and $V_j = V(\mathbf{r}_j) - V(\mathbf{r}_{\text{ref}})$ is the potential difference between electrode j located at $\mathbf{r}_j$ and reference electrode located at $\mathbf{r}_{\text{ref}}$.

Finally, due to the linearity of the field equations and equation (1.2), the EEG or MEG field due to the joint primary current $\mathbf{J}^{\text{p}}(\mathbf{r})$ is a weighted sum of the fields due to component source currents $\mathbf{J}_s^{\text{p}}(\mathbf{r})$. Each sensor recording $y_i(t)$, $i = 1 \ldots N_{\text{ch}}$, is thus the sum of the matrix elements A_{is} weighted by the amplitudes $x_{st} = x_s(t)$,

$$y_i(t) = A_{i1}x_1(t) + A_{i2}x_2(t) + \cdots + A_{iN_{\text{s}}}x_{N_{\text{s}}}(t) + \eta_i(t)\,. \tag{1.5}$$

Writing this equation in matrix form with $[y_1(t), \ldots, y_{N_{\text{ch}}}(t)]^{\text{T}}$ and $[\eta_1(t), \ldots, \eta_{N_{\text{ch}}}(t)]^{\text{T}}$ being the tth column of matrices $\mathbf{Y}$ and η for $t = 1, \ldots, N_{\text{t}}$, we arrive at equation (1.1).

Everything in this book refers to the fundamental equation (1.1): how $\mathbf{Y}$ is measured and how η or its effects are minimized, how $\mathbf{A}$ is determined, and, finally, how the amplitudes in $\mathbf{X}$ and the source locations are estimated.

Before going to the intricacies of solving equation (1.1), it is important to realize and understand clearly what it means. The interpretation of equation (1.1) is as simple as the equation: MEG and EEG measure brain electrical activity *directly*; see figure 1.5. Although the weights (matrix elements of $\mathbf{A}$), with which sources are seen by different MEG or EEG sensors, are different, each source waveform (i.e., $x_s(t)$ as a function of t) is detected almost identically (apart for scaling by the weights) by the different channels; see figure 1.5. However, when multiple sources are active at the same time, the signals from different sources are superposed or mixed as described by the elements of $\mathbf{A}$; see figure 1.6. Therefore, $\mathbf{A}$ in the multisource case is also called the mixing matrix.

The challenge is to determine $\mathbf{A}$, minimize the effect of η, and solve equation (1.1) for $\mathbf{X}$. Because of the noise and, in particular, the nonuniqueness of the inverse problem, $\mathbf{X}$ cannot generally be determined with certainty even if $\mathbf{A}$ were known exactly. If the a priori probability density $p(\mathbf{X})$ for $\mathbf{X}$ and the conditional probability density $p(\mathbf{Y}|\mathbf{X})$ of $\mathbf{Y}$ given $\mathbf{X}$ are known, then the a posteriori probability distribution for $\mathbf{X}$ is, according to Bayes's rule, $p(\mathbf{X}|\mathbf{Y}) = p(\mathbf{X})p(\mathbf{Y}|\mathbf{X})/p(\mathbf{Y})$. This can be considered the (statistical) solution for equation (1.1); see figure 1.7. Unfortunately, the Bayesian approach is very often impractical because of computational complexities and because of the difficulty of describing a priori information adequately as a probability density. However, it is good to understand Bayesian thinking and, in particular, to keep in mind that the selection of the method to solve the inverse problem should be based on a priori or supplementary knowledge about the source-current distribution. In some cases such as in Wiener estimation (see section 4.5), the optimal estimator is either the average of the a posteriori distribution or the maximum a posteriori estimator if the data equation is linear and the relevant probability distributions are Gaussian.

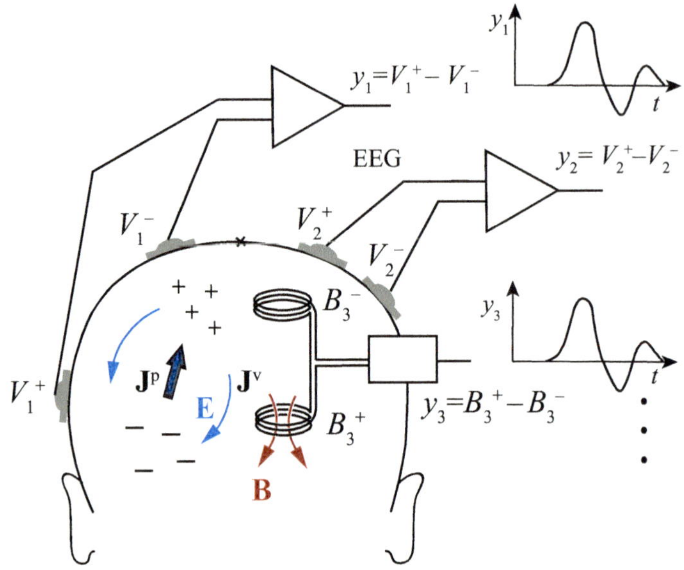

Figure 1.5
EEG electrodes record voltages and magnetometer coils measure magnetic fields, $\mathbf{Y}(t) = [y_1(t), \ldots, y_{N_\mathrm{ch}}(t)]^\mathrm{T}$, that depend linearly on sources in the brain ($\mathbf{X}$, see figure 1.4). One spatially fixed source current $\mathbf{J}^\mathrm{p}$ with a time-varying source waveform $x(t)$ produces identical waveshapes in all MEG and EEG channels, assuming that tissue conductivities do not depend on frequency. This may be approximately true for typical signals, but frequency dependency of tissue conductivity has been reported to influence high-frequency components of EEG more than those of MEG [67]; this may be of importance at least when locating so-called 600-Hz signals [66] or signals of even higher frequency. See also [27].

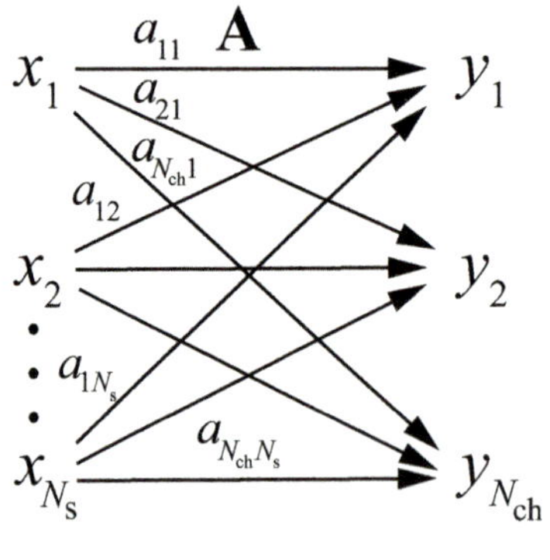

Figure 1.6
Mixing of signal components from different sources x_s. Each measured waveform $y_i(t)$ is a linear combination of the source waveforms.

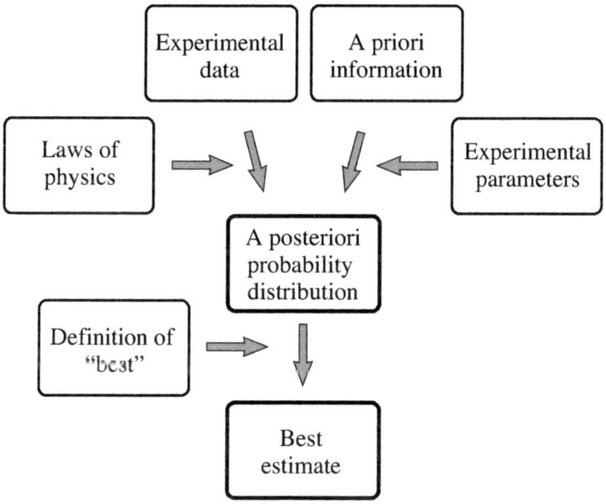

Figure 1.7
The a posteriori probability distribution can be considered the solution of the inverse problem. From it, different distinct solutions can be derived. For example, if one has defined what is meant by "best estimate," one can obtain the best estimate from the a posteriori probability distribution (e.g., the solution that minimizes the expected squared error).

Thus, depending on what is known or can be assumed a priori about **A** or **X** and what kind of analytical or computational machinery is assumed to be available, various other approaches to solve equation (1.1) have been proposed. We will discuss these issues in detail in subsequent chapters. We will describe and analyze the events in the brain that lead to the neuronal source currents $\mathbf{J}_s^p$. We will discuss how **A** depends on the geometry (conductivity distribution) of the head and on the characteristics (locations, orientations, shapes, frequency response, calibration) of the measurement sensors. We will describe ways to measure the bioelectromagnetic field, yielding **Y**. We will analyze the noise η and ways to reduce its effects. We will give examples of how to estimate **X** on the basis of **Y**. Starting from the physics and physiology of the brain and from the laws of electromagnetism, we will describe the nature and properties of the brain's electric and magnetic fields. Although not going into great detail about the art of instrument development, we will provide the basic principles of measurement. The emphasis of the book will be on the understanding of how information is conveyed by the measured MEG/EEG signals and how this information is used and transformed into a form that helps the neuroscientist or clinician learn what the brain is doing.

1.5 Historical and Technical Background of EEG

Nearly a century after Galvani's demonstrations of animal electricity [96, 250], Eduard Hitzig and Gustav Fritsch in Berlin, studying a dog [95], and David Ferrier of King's College in London, working with rabbits, cats, dogs, and monkeys [93], observed muscle movements after electrical stimulation of the animals' cortex. Then, on August 4, 1875, Richard Caton of Liverpool reported at the 43rd Annual Meeting of the British Medical Association in Edinburgh the first electrical recordings of brain activity [50]. He had detected signals from the cortex of the rabbit and monkey using a galvanometer. This was the beginning of electrical brain recordings, although it took another half-century until Hans Berger in Jena, after his first attempt in 1920, successfully recorded in 1924 the first human electroencephalogram [30, 207]. Berger's discovery of the alpha and beta rhythms and the subsequent confirmation of his results by Adrian and Matthews [2] aroused the interest of many investigators and soon led to clinical applications. The first "unambiguous" sensory evoked responses were recorded (according to Luck [194]) by Pauline and Hallowell Davis in 1935–1936 [71, 72]. Two other subsequent findings may be worth mentioning here as they demonstrated the potential value of EEG in revealing brain mechanisms in mental tasks.

William Grey Walter in Bristol was one of the early developers of the technique, investigating brain rhythms and their relationship with epilepsy and other clinical conditions. Studying evoked potentials, Grey Walter discovered a component called *contingent negative variation* or CNV [321]. The CNV develops in the brain if a subject is instructed to perform a motor task in response to a sensory stimulus presented about a second after a first stimulus that indicates the beginning of a trial. A related EEG deflection, the readiness potential or *Bereitschaftspotential*, was discovered by Kornhuber and Deecke [177, 178]; it is observed over the premotor regions up to one second before a voluntary movement is performed.

After the pioneers' findings, a large number of event-related EEG components have been identified in different sensory and motor modalities. These include auditory brainstem potentials, which can reveal abnormalities in the relevant neuronal pathways; the auditory, visual, and somatosensory cortically generated responses with multiple peaks and troughs are often named by their polarity: for example, P50, N100, and so on are common auditory components where the number refers to the time after the stimulus in milliseconds and P or N indicates whether the potential over the region of interest is positive or negative with respect to a reference electrode. A quite remarkable and widely used auditory evoked potential is the mismatch negativity or MMN, discovered by Näätänen in Helsinki in 1978 [216]. The MMN is a negative deflection in the EEG signal evoked by a deviant auditory stimulus embedded in a sequence of repeated or predictable stimulus presentations. As an automatic response, it can be elicited without the subject paying attention. The MMN reflects the formation of a memory trace by the standard stimuli.

The time from the 1930s until 1970s was also a period of active research in the study of spontaneous brain activity. The alpha rhythm and other brain rhythms were characterized in healthy people and in various patient groups. One culmination of these studies was the landmark study, published in *Science* in 1977 [155], on *neurometrics* by E. R. John of New York City with local and Cuban collaborators. The goal in neurometrics was to use computerized analysis of EEG data to classifify patients' conditions and predict the outcome of disease or the development of a child. With the subsequent development of EEG and computer technology, so-called *quantitative EEG* or qEEG emerged and has been used for diagnostics and biofeedback therapy [90]. With recent developments in data analysis such as machine learning, neurometrics, to use the old term, based on EEG or MEG, is advancing as a very powerful approach. In parallel, efforts intensified to develop methods for solving the inverse problem (i.e., for determining the spatiotemporal activation pattern in the brain). We will discuss these methods in later chapters of the book.

EEG is now a widely used tool, with established electrode-placement standards such as the 10–20 system for originally nineteen scalp electrodes [170, 280] (see figure 1.8),

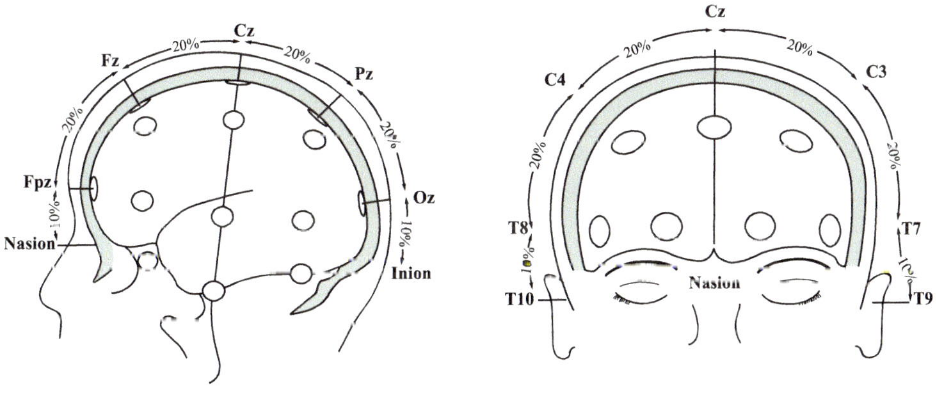

Figure 1.8
Placement of EEG electrodes according to the International 10–20 system (modified from [170, 280]). The symmetry line along the scalp from inion to nasion is divided into segments that cover 10 or 20 percent of this distance; the midpoint of the line is the vertex. Similarly, the shortest line along the scalp between preauricular points that goes via the vertex is divided in similar segments. These lines and divisions are used to define electrode locations.

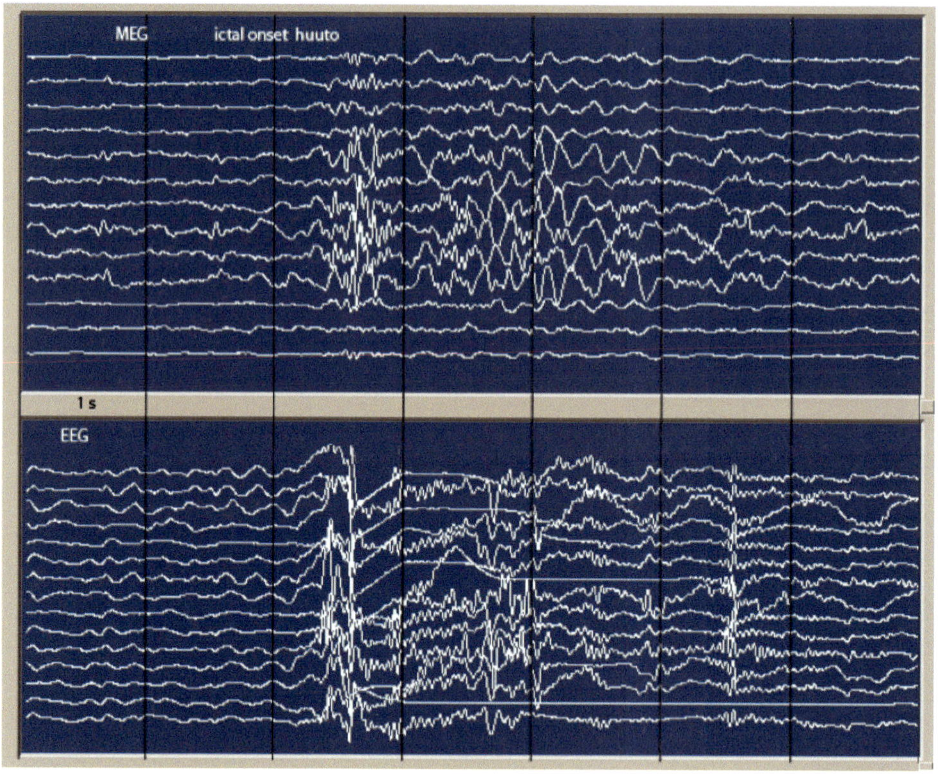

Figure 1.9
MEG and EEG signals showing, after low-amplitude signals, the onset of a seizure in an epileptic patient. The vertical bars show the timescale of one second. Notice that the MEG signals are prominent only in the channels over the suspected source area whereas EEG signals are more widespread. The MEG signals were recorded with planar gradiometers of the Elekta Neuromag system at the BioMag Laboratory of Helsinki University Hospital. Courtesy of Ritva Paetau and Jyrki Mäkelä.

subsequently extended [236]. Standard electrode placement, however, is not a necessity or not even possible because of differences in head shapes and sizes. What is important is that electrode locations are *known*, so that the signals recorded with them can be properly analyzed. The number of electrodes should be sufficient, perhaps sixty or more, for inverse solvers to work adequately [291]. For an example of EEG signals, see figure 1.9.

For detailed accounts on the history of electrophysiology and the development of EEG instrumentation, several excellent texts are available [37–39, 62]. The fundamentals of EEG and related techniques are described in many books and journal articles [194, 256, 278].

1.6 Historical and Technical Background of MEG

Three Nobel prize-winning discoveries paved the way for the development of the SQUID magnetometer, which made magnetoencephalography possible. Half a century after the discovery of superconductivity by Kamerlingh Onnes in Leiden [234] and soon after the publication of the BCS theory of superconductivity by Bardeen, Cooper, and Schrieffer [24], Cambridge graduate student Brian Josephson theoretically predicted [158, 159] the quantum mechanical behavior of a weak link between superconductors that is now called the *Josephson effect*. This discovery enabled the development of extremely sensitive magnetometers and thus the measurement of cerebral magnetic fields.

Josephson's equations relate the voltage $V(t)$ and current $I(t)$ through a weak link (a narrow constriction or thin insulating barrier that allows quantum-mechanical tunneling) with the phase difference $\phi(t)$ across the junction of the order parameter that describes the *Cooper pairs* of electrons in the conductor, $V(t) = (\hbar/2e)\partial\phi/\partial t$ and $I(t) = I_c \sin\phi(t)$, where $\hbar$ is Planck's constant and I_c is the maximum current the junction can sustain without a voltage.

If one or two weak links or *Josephson junctions* are arranged to interrupt a superconducting ring, a very interesting situation arises. Because magnetic flux is quantized in a superconducting loop (in multiples of the flux quantum $\phi_0 = h/2e$) [74, 85], shielding currents have to flow around the ring to compensate for the flux due to external magnetic fields. As Josephson junctions have a maximum current (when $\phi = \pi/2$), the current–voltage or I–V characteristics of the ring change with changing external magnetic field. Such a ring is called a superconducting quantum interference device or SQUID [54, 169]. If constant current is passed through the SQUID as in figure 1.10, the voltage measured over the SQUID is a periodic function of the applied external flux, the period being ϕ_0. By measuring the voltage, one then measures the external magnetic field.

In practice, the operation of the SQUID is linearized by applying a feedback magnetic field to the field-detecting loop, keeping the flux in the loop constant. Then, the feedback current is proportional to the measured field. James (Jim) Zimmerman and his coworkers at the Ford Motor Company in Dearborn, Michigan, developed the first practical SQUIDs for biomagnetism in the late 1960s [287]. Having first been used to record cardiac magnetic signals (i.e., the magnetocardiogram [59]), SQUIDs soon

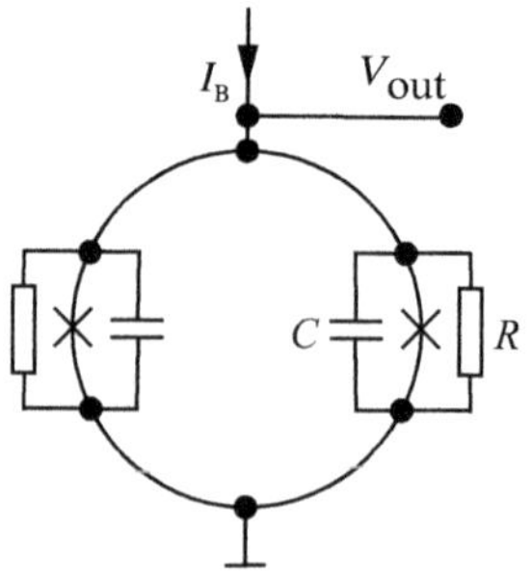

Figure 1.10
A DC SQUID has two Josephson junctions, depicted with the crosses. The physical junctions also have capacitance C and resistance R, the proper values of which are crucial for SQUID performance. Bias current I_B that exceeds the sum of the critical currents of the junctions is fed through the SQUID, producing a voltage that depends periodically on the external flux through the SQUID ring. Often, as in figure 1.11, R and C are omitted and the Josephson junction is represented just with a cross.

revolutionized magnetic brain measurements as well. The Zimmerman SQUID had one Josephson junction that could be adjusted mechanically; it was probed inductively with radio frequencies and therefore called RF SQUID. Modern SQUID magnetometers (see figures 1.11 and 1.12) use two junctions. Because they can be probed by observing the voltage due to a DC bias current, they are called DC SQUIDs.

The first successful biomagnetic measurements had been performed in 1963, when Baule and McFee, far away from electric disturbances on a meadow in Syracuse, New York, used an induction-coil magnetometer to measure the human magnetocardiogram [26]. An induction-coil magnetometer was also used for the first MEG recordings, when David Cohen in 1968 observed the magnetic alpha rhythm by averaging the magnetic signal by phase-locking with simultaneously recorded EEG [55]. As mentioned earlier, SQUIDs revolutionized biomagnetic measurements. The pioneer in both MCG and MEG measurements using SQUIDs was Cohen, who at MIT was the first to measure the magnetic alpha rhythm of the brain in 1971 [57]. After his recordings of the spontaneous brain rhythms, much of the emphasis shifted to evoked responses. There was a lot of literature on electric evoked potentials, and many investigators simply repeated with MEG what had been done earlier with EEG. Progress was essentially limited only by the technological development. Early SQUID magnetometers were still quite noisy and unreliable, and external magnetic disturbances were often a problem.

The first magnetic evoked responses, visually evoked fields or VEFs, were recorded in 1975 at New York University (NYU) by Brenner et al. [41] and at MIT by Teyler et al.

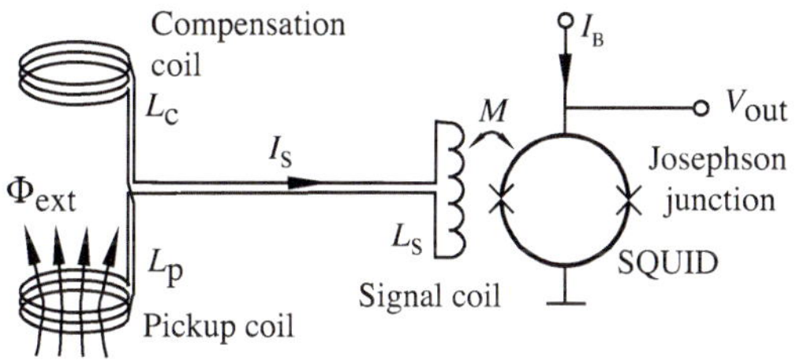

Figure 1.11
A flux transformer circuit is a closed superconducting loop consisting of a pickup coil, possible compensation coil(s), a signal coil, and connecting wires. The pickup coil is meant to measure the magnetic field from the brain while the compensation coil, being far away from the head, is meant to measure external disturbances. When the signal and compensation coils have the same area and the winding direction is opposite, a gradiometer is formed. This makes the sensor insensitive to homogeneous magnetic fields, reducing interference from distant sources. The purpose of the signal coil is to couple the signal to the SQUID; when the magnetic flux in the pickup loop changes, a shielding current I_S is formed so as to keep the total flux in the flux transformer constant. This current then produces a magnetic field in the SQUID and a signal V_{out} can be measured when a bias current I_B is fed through the SQUID. Other types of pickup and compensation coil arrangements are depicted in figure 1.12.

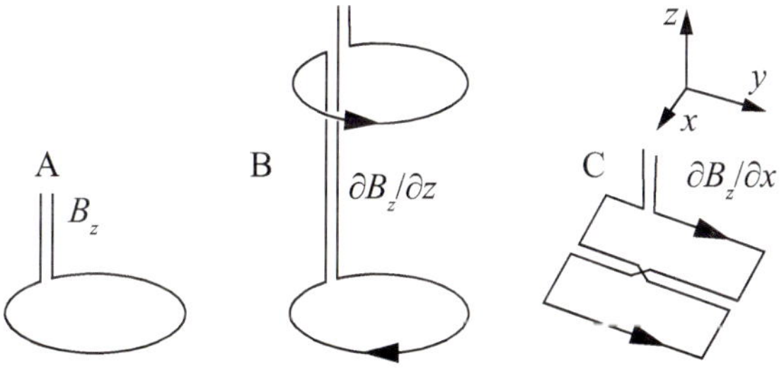

Figure 1.12
Sensor coil types. (A) Magnetometer, (B) axial gradiometer, (C) planar gradiometer.

[308]. In 1978, somatosensory evoked magnetic fields (SEFs) were reported by Brenner et al. [40] and auditory evoked magnetic fields (AEFs) by Reite et al. [257], the latter at the National Bureau of Standards in Boulder, Colorado. Soon thereafter, high-quality AEF recordings were made by Aittoniemi et al. [11] and Hari et al. [118] at the Helsinki University of Technology (HUT), Farrell et al. [91] at Case Western University, and Romani et al. [261] at NYU. The first MEG recordings of epileptic activity were published by Barth et al. [25] in 1982. These early studies, performed with single-sensor magnetometers,

had demonstrated the potential of MEG in detecting and locating electrical brain activity in both healthy subjects and in patients; the field only needed better techniques for measurement and for data analysis. Epilepsy would become the prime clinical application of MEG [19, 63, 73, 97, 167, 292]. For the reader interested in the large number of MEG studies during and after the 1980s when the instrumentation was developed from single sensors to large arrays of SQUID magnetometers, we refer to books and review articles and indirectly to original papers they are based on [10, 111, 120–122, 217].

After pioneering biomagnetic studies by Toivo Katila and his group at HUT [12, 162, 163, 264], it was realized that we need multichannel magnetometers to speed up the mapping of the spatial form of the cerebral magnetic field. In Helsinki, this development started at the Low Temperature Laboratory under the leadership of Prof. Olli V. Lounasmaa. After the first attempt to build a seven-channel instrument with a novel flux-guide method [88], more successful multichannel designs were introduced by Ilmoniemi et al. [143], Knuutila et al. [171], and Kajola et al. [4, 160]. A fundamental advance was made by the team at the newly founded startup Neuromag Ltd., where the team led by Antti Ahonen introduced the first whole-head MEG instrument in 1992 [7, 8]. Whole-head devices revolutionized MEG, allowing one to obtain the complete magnetic field pattern over the head at once. This technical breakthrough enabled investigators to bring the field to a completely new level of sophistication [218, 265, 266]. The amount of data increased and their quality improved vastly. The limiting factor was no longer the instrumentation; the bottleneck was now data analysis.

1.7 State of the Art and Future Prospects for MEG Technology

At the same time when whole-head MEG systems with over 100 channels were made available by Neuromag and other manufacturers [184, 320], the question was raised as to how many channels are needed or are useful. Since the superconducting sensors are within a liquid-helium dewar and must therefore be quite distant from the scalp, the spatial frequencies in the magnetic field pattern are relatively low. The problem of sufficient sampling was solved by generalizing the Nyquist sampling theory to three-dimensional vector fields [6]. The analysis implies that the right sampling density had been reached. However, a new system with combined planar gradiometers and magnetometers was introduced later: the current state of the art is represented by the 306-channel MEG system of MEGIN Ltd. (originally Neuromag Oy), in which the field distribution is oversampled to a considerable extent.

MEG technologies reached essentially the present level with hundreds of recording channels in the 1990s, after which no dramatic changes have taken place from the regular user's point of view. Noise levels have become lower, reliability and noise rejection have improved, and closed-cycle helium systems have been developed, but the approach has remained the same. Similarly, multichannel EEG systems have been available and in common use for decades.

However, we are now entering an era in which any of several new technologies or methodologies may revolutionize the way MEG and EEG measurements are performed and analyzed. In particular, MEG technology is pushed forward on several fronts while MEG and EEG data-analysis methods are refined with the help of machine learning, Bayesian analysis, and other advanced approaches; new signal-separation and inverse-solution methods are making source localization more powerful and more reliable. The new MEG technologies include *hybrid MEG–MRI* systems [176, 188, 316], *optically pumped magnetometers* (OPMs, also called atomic magnetometers [34–36, 138, 174, 285]), *hybrid quantum interference devices* or hyQUIDs, and sensors based on *high-temperature superconductors* (high-T_C) [229], *giant magnetoresistance* [49, 241, 242], and *nitrogen vacancies in diamonds* [161]. The latter two can be miniturized and have been demonstrated to enable the recording of magnetic fields from brain slices or even single neurons. All mentioned technologies involve quantum sensing; for a review of developments in this field, see Degen et al. [75]. The main advantages of OPMs and high-T_C and magnetoresistance sensors are lower cost than present technology and the possibility of placing the sensors closer to the scalp [138] than is possible with traditional low-T_C SQUID sensors, which have to be cooled down to 4 degrees K. Brief descriptions of high-T_C sensors, OPMs, hyQUIDs, and hybrid MEG–MRI systems and their advantages are given next.

1.7.1 High-T_C SQUID Magnetometers

The discovery and development of materials that are superconductors at the temperature of liquid nitrogen (77 degrees K at atmospheric pressure) provided the possibility of developing SQUIDs that need not be cooled with the more expensive helium. Furthermore, because of the higher latent heat, requirements for thermal insulation for liquid-nitrogen-cooled dewars are much smaller. Because no superinsulator layers are needed in the dewar's vacuum space, high-T_C sensors can be at a distance of only 1 mm from the room-temperature surface, whereas the sensor–scalp distance in helium dewars is typically 3 cm or more. Although the noise level of high-T_C SQUIDs is higher than that of low-T_C SQUIDs, the stronger neuromagnetic field near the scalp partially compensates for the higher noise. Also, measurements close to the scalp provide a better spatial resolution than those further away. In fact, large high-T_C SQUID arrays close to the scalp would convey (depending on sensor parameters) more information from the brain than low-T_C SQUIDs at a larger distance [15, 258, 277] (i.e., they can have a higher channel capacity [166]).

Although the ability of high-T_c SQUIDs to measure brain signals has been demonstrated [229] and although these sensors offer a large bandwidth and operation in less well-shielded rooms compared to optically pumped magnetometers, it remains to be seen what their preferred applications will be.

1.7.2 Optically Pumped Magnetometers

After the demonstration by Kominis et al. [174] that OPMs (originally called atomic magnetometers) can provide femtotesla-level sensitivity similar to that of typical SQUID magnetometers, practical sensors have been developed and applied to the measurement of both cardiac [32, 44] and neuromagnetic signals [34, 36]. Advantages of the OPM technology include compact sensors, no need for refrigeration, and the ability to place the sensors close to the head, which results, as is the case for high-T_c SQUIDs, in increased rate of information from the brain [138].

The operation of OPMs is based on the influence of the magnetic field on optically polarized spins of the valence electrons of vaporized alkali metals such as potassium [328] or rubidium [157, 267]. To operate these devices, the alkali metal housed in a transparent probe has to be heated to 120 degrees Centigrade or higher to reach a sufficient vapor pressure. The readout is performed by measuring the polarization rotation or transmissivity of light through the vapor cavity, which depends on how much the spin polarization has been tilted by the magnetic field.

Current OPM systems have only a small dynamic range and small bandwidth, and they require a very low (on the order of nanoteslas) ambient field; therefore, they must be used within magnetically well-shielded enclosures. Several MEG research groups use the commercially available general-purpose OPM sensors of QuSpin, Inc. (Louisville, CO). For a discussion of microfabrication techniques of OPMs and their use, see Alem et al. [14] and Jiménez-Martinez and Knappe [154].

The first clinically useful application of OPMs is the measurement of the fetal magnetocardiogram [327]. OPMs also show great promise for MEG as their noise level is similar to that of typical SQUIDs and as they can be placed closer to the head than low-T_c SQUIDs sensors. Boto et al. [35] demonstrated that a small number of OPMs can be mounted in a head cast, allowing MEG studies where the head is allowed to move. For an OPM array, see figure 1.13.

1.7.3 hyQUIDs

The hyQUID is a superconducting ring interrupted by one or more short sections of normal metal such as silver, with such sections forming superconductor–normal–superconductor (SNS) structures. The device is operated as a magnetometer by

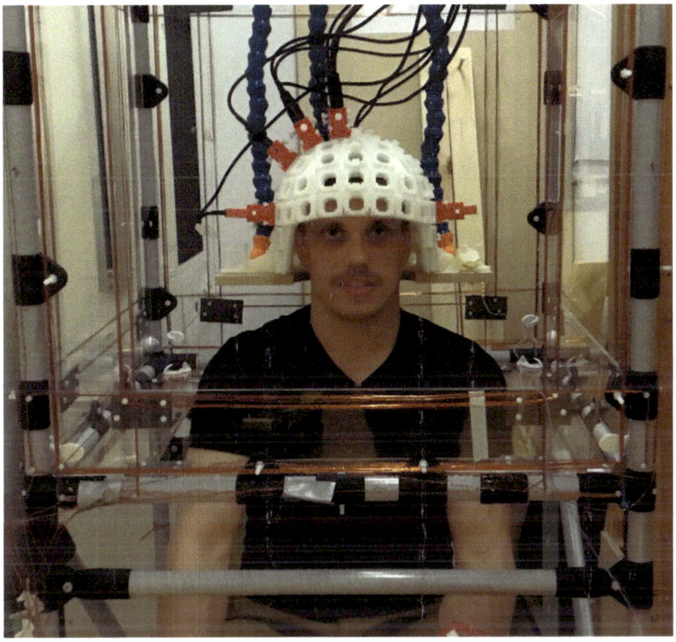

Figure 1.13
Optically pumped magnetometers (OPMs) mounted on a whole-head sensor frame at Aalto University; the device has been successfully used within a magnetically shielded room. The helmet-shaped frame and the subject are surrounded by a set of coils that are used to ensure that the static background magnetic field is close to zero. The sensors have been fabricated by QuSpin (Boulder, Colorado). Courtesy of Lauri Parkkonen.

measuring changes in the resistivity of the normal metal. The densities of charge carriers and holes in the metal change because of the so-called Andreev reflection [16] of electrons at superconductor–normal-metal boundaries depending on the supercurrent through the junction [247], which, in turn, depends on the phase difference over the junction. In the Andreev reflection, an electron with spin up (along an arbitrary direction) entering a superconductor becomes a Cooper pair in the superconductor and a hole with spin down in the normal metal. The hyQUID has been developed for MEG applications by York Instruments Ltd. (York, England), which claims the device to be more sensitive than the SQUID. At the time of this writing, this claim remains to be publicly verified. It should also be mentioned that other new types of superconducting quantum sensors have emerged such as the *superconducting quantum interference proximity transistor* or SQUIPT [101, 152] and the *kinetic inductance magnetometer* [195], the former being based on SNS junctions not unlike those in hyQUIDs.

1.7.4 Hybrid MEG–MRI Systems

As will be elaborated in chapters 5 to 9, sophisticated signal-analysis methods such as beamforming and MUSIC algorithms to locate multiple sources in the brain have been developed. However, the performance of these techniques, where the mixing matrix has to be untangled and/or a priori information has to be relied on, depends, in addition to the signal-to-noise ratio and calibration of the sensors, on knowing the precise locations and orientations of the sensors as well as the conductivity geometry of the subject's head. In particular, a priori information about the source distribution is of crucial importance in making the solution of the inverse problem unique and thereby usable.

The demonstration by the group of John Clarke at Berkeley University in 2004 [203] that MRI signals at 100-microtesla fields can be measured with SQUIDs opened the way to the development of hybrid MEG and MRI. Zotev et al. [335] were the first to show that this ultra-low-field MRI (ULF MRI) methodology can be combined with MEG. One key advantage of this combination is that one will be able to obtain MEG and MRI data in the same coordinate system, avoiding registration errors [279] as demonstrated theoretically by Mäkinen et al. [199]. Furthermore, ULF MRI may allow one to measure tissue conductivity values as soon as sufficient sensitivity is reached in current-density imaging (CDI [225, 317]). With improved knowledge of the conductivities, one can expect a major improvement in MEG source localization accuracy and reliability. Furthermore, if conductivity values can also be sufficiently well determined for the skull and the scalp (this is not crucial for MEG), then EEG can be made far more accurate in source localization than what it is today. The present state of the art of the techniques for MEG–MRI [188, 316] (see figure 1.14) is expected to be soon replaced by systems with far better signal-to-noise ratios [92, 298], allowing high-contrast MRI to be obtained with better than 2-mm resolution in twenty minutes so that the combined MEG–MRI measurement would not take longer than the MEG measurement alone.

1.7.5 MEG Systems for Measuring Infants and Fetuses

MEG offers distinct advantages when measuring brain activity in babies. Because radial conductivity anisotropy of the (roughly spherical) head does not affect MEG signals but can affect EEG dramatically, the infant with fontanelles is far easier to study with MEG than with EEG [144, 145, 185] (however, see Lau et al. [182]). Furthermore, it may be tedious to attach a large number of electrodes on an infant's scalp while MEG can be started with a less time-consuming preparation. Early MEG experiments on babies were performed exclusively with devices built for adult subjects [135, 150, 180]. Since children's heads are small compared to standard MEG helmets, one can effectively measure

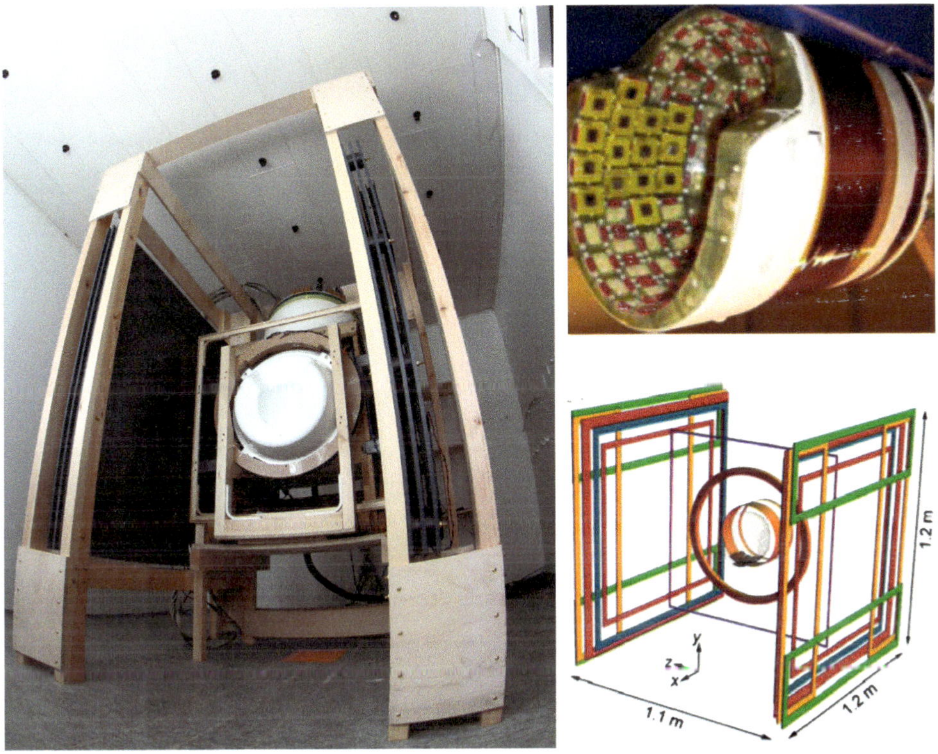

Figure 1.14
Prototype of the hybrid MEG–MRI instrument at Aalto University. The dewar with the helmet-shaped bottom is surrounded by several coils that produce the rf field, measurement field, and the gradient fields required for ultra-low-field MRI. *Upper right*: The partially filled helmet-shaped sensor-array framework for 102 sensor units. The superconducting magnet to produce prepolarization of the tissue surrounds the sensor array. *Lower right*: The arrangement of the various MRI coils.

signals only on one side of the head at a time. To improve on this situation, arrays that better fit the size of the infant's head have been developed [1, 156, 232, 259]. As the latest development, Okada et al. [230] introduced what is called *babyMEG* (figure 1.15): an array of 270 sensors in a helmet-shaped arrangement so that the sensors are on average only 8.5 mm from the helmet's room-temperature surface. The helmet is designed to accommodate 95 percent of thirty-six-month-old babies in the United States. The small size of the helmet, the proximity of the sensors to the child's head, the 15-mm sensor-coil spacing, and the low noise level provide an unprecedented spatial resolution in mapping brain activity. BabyMEG is located in a single-layer shielded room

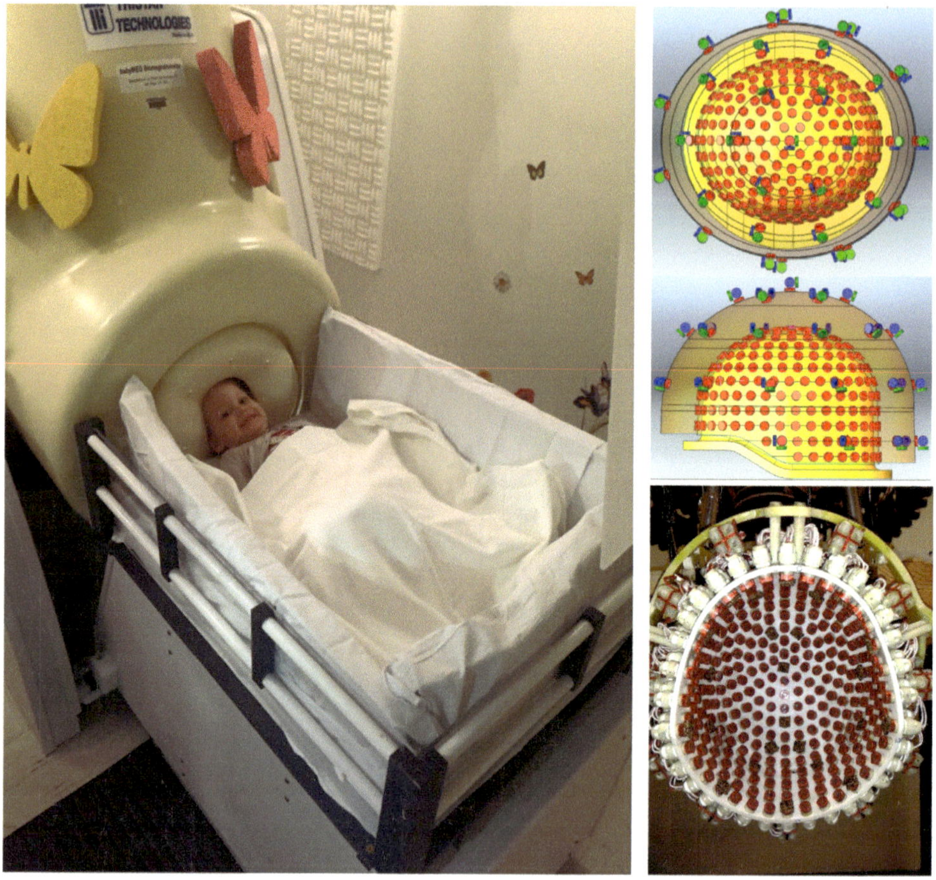

Figure 1.15
The "BabyMEG" at Boston Children's Hospital. The helmet-shaped sensor array within the liquid-helium dewar contains 270 magnetomers (the dense array in the illustrations and photograph on the right) and thirty-five sensor units measuring all three components of the magnetic field in an outer layer 4 cm from the dense array. This arrangement allows a very effective rejection of external magnetic disturbances using signal-space projection. Courtesy of Yoshio Okada.

in Boston Children's Hospital; external disturbances and those from the closed-cycle helium refrigeration system are very effectively eliminated by the signal-space projection technique, also thanks to additional sensors placed farther away from the head [312, 322].

Recently, Compumedics (Victoria, Australia) announced an MEG system that includes helmet-shaped arrays for both adult and pediatric subjects in the same dewar (figure 1.16). Also, brain activity of fetuses can be detected magnetically

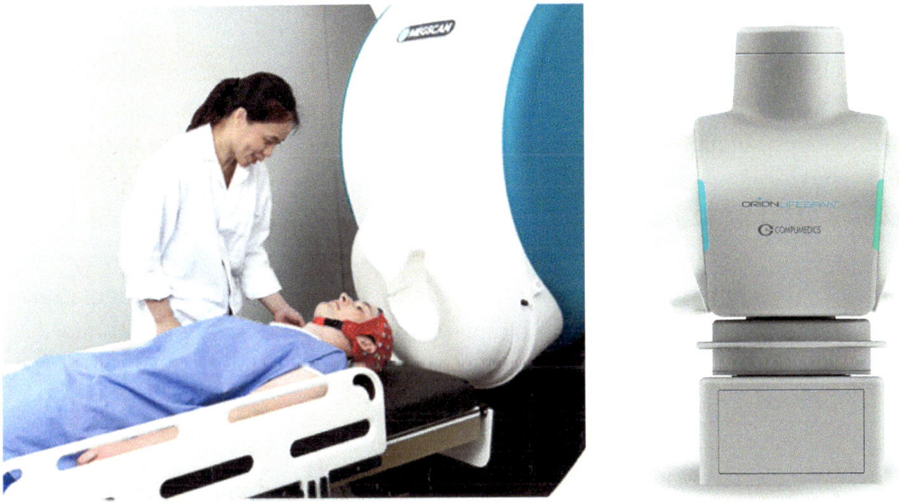

Figure 1.16
Left: The MEGSCAN system (York Instruments, York, UK); the sensor array with HyQUID sensors are cooled without the need for liquid helium. Photo: York Instruments. *Right*: Dual-head MEG system "Orion Lifespan MEG" (Compumedics, Victoria, Australia), with a pediatric helmet on one side and a helmet for adult subjects on the other side. The system can be rotated so that adults and children can be studied in succession. Photo: Compumedics.

[86, 87, 134, 192, 252, 286], although the signals outside the mother's abdomen are quite weak and their interpretation is hard because the fetus can be in different positions and orientations and can move during the measurement. Furthermore, the modeling of signal generation is complicated because of the various tissue and fluid compartments that shape the electromagnetic field and because it is hard or impossible to know accurately the conductivity geometry or even the position or orientation of the fetus's head. Nevertheless, interesting information on the development of the fetus can be obtained, as demonstrated, for example, in a study where numerosity discrimination appeared to exist in fetuses in the last trimester of pregnancy [275].

1.7.6 Methods to Remove Unwanted Components in the Data

Since magnetic fields from the brain are very weak, in the femtotesla (10^{-15} T) range, it is a great challenge to measure them reliably when external disturbances can be six to nine orders of magnitude higher. Several methods are used to shield against the external magnetic fields or to minimize their effect on experimental results. The first method is to perform the measurements far away from electric motors, elevators, cars and trucks, tram and subway lines, and power lines. However, this is not sufficient or usually not even possible, so further measures are needed.

Typically, MEG measurements are performed in a magnetically shielded room (MSR) [56, 165]. Conducting rooms made of aluminum are effective only at frequencies of several tens of hertz or higher. In conducting rooms, the shielding is based on eddy currents induced by changing magnetic fields: according to Lenz's law, induced currents oppose the change that induce them [126, 200, 331]. For proper shielding at DC and low frequencies, shielded rooms are usually made of *mu-metal* with a magnetic permeability of at least 30,000 times that of vacuum. In a mu-metal enclosure, the external magnetic field lines tend to follow the path of least magnetic reluctance (which is inversely proportional to permeability), leaving the interior of the room free from most of the external field. Best shielding is obtained by constructing multiple shells of mu-metal. In an eight-shell room and active shielding, a shielding factor of 2×10^8 at 5 Hz has been reported [33]. However, such heavy rooms are too expensive for most applications. Typical MSRs attenuate the external field by two to three orders of magnitude at DC and below 1 Hz and four to five orders of magnitude above 10 Hz. This makes the recording of MEG easier, but additional measures are needed to shield against external magnetic interference.

An effective and straightforward method to attenuate external disturbances is to measure the gradient or first derivative of one component of the magnetic field such as $\partial B_z/\partial z$ or $\partial B_z/\partial x$. This can be done by forming the flux transformer so that it measures the difference of the same field component at two separate locations; see figure 1.11. More effective attenuation is achieved by second-order or even third-order gradiometers [58, 319]. This method works because sources of external magnetic fields are generally far away and produce a relatively uniform magnetic field or uniform gradient near the subject.

It was realized by Samu Taulu and his colleagues that a suitable (over)sampling of the field (including information about the radial dependence of the field from a chosen origin of the coordinate system within the measurement helmet) can be used to divide the multidimensional signal in two parts, one arising from sources within the measurement helmet and one arising from outside disturbances [304, 305]. This division, named signal-space separation or SSS, proved to be able to eliminate much of the external disturbances even in cases when the spatial patterns of the disturbances are unknown a priori. This method is used in the MEGIN (former Neuromag) system and is very effective, but one has to keep in mind in further data analysis that the rank of the original 306-channel data is reduced by SSS to less than 100, meaning that care has to be exercised when using analysis techniques where this may have significance.

If the spatial patterns of the external magnetic field are known a priori or can be determined (as is often done by performing *empty-room measurements*), the magnetic disturbances can very effectively be eliminated without a large reduction of signal-space rank. The method used is called *signal-space projection* or *SSP* [141, 312]; see section 5.4. SSP is also effective for suppressing artifacts in MEG and EEG signals that originate in the subject such as those from eye movements or blinks [133], cardiac activity [226], and muscle activity [198, 214]. It can also be used as a signal-separation method to simplify the analysis of brain activity [306].

Finally, unwanted components in the data can be removed by any data-analysis method that can separate external interferences, artifacts, or noise from the signal produced by the sources of interest. An ingenious blind source-separation method for sufficiently stationary multidimensional data is the *independent component analysis* or *ICA* [136, 137, 219], which is very effective, assuming that the underlying component values have suitably non-Gaussian probability distributions (except for one) and if they are independent of each other (meaning that one cannot gain any information of any component by learning the values of the other components); see chapter 8. In MEG and EEG, because of causal inter-area connections in the brain, source currents often fail to be sufficiently independent for ICA to separate the underlying components correctly. Such cases can be hard to notice, because ICA always gives a result based on the assumption of independence. In particular, source activities elicited by the same sensory or magnetic stimulus are not independent.

2 Genesis of MEG and EEG

The relationship between currents in the brain and the electric and magnetic fields outside the head is described mathematically by Maxwell's equations. In the frequency range of interest (0 Hz to about 1 kHz), the relationship is quite independent of frequency (but see papers by Stinstra and Peters [297] and by Pfurtscheller and Cooper [249], who point out that the generally higher conductivity of tissues at higher frequencies leads to lower inter-electrode voltages at those frequencies). Therefore, all EEG and MEG signal waveforms for each source in the brain are essentially identical: they are the same as the source waveform. Thus, for any single source or for any synchronous set of sources, the waveforms and, consequently, the frequency content of EEG and MEG are almost identical. However, the situation becomes much more complicated when there are many simultaneous sources: the waveforms and frequency spectra obtained from different channels (EEG or MEG) are generally not identical, because the signals originating from multiple sources are mixed in different proportions in different sensors. We will deal with this mixing problem in later chapters of the book, explaining how the source-level signals can be unmixed.[1]

2.1 Maxwell's Equations

Maxwell's equations describe how the *electric field* **E** and *magnetic flux density* **B** (in this book referred to as *magnetic field*) are produced by the charge density ρ and current density **J**, respectively:

$$\nabla \cdot \mathbf{E}(\mathbf{r}, t) = \rho(\mathbf{r}, t)/\epsilon_0 \tag{2.1}$$

$$\nabla \times \mathbf{E}(\mathbf{r}, t) = -\frac{\partial \mathbf{B}(\mathbf{r}, t)}{\partial t} \tag{2.2}$$

1. Much of the text in this section is based on Ilmoniemi [139].

$$\nabla \cdot \mathbf{B}(\mathbf{r}, t) = 0 \tag{2.3}$$

$$\nabla \times \mathbf{B}(\mathbf{r}, t) = \mu_0 \left(\mathbf{J}(\mathbf{r}, t) + \epsilon_0 \frac{\partial \mathbf{E}(\mathbf{r}, t)}{\partial t} \right). \tag{2.4}$$

The *quasi-static approximation* for time-dependent electromagnetic phenomena means that the time derivatives of **B** and **E** are omitted when electric or magnetic fields are calculated from charges and currents. In neuromagnetism or neuroelectricity, this is an excellent approximation. The magnetic field is then due to the distribution of the (total) electric current $\mathbf{J}(\mathbf{r})$ only, and the electric field arises only from the charge density $\rho(\mathbf{r})$. We then have

$$\nabla \cdot \mathbf{E}(\mathbf{r}, t) = \rho(\mathbf{r}, t)/\epsilon_0 \tag{2.5}$$

$$\nabla \times \mathbf{E}(\mathbf{r}, t) = 0 \tag{2.6}$$

$$\nabla \cdot \mathbf{B}(\mathbf{r}, t) = 0 \tag{2.7}$$

$$\nabla \times \mathbf{B}(\mathbf{r}, t) = \mu_0 \mathbf{J}(\mathbf{r}, t) \ . \tag{2.8}$$

In this quasi-static approximation, the electric field can be described by the *scalar potential* V by equation $\mathbf{E}(\mathbf{r}) = -\nabla V(\mathbf{r})$ while the magnetic field can be described by the vector potential **A**: $\mathbf{B}(\mathbf{r}) = \nabla \times \mathbf{A}(\mathbf{r})$. Both **J** and $\rho(\mathbf{r})$ result from the source (or primary) current density $\mathbf{J}^{\mathrm{p}}(\mathbf{r})$ (see figure 2.1); therefore, the magnetic field **B**, the electric

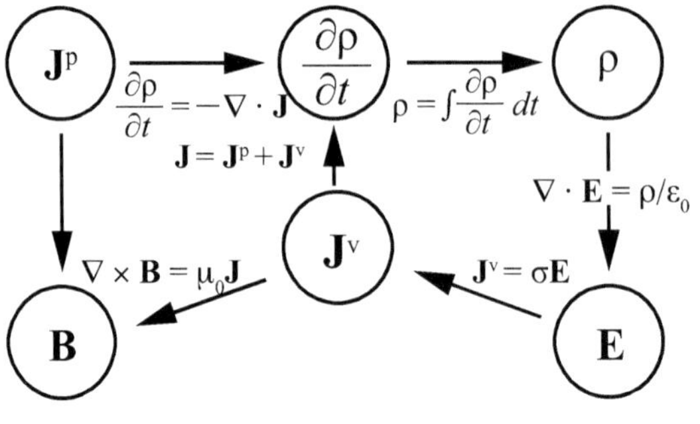

Figure 2.1
Fields generated by the primary current $\mathbf{J}^{\mathrm{p}}$. Since $\mathbf{J}^{\mathrm{p}}$ (as well as the volume current $\mathbf{J}^{\mathrm{v}}$) is the flow of electric charge, it causes the charge distribution $\rho(\mathbf{r})$ to change according to the continuity equation $\partial\rho/\partial t = -\nabla \cdot \mathbf{J} = -\nabla \cdot (\mathbf{J}^{\mathrm{p}} + \mathbf{J}^{\mathrm{v}})$. The resulting charge distribution, according to Maxwell's first equation, gives rise to the electric field $\mathbf{E}(\mathbf{r})$, which drives the volume current $\mathbf{J}^{\mathrm{v}} = \sigma \mathbf{E}$. $\mathbf{J}^{\mathrm{v}}$ and $\mathbf{J}^{\mathrm{p}}$ together produce the magnetic field according to Maxwell's fourth equation. In MEG and EEG applications, the time derivatives of **E** or **B** need not be taken into account in computing the fields: the quasistatic approximation is adequate. Although Maxwell's equations may look complicated, please note that all quantitative dependencies illustrated in the figure are linear. For example, $\mathbf{E}(\mathbf{r}) = (4\pi\epsilon_0)^{-1} \int \mathbf{g}(\mathbf{r}, \mathbf{r}')\rho(\mathbf{r}')dV'$ (equation (3.22)).

field $\mathbf{E}$, and V are proportional to $\mathbf{J}^p(\mathbf{r})$ (for the definition of primary current, see section 2.1.1).

2.1.1 Impressed and Primary Currents

As discussed in section 1.4, the total current $\mathbf{J}$ can be divided as

$$\mathbf{J}(\mathbf{r}) = \mathbf{J}^p(\mathbf{r}) + \mathbf{J}^v(\mathbf{r}) = \mathbf{J}^p(\mathbf{r}) + \sigma(\mathbf{r})\mathbf{E}(\mathbf{r}), \tag{2.9}$$

where $\mathbf{J}^p$ is the primary, or source, current, σ is the conductivity, and $\mathbf{J}^v$ is the passive ohmic or volume current. Here $\mathbf{J}^p$ is nonzero only at the site of neuronal activity. The conductivity $\sigma(\mathbf{r})$ depends on the location $\mathbf{r}$. In fact, the conductivity can also be a function of frequency and it can also depend on time (e.g., because of neuronal activity or hemodynamic changes). However, in most cases, the conductivity can be assumed to be unchanging and independent of frequency without causing significant errors in the analysis of MEG or EEG signals.

It is noteworthy that equation (2.9) depends critically on what length scale $\mathbf{E}(\mathbf{r})$, $\sigma(\mathbf{r})$, and $\mathbf{J}(\mathbf{r})$ are described. If $\sigma(\mathbf{r})$ is described on the fine scale of cells and membranes, $\mathbf{J}^p(\mathbf{r})$, to a large extent, is transmembrane current, often also called *impressed current*. On the more macroscopic level where $\sigma(\mathbf{r})$ in equation (2.9) is defined on a length scale of about a millimeter, primary current is the current that is impressed by microscopic phenomena and guided locally by the microscopic geometry of brain tissue [46, 212, 231, 233]. When MEG or EEG is interpreted in terms of sources, the primary current is estimated or interpreted. There is still confusion in the literature regarding primary current: $\mathbf{J}^p$ is not purely intracellular, and not all intracellular current is primary current. Both EEG and MEG are produced by the same intra- and extracellular primary currents, although different EEG or MEG sensors measure the primary current distribution with different sensitivity profiles or lead fields.

2.2 The Current Dipole and Lead Fields

A central concept in bioelectromagnetism is the *current dipole*, which is used to approximate a concentration of primary current in a small volume. Ideally, a current dipole with *moment* $\mathbf{Q} = Q\,\mathbf{e}_Q$ at *location* $\mathbf{r}_Q$ is a concentration of primary current at a single point:

$$\mathbf{J}^p_Q(\mathbf{r}) = Q\delta(\mathbf{r} - \mathbf{r}_Q)\mathbf{e}_Q, \tag{2.10}$$

where $\delta(\mathbf{r})$ is the Dirac delta function, the scalar Q is the *amplitude*, and the unit vector $\mathbf{e}_Q$ is the *orientation* of the dipole. Explicitly, if $\mathbf{J}^p(\mathbf{r})$ is a primary current distribution in a small volume containing location $\mathbf{r}_Q$, then $\mathbf{J}^p(\mathbf{r})$ can be approximated by a current dipole $\mathbf{J}^{\mathrm{p}}_{\mathrm{Q}}(\mathbf{r}) = \delta(\mathbf{r} - \mathbf{r}_Q)\,\mathbf{Q}$ with moment

$$\mathbf{Q} = \int \mathbf{J}^{\mathrm{p}}(\mathbf{r})\,dV\,, \tag{2.11}$$

where the integral is at least over the volume where $\mathbf{J}^{\mathrm{p}}(\mathbf{r}) \neq 0$.

Generally, any primary current distribution $\mathbf{J}^{\mathrm{p}}(\mathbf{r})$ can be approximated by a sum of current dipoles by dividing the relevant volume (such as the brain) to small-volume elements (voxels) and placing a current dipole with moment $\mathbf{Q} = \int_{\mathrm{voxel}} \mathbf{J}^{\mathrm{p}}(\mathbf{r})\,dV$ to the center of each element.

The center of gravity of the primary current $\mathbf{J}^{\mathrm{p}}$ is

$$\mathbf{r}_{\mathrm{CoG}} = \int \|\mathbf{J}^{\mathrm{p}}(\mathbf{r})\|\,\mathbf{r}\,dV \bigg/ \int \|\mathbf{J}^{\mathrm{p}}(\mathbf{r})\|\,dV\,. \tag{2.12}$$

In EEG and MEG applications, a current dipole is often used as an equivalent source for primary currents that may extend even over several centimeters. If an extended source is modeled with an *equivalent current dipole* (one that reproduces MEG or EEG data as closely as possible), one should keep in mind that the location of this dipole does not generally need to coincide with the center of gravity of the primary current distribution. For example, the single equivalent dipole that best accounts for the measured magnetic field pattern due to a pair of parallel superficial sources is deeper than the two sources [51]. Also, the amplitude or orientation of an equivalent dipole is not necessarily the same as those of the dipole moments that constitute the distributed source.

The current dipole is often considered a line element of primary current; however, it equally well approximates a layer or any other shape of primary current (see figure 2.2). Although there are no actual current dipoles in the brain, this concept is useful in locating current sources clustered in small volumes. As already mentioned, one can also approximate an extended distribution of primary current with a collection of discrete current dipoles.

By making use of the concept of a current dipole, the (measured) sensor signals, due to a primary current distribution $\mathbf{J}^{\mathrm{p}}(\mathbf{r})$, can be presented in terms of the *lead fields* $\mathbf{L}_i(\mathbf{r})$ as follows.

The part of $\mathbf{J}^{\mathrm{p}}(\mathbf{r})$ in an infinitesimal volume dV' at the point $\mathbf{r}'$ can be represented as a current dipole $\mathbf{J}^{\mathrm{p}}(\mathbf{r}')\delta(\mathbf{r} - \mathbf{r}')\,dV'$. By writing $\mathbf{J}^{\mathrm{p}}(\mathbf{r}')$ componentwise,

$$\mathbf{J}^{\mathrm{p}}(\mathbf{r}') = J^{\mathrm{p}}_1(\mathbf{r}')\mathbf{e}_1 + J^{\mathrm{p}}_2(\mathbf{r}')\mathbf{e}_2 + J^{\mathrm{p}}_3(\mathbf{r}')\mathbf{e}_3\,, \tag{2.13}$$

where $\mathbf{e}_1$, $\mathbf{e}_2$, and $\mathbf{e}_3$ are the cartesian coordinate unit vectors, we can write

$$\mathbf{J}^{\mathrm{p}}(\mathbf{r}')\,\delta(\mathbf{r} - \mathbf{r}')dV' = \sum_{j=1}^{3} J^{\mathrm{p}}_j(\mathbf{r}')\delta(\mathbf{r} - \mathbf{r}')\,dV'\,\mathbf{e}_j\,. \tag{2.14}$$

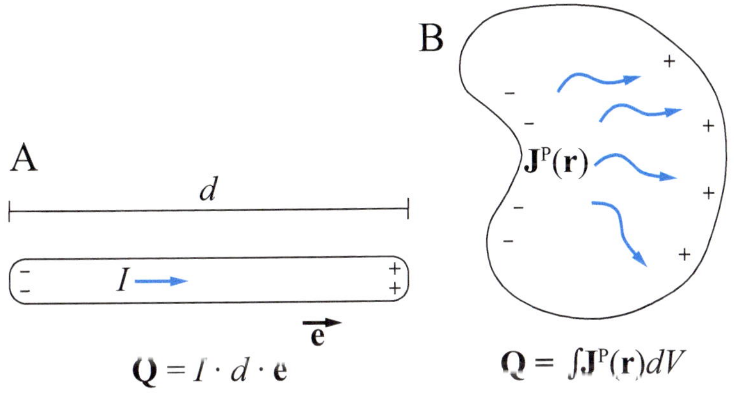

Figure 2.2
Current dipole. (A) Often depicted as a short line element of current, the current dipole consists of a sink and a source in conducting medium, and the current flowing from sink to source. The sink and source are the negative and positive poles of the dipole, giving rise to volume current in the conducting medium. (B) The current dipole describes the leading term in the current multipole expansion of any shape of current distribution.

Let $L_{ij}(\mathbf{r}')$ be the reading in sensor i due to the unit source dipole $\delta(\mathbf{r}-\mathbf{r}')\mathbf{e}_j$ at $\mathbf{r}'$, $j = 1, 2, 3$. The three-vector

$$\mathbf{L}_i(\mathbf{r}') = [L_{i1}(\mathbf{r}'),\ L_{i2}(\mathbf{r}'),\ L_{i3}(\mathbf{r}')]^{\mathrm{T}} \tag{2.15}$$

is called the *lead field* of sensor i at location $\mathbf{r}'$. It follows that if we denote by dy_i the signal in sensor i due to the dipole $\mathbf{J}^{\mathrm{p}}(\mathbf{r}')\delta(\mathbf{r}-\mathbf{r}')dV'$, then by equation (2.14) and the linearity of the field equations, we get

$$dy_i = \sum_{j=1}^{3} L_{ij}(\mathbf{r}')\, J_j^{\mathrm{p}}(\mathbf{r}')\, dV' = \mathbf{L}_i(\mathbf{r}') \cdot \mathbf{J}^{\mathrm{p}}(\mathbf{r}')\, dV'\,. \tag{2.16}$$

By integrating this equation over the volume containing $\mathbf{J}^{\mathrm{p}}$, we get the wanted representation of the signal y_i in sensor i, due to primary current $\mathbf{J}^{\mathrm{p}}$, in terms of the lead field $\mathbf{L}_i(\mathbf{r}')$,

$$y_i = \int \mathbf{L}_i(\mathbf{r}') \cdot \mathbf{J}^{\mathrm{p}}(\mathbf{r}')\, dV'\,. \tag{2.17}$$

In numerical simulations of MEG or EEG, one wants to compute the sensor signal for a given head model due to any primary current distribution. To that end, the lead fields $\mathbf{L}_i(\mathbf{r}')$ are computed and saved for a sufficiently dense grid of locations $\mathbf{r}'$, and then the integrals (2.17) are approximated with appropriate sums. This enables fast computing of the sensor signals for arbitrary primary current distributions.

2.3 Cellular Basis for Electromagnetic Fields

The electric currents and charge accumulation associated with action potentials and postsynaptic potentials give rise to MEG and EEG signals.

Information from one neuron to another is transmitted along the axon in the form of electrical pulses called action potentials (see figure 2.3). When a pulse reaches a synapse, transmitter molecules are released from the presynaptic cell to the synaptic cleft that separates the two neurons (see figure 2.4). Some of the transmitter molecules diffuse to the postsynaptic membrane, where they increase its permeability to ions. Because of the transmembrane voltage and the ionic concentration gradients over the membrane, two things happen: (1) the increased conductivity of the membrane increases ohmic currents into the cell, and (2) the concomitant increase in the diffusion coefficients for the ions facilitates diffusion across the membrane, with sodium diffusing in and potassium diffusing out (in an excitatory synapse). There will be a net current flow through the membrane (this is the "impressed current"); because of the consequent charge distribution changes ($\partial\rho/\partial t = -\nabla\cdot\mathbf{J}$), the electric field is changed accordingly ($\nabla\cdot\mathbf{E} = \rho/\epsilon_0$), and return currents $\mathbf{J}^{\mathrm{v}} = \sigma\mathbf{E}$ flow in the tissue (see figure 2.5). EEG and MEG signals are believed to be largely due to these three-dimensional current loops that are of synaptic origin. MEG is generated by both the primary currents and volume currents, whereas EEG signals arise from the electric field produced by the changing distribution of

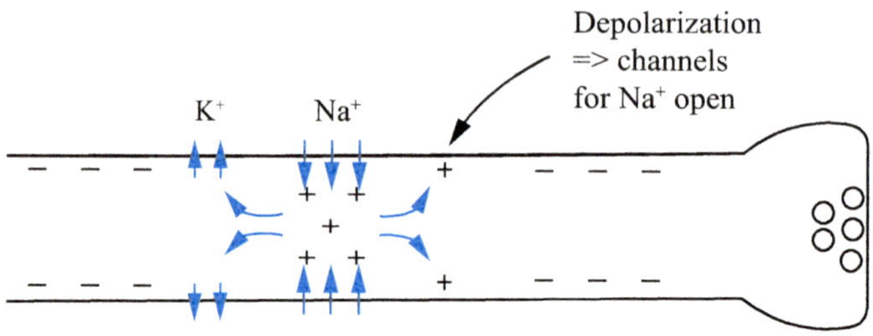

Figure 2.3
Action potential moving towards a synapse on the right. As positive sodium ions (Na^+) enter the cell, the interior becomes more positive and the resulting electric field, directed to both right and left, drives electric current flow. On the right, it depolarizes the membrane, causing new sodium channels to open and the action potential to move to the right. On the left, potassium channels have opened just after the opening of the Na channels, resulting in positive impressed-current ion flow out from the cell, repolarizing the membrane.

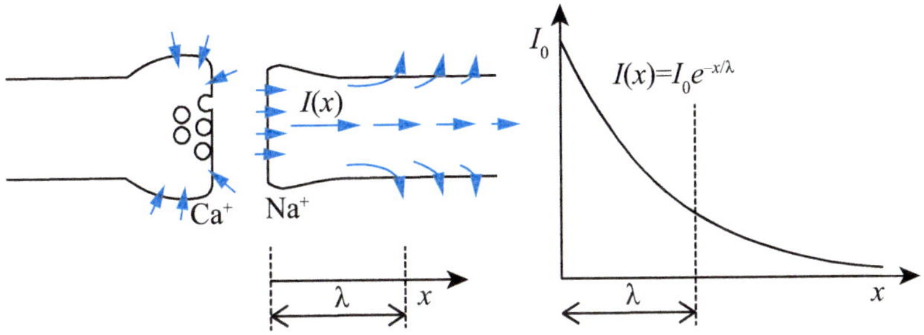

Figure 2.4
Calcium ions enter the cell due to the opening of voltage-sensitive Ca^+ channels when an action potential arrives at a synapse. This triggers the fusion of transmitter-molecule-containing vesicles to the presynaptic cell membrane; the vesicles then release transmitter molecules to the synaptic cleft, causing opening of ion channels postsynaptically. This causes a postsynaptic electric current, which is assumed to be the main primary-current component to which MEG and EEG are sensitive. The postsynaptic dendritic current $I(x)$ diminishes (in a simplified model of the dendrite) exponentially as a function of distance from the synapse, with the length constant λ defining the rate of change of the current.

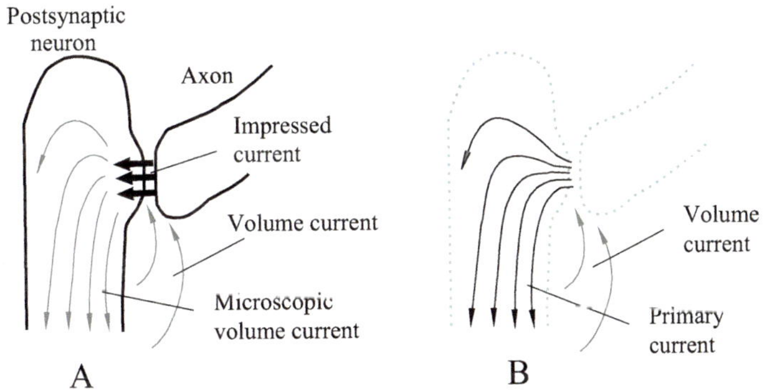

Figure 2.5
Impressed versus primary currents. (A) The opening of ion channels in the postsynaptic membrane causes ions to move into the neuron, making the interior of the cell more positive and the exterior more negative. The resulting electric field gives rise to electric currents that constitute the microscopic volume current. (B) The macroscopic primary current $\mathbf{J}^p(\mathbf{r}) = \mathbf{J}^{total}(\mathbf{r}) - \sigma(\mathbf{r})\mathbf{E}(\mathbf{r})$ is defined to be the current that is not explained by the volume current $\mathbf{J}^v(\mathbf{r}) = \sigma(\mathbf{r})\mathbf{E}(\mathbf{r})$ caused by the macroscopic electric field $\mathbf{E}(\mathbf{r})$ and the macroscopic conductivity $\sigma(\mathbf{r})$.

electric charge. However, since $\mathbf{J}^{\mathrm{p}}$ produces $\mathbf{J}^{\mathrm{v}}$ and $\mathbf{J} = \mathbf{J}^{\mathrm{v}} + \mathbf{J}^{\mathrm{p}}$ produces any changes in ρ, both EEG and MEG result from $\mathbf{J}^{\mathrm{p}}$ (see figure 2.1).

The primary currents in the neurons, although resulting mainly from synaptic transmembrane currents, are guided by the dendritic tree and other structures of the cell. In pyramidal cells, which are believed to be the main sources of electromagnetic fields detected by EEG or MEG, the dendritic trees are oriented predominantly perpendicular to the cortical surface. Therefore, it appears natural that the preferred direction of the primary current is also perpendicular to the cortical surface. A large amount of experimental evidence has been obtained where the primary current has been concluded to be approximately perpendicular to the cortex, supporting this interpretation. However, as was already pointed out, not all primary current is intracellular and not all intracellular current is primary current since the volume current also flows to a large extent via the intracellular compartments.

2.4 The Action Potential

The propagating action potential in an idealized straight axon in a uniform bath of conducting extracellular fluid is described by the cable equation and by the Hodgkin–Huxley model [130]. The cable equation in the absence of the action potential describes how the membrane potential V_{m} behaves:

$$\lambda^2 \frac{\partial^2 V_{\mathrm{m}}}{\partial x^2} - V_{\mathrm{m}} = \tau \frac{\partial V_{\mathrm{m}}}{\partial t}, \tag{2.18}$$

where $\lambda = \sqrt{r_{\mathrm{m}}/r_{\mathrm{i}}}$ is the length constant and $\tau = c_{\mathrm{m}} r_{\mathrm{m}}$ is the time constant; r_{m} is the resistance of a segment of the membrane times the segment length, r_{i} is the intracellular resistance per unit length, and c_{m} is the capacitance of the membrane per unit length of the axon.

Action potentials propagating in axons produce only very small electromagnetic fields at a distance. However, when arriving at a synapse, the symmetry of the axonal current pattern is broken, and instead of the quadrupolar pattern, there will be a dipolar current distribution, which contributes to high-frequency MEG and EEG.

2.5 Postsynaptic Potential (PSP)

When transmitter molecules attach to ligand-gated (molecule-activated) ion channels on the postsynaptic membrane, the channels open and, depending on the type of channel, electric current flows into the postsynaptic cell or out from it, causing an excitatory or inhibitory postsynaptic potential (EPSP or IPSP), respectively. The current flow into the cell lasts typically tens of milliseconds, forming an element of primary current. The current is guided along the dendrite (see figure 2.5). This miniature current element can be modeled as a current dipole with amplitude $Q(t) = I(t) \cdot d$ where $I(t)$ is the current and d is the characteristic length of the current path within the cell, which in a simple case is the length constant of the dendritic branch in question: $d = \lambda$ (see figure 2.4). For an order-of-magnitude estimate of Q, assume a postsynaptic potential $\Delta V_\mathrm{m} = 50\,\mathrm{mV}$, postsynaptic dendrite diameter of $D = 10^{-6}$ m, length constant $\lambda = 0.5$ mm and resistivity of the cytoplasm $\rho_\mathrm{int} = 1\,\Omega\mathrm{m}$. Then, the postsynaptic resistance $R_\mathrm{PSP} = \lambda\rho/(\pi D^2/4)$ and $I = \Delta V_\mathrm{m}/R_\mathrm{PSP}$, whereby $Q = \lambda I = (\pi/4\rho)D^2 \Delta V_\mathrm{m}$ is independent of λ. Inserting the assumed values, we obtain $Q = 4 \times 10^{-14}$Am. If one assumes that I, instead of ΔV_m, is fixed, then Q is proportional to λ.

2.6 Description of Synaptic Activity

The most common excitatory synapse in the brain is glutamatergic, that is, it uses glutamate as transmitter molecule. The AMPA-type glutamate receptor channels are permeable to both Na^+ and K^+ ions. This will cause sodium to diffuse into the cell and potassium to diffuse out after an action potential has released glutamate to the synaptic cleft. However, the net effect is current flow into the cell and depolarization (i.e., an EPSP). The positive charge injected into the cell causes an intracellular electric field, which drives current along the postsynaptic neuron. This, in turn, changes the membrane potential along the dendrite and causes the flow of current through it. Current in the surrounding tissue then closes the current loop. The most common inhibitory neurotransmitter is gamma-aminobutyric acid or GABA, which, by opening channels for either Cl^- or K^+ ions, causes current flow out from the postsynaptic cell. Although the polarity of the transmembrane current is opposite to that of EPSPs, the overall IPSP current pattern may be similar because of the different locations of the synapses: excitatory synapses tend to be located in apical dendrites of pyramidal neurons while inhibitory synapses are more often in basal dendrites or soma. Thus, the current direction in both cases can be the same: toward soma.

3 Forward Problem

Starting from Maxwell's equations (2.1) – (2.4), we consider in this chapter the electromagnetic forward problem, that is, calculating the magnetic and electric fields $\mathbf{B}(\mathbf{r})$ and $\mathbf{E}(\mathbf{r})$ due to a known distribution of primary current $\mathbf{J}^{\mathrm{p}}(\mathbf{r})$, given the conductivity distribution $\sigma(\mathbf{r})$ of the relevant volume such as the head. Unfortunately, the conductivity distribution is in practice incompletely and inaccurately known. Therefore, we must use estimated conductivity values and approximate representations of tissue shapes. We will discuss different categories of approximations of conductivity distributions by defining conductor models with different degrees of detail and complexity. We describe how we can use these models to calculate the electromagnetic fields due to neuronal currents in the brain. Both analytical and numerical methods will be explained with their relative merits and limitations.

Studies of the forward problem have been reported in a multitude of papers. The forward-problem solutions by Grynzpan and Geselowitz [99, 107] and by Cuffin and Cohen [65] laid the foundation for the proper analysis of the inverse problem. Subsequently, a large number of papers have been published on the topic. In this chapter, we will derive solutions for several conductor models such as the homogeneous infinite conductor, semi-infinite homogeneous conductor, spherically symmetric conductor, multi-shelled models [294, 303], and the general case.

For the interpretation of EEG and MEG signals, we wish to determine the sensitivity pattern for the primary current of each sensor i, $i = 1, 2, \ldots, N_{\mathrm{ch}}$, where N_{ch} is the number of channels, or sensors. The signal $y_i(t)$ recorded by sensor i, as explained in section 2.2, can be given by

$$y_i(t) = \int_V \mathbf{L}_i(\mathbf{r}) \cdot \mathbf{J}^{\mathrm{p}}(\mathbf{r}, t)\, dV\,, \tag{3.1}$$

where $\mathbf{L}_i(\mathbf{r})$ is the lead field of sensor i at location $\mathbf{r}$. This crucial relationship is illustrated in figure 3.1. If considered a function of $\mathbf{r}$, then $\mathbf{L}_i(\mathbf{r})$ is called the *sensitivity*

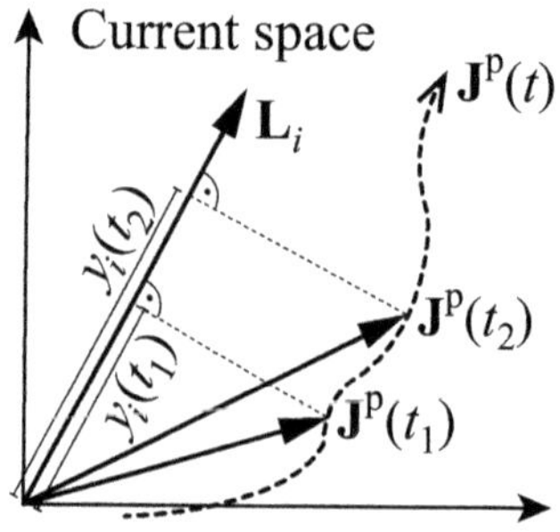

Figure 3.1
The evolution of the brain state, as reflected by the changing primary-current distribution $\mathbf{J}^p = \mathbf{J}^p(t)$, is projected on one of the lead-field vectors (the ith), $\mathbf{L}_i$, resulting in the time evolution of the measured signal $y_i(t)$, which is the projection of $\mathbf{J}^p(t)$ onto $\mathbf{L}_i$. The values of y_i are depicted at time instants t_1 and t_2. Note that the components of $\mathbf{J}^p$ that are orthogonal to $\mathbf{L}_i$ are silent to the measurement with sensor i, that is, only the projection $y_i(t) = \langle \mathbf{L}_i, \mathbf{J}^p(t) \rangle = \int \mathbf{L}_i(\mathbf{r}) \cdot \mathbf{J}^p(\mathbf{r}, t) dV$ is detected.

pattern of the sensor i. The function $\mathbf{L}_i(\mathbf{r})$ is a vector field: its direction at $\mathbf{r}$ defines the direction for which the sensor i is most sensitive with respect to the primary current at $\mathbf{r}$. Any element of primary current perpendicular to the local lead field is invisible to this sensor. The norm $||\mathbf{L}_i(\mathbf{r})||$ is the sensitivity of sensor i to the primary current at $\mathbf{r}$.

Both the lead field $\mathbf{L}_i(\mathbf{r})$ and the primary-current distribution $\mathbf{J}^p(\mathbf{r})$ can be considered vectors in a space of (quadratically integrable) vector-valued functions. In that function space, the inner product between vectors $\mathbf{x}_1$ and $\mathbf{x}_2$ is defined by

$$\langle \mathbf{x}_1, \mathbf{x}_2 \rangle = \int_V \mathbf{x}_1(\mathbf{r}) \cdot \mathbf{x}_2(\mathbf{r}) \, dV \, . \tag{3.2}$$

Thus,

$$y_i(t) = \langle \mathbf{L}_i, \mathbf{J}^p(t) \rangle \tag{3.3}$$

is a projection of $\mathbf{J}^p(t)$ on the lead field $\mathbf{L}_i$ (multiplied by the norm of the lead field).

If the measured signals $y_i(t)$, $i = 1, 2, \ldots, N_{ch}$, are collected to a vector $\mathbf{y}(t) = [y_1(t), y_2(t), \ldots, y_{N_{ch}}(t)]^T$, we have

$$\mathbf{y}(t) = [\langle \mathbf{L}_1, \mathbf{J}^p(t) \rangle, \ldots, \langle \mathbf{L}_{N_{ch}}, \mathbf{J}^p(t) \rangle]^T \, . \tag{3.4}$$

EEG and MEG measurements thus provide an N_{ch}-dimensional *projection* of the infinite-dimensional primary-current vector. One dimension of this projection (equation (3.3)) is illustrated in figure 3.1. It is important to realize that the measurements provide the projected components of $\mathbf{J}^{\mathrm{p}}(t)$ directly, without any distortion (apart from additive noise). However, no information about the primary-current components orthogonal to the subspace spanned by the lead fields is obtained. If we have no a priori information about $\mathbf{J}^{\mathrm{p}}(t)$, the best estimate we can make on the basis of the measured data for the primary current is the orthogonal projection $\widetilde{\mathbf{J}}^{\mathrm{p}}(t)$ of $\mathbf{J}^{\mathrm{p}}(t)$ onto the function subspace spanned by $\mathbf{L}_1, \ldots, \mathbf{L}_{N_{\mathrm{ch}}}$. This estimate is called the *minimum-norm estimate*, because its norm $||\widetilde{\mathbf{J}}^{\mathrm{p}}|| = \langle \widetilde{\mathbf{J}}^{\mathrm{p}}, \widetilde{\mathbf{J}}^{\mathrm{p}} \rangle^{1/2}$ is least among all current distributions $\mathbf{K}$ with $\langle \mathbf{L}_i, \mathbf{K} \rangle = y_i,\ i = 1 \ldots N_{\mathrm{ch}}$; see also chapter 5. An explicit representation of $\widetilde{\mathbf{J}}^{\mathrm{p}}(t)$ is given by

$$\widetilde{\mathbf{J}}^{\mathrm{p}}(t) = \sum_{i=1}^{N_{\mathrm{ch}}} \alpha_i(t)\, \mathbf{L}_i\,, \tag{3.5}$$

where

$$[\alpha_1(t), \ldots, \alpha_{N_{\mathrm{ch}}}(t)]^{\mathrm{T}} = \mathbf{A}^{-1}\mathbf{y}(t) \ \text{ with } \ \mathbf{A}(i,j) = \langle \mathbf{L}_i, \mathbf{L}_j \rangle \tag{3.6}$$

for $1 \le i, j \le N_{\mathrm{ch}}$; $\mathbf{A}^{-1}$ is the inverse of the Gram matrix $\mathbf{A}$.

3.1 Conductor Models

A conductor model is a description of the conductivity distribution of a region of space such as the human body or head. Always an approximate representation of reality, a conductivity model enables one to calculate the electromagnetic field due to primary currents. In general, the more detailed and accurate the conductivity model is, the more accurately the electromagnetic fields can be calculated. However, detailed conductor models are difficult to construct, and their use requires more computer time than simple models. After some basic considerations in the next section, we will discuss the use and characteristics of several conductor models.

3.2 The General Case

We start with the *quasi-static* approximation of Maxwell's equations, which describes how the electromagnetic field is produced in situations where the electric and magnetic fields change so slowly that their time derivatives do not need to be taken into account in calculating the electromagnetic fields. The resulting quasi-static approximation is valid at the frequencies relevant for MEG and EEG, which are at most on the order of 1 kHz [310]. The *first Maxwell equation*, also called Gauss's law, describes how electric charges $\rho(\mathbf{r})$ give rise to the electric field in a volume with constant electric permittivity ϵ_0 (the permittivity of vacuum):

$$\nabla \cdot \mathbf{E}(\mathbf{r}, t) = \rho(\mathbf{r}, t)/\epsilon_0 \,. \tag{3.7}$$

The integral form of Gauss's law asserts explicitly that the electric flux through the surface ∂V of volume V is the total charge within the volume divided by ϵ_0:

$$\int_{\partial V} \mathbf{E} \cdot d\mathbf{S} = \frac{1}{\epsilon_0} \int_V \rho \, dV \,. \tag{3.8}$$

Maxwell's second equation, which would also be called Faraday's law of induction if the time derivatives were included, in the quasi-static approximation states that

$$\nabla \times \mathbf{E}(\mathbf{r}, t) = 0 \,, \tag{3.9}$$

meaning that the electric field can be derived from an *electric potential* V as

$$\mathbf{E} = -\nabla V \,. \tag{3.10}$$

According to *Maxwell's third equation*, Gauss's law for magnetic fields, when no magnetic charges (magnetic monopoles) are assumed to exist,

$$\nabla \cdot \mathbf{B}(\mathbf{r}, t) = 0 \,. \tag{3.11}$$

The *fourth Maxwell equation*, Ampère's circuital law, expresses in the quasi-static case that the curl of the magnetic field is proportional to the total current density:

$$\nabla \times \mathbf{B}(\mathbf{r}, t) = \mu_0 \mathbf{J}(\mathbf{r}, t) \,, \tag{3.12}$$

where μ_0 is the *magnetic permeability* of vacuum. Note that here

$$\mathbf{J} = \mathbf{J}^{\mathrm{p}} + \sigma \mathbf{E} = \mathbf{J}^{\mathrm{p}} - \sigma \nabla \mathbf{V} \,, \tag{3.13}$$

where $\sigma \mathbf{E} = -\sigma \nabla \mathbf{V}$ is the *volume*, or *ohmic, current* density and $\sigma = \sigma(\mathbf{r})$ is the conductivity at $\mathbf{r}$, $\mathbf{J}$ is the *total current* and $\mathbf{J}^{\mathrm{p}}$ the *primary* or *source current*.

In its integral form, Ampère's circuital law states that the line integral of the magnetic field around path C forming a loop is proportional to the flux of the electric current through the loop:

$$\int_C \mathbf{B} \cdot d\mathbf{l} = \mu_0 \int_S \mathbf{J} \cdot d\mathbf{S} \,, \tag{3.14}$$

where S is a surface patch circumvented by C; the equation follows from (3.12) by Stokes's theorem. This means that magnetic field lines curl around the flow paths of the current (see figure 3.2).

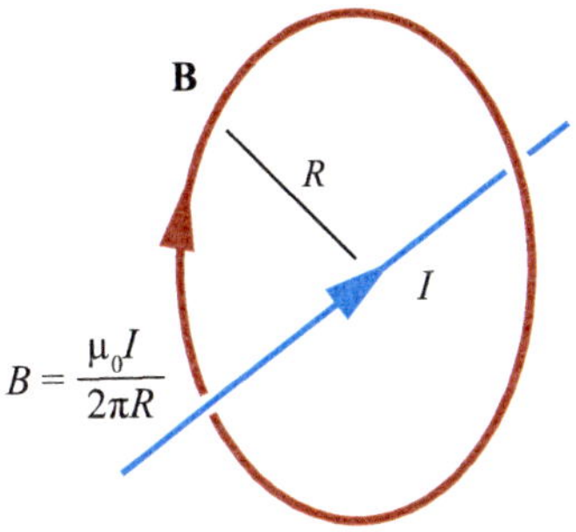

Figure 3.2
Illustration of Maxwell's fourth equation. Magnetic flux lines circle about current-flow lines. The direction of the magnetic field is obtained from the right-hand rule: if the right thumb is oriented along the current, the other fingers show the direction of the magnetic field.

The solution to equations (3.11) and (3.12), with the requirement that the magnetic field vanishes at infinity, is given by the Ampère–Laplace (or Biot–Savart) law[1]:

$$\mathbf{B}(\mathbf{r}) = \frac{\mu_0}{4\pi} \int \mathbf{J}(\mathbf{r}') \times \frac{\mathbf{r} - \mathbf{r}'}{||\mathbf{r} - \mathbf{r}'||^3} dV' . \tag{3.15}$$

This equation also yields the vector potential $\mathbf{A}(\mathbf{r})$, which is related to $\mathbf{B}(\mathbf{r})$ by $\mathbf{B}(\mathbf{r}) = \nabla \times \mathbf{A}(\mathbf{r})$. Namely, by using the identities $\nabla \times (\mathbf{J}(\mathbf{r}')/||\mathbf{r} - \mathbf{r}'||) = -\mathbf{J}(\mathbf{r}') \times \nabla(1/||\mathbf{r} - \mathbf{r}'||)$ and $\nabla(1/||\mathbf{r} - \mathbf{r}'||) = -(\mathbf{r} - \mathbf{r}')/||\mathbf{r} - \mathbf{r}'||^3$ and moving ∇ to the front of the integral, we get

$$\mathbf{B}(\mathbf{r}) = \frac{\mu_0}{4\pi} \nabla \times \int \frac{\mathbf{J}(\mathbf{r}')}{||\mathbf{r} - \mathbf{r}'||} dV' , \tag{3.16}$$

which yields the vector potential $\mathbf{A}(\mathbf{r})$ of $\mathbf{B}(\mathbf{r})$ as

$$\mathbf{A}(\mathbf{r}) = \frac{\mu_0}{4\pi} \int \frac{\mathbf{J}(\mathbf{r}')}{||\mathbf{r} - \mathbf{r}'||} dV' . \tag{3.17}$$

To compute $\mathbf{B}$, $\mathbf{E}$, and $\mathbf{J}$ due to a given source-current density $\mathbf{J}^p$, we still need to find V. For practical purposes, we may assume here that the conductivity is piecewise constant. By equation (3.13), $\mathbf{J} = \mathbf{J}^p - \sigma \nabla V$. By equation (3.12), $\nabla \cdot \mathbf{J} = 0$, and we have $\nabla \cdot (\sigma \nabla V) = \nabla \cdot \mathbf{J}^p$, which in a region of constant $\sigma > 0$ yields

1. Equation (3.15) is also called the *Biot–Savart* law, although this term should perhaps be reserved for the case when the current flow is confined to one-dimensional wires and the volume integral can be replaced by a line integral over a closed loop of wire.

$$\nabla^2 V = \frac{1}{\sigma}\nabla \cdot \mathbf{J}^{\mathrm{p}}, \tag{3.18}$$

where we have denoted $\nabla \cdot \nabla = \nabla^2$, which is the Laplace operator. The potential V is the unique solution of the (Poisson) equation (3.18) with the requirement that $V(\mathbf{r}) \to 0$ as $||\mathbf{r}|| \to \infty$ and with the boundary conditions

$$V' = V'' \text{ and } \sigma' \frac{\partial V'}{\partial n} = \sigma'' \frac{\partial V''}{\partial n} \tag{3.19}$$

on interfaces between regions of constant conductivities σ' and σ''. Here $\partial/\partial n$ denotes the normal derivative on an interface.

Computing V numerically from equation (3.18) is usually carried out by the *finite element method* (FEM) or by the *boundary element method* (BEM). Later, we derive the surface integral equation for BEM. If an anatomically realistically shaped head model is approximated by some simpler head models, solving equation (3.18) becomes much easier and faster. In the case of MEG, it was realized early that since the skull is a poor conductor of electricity, the contribution to the magnetic field of currents in it and in the scalp is negligible; thus, one can use a conductivity model that takes into account currents only in the intracranial volume [115, 142]. Examples of simple conductivity models are the infinite homogeneous conductor, the semi-infinite homogeneous conductor, and the spherically symmetric model.

3.3 Fields in an Infinite Homogeneous Medium

Here we assume an infinite homogeneous medium where $\sigma \geq 0$, $\epsilon = \epsilon_0$ and $\mu = \mu_0$ are constant everywhere. We first consider V and $\mathbf{E}$ in the simple case that the medium is vacuum (i.e., $\sigma = 0$), and the source is a single point charge q at point $\mathbf{r}_q$, that is, $\rho(\mathbf{r}) = q\delta(\mathbf{r} - \mathbf{r}_q)$. Equations (3.7) and (3.10) imply that $\nabla^2 V = -\rho/\epsilon_0$. This is a Poisson equation for V in *vacuum* and the solution (vanishing at infinity) is

$$V(\mathbf{r}) = \frac{1}{4\pi\epsilon_0}\frac{q}{||\mathbf{r} - \mathbf{r}_q||}, \tag{3.20}$$

which, with equation (3.10) and the identity $\nabla(1/||\mathbf{r} - \mathbf{r}_q||) = -||\mathbf{r} - \mathbf{r}_q||^{-3}(\mathbf{r} - \mathbf{r}_q)$, implies that

$$\mathbf{E}(\mathbf{r}) = \frac{q}{4\pi\epsilon_0}\frac{\mathbf{r} - \mathbf{r}_q}{||\mathbf{r} - \mathbf{r}_q||^3}. \tag{3.21}$$

Consequently, the electric field of a charge distribution $\rho = \rho(\mathbf{r}')$ is, because of the linear summation of electromagnetic fields,

$$\mathbf{E}(\mathbf{r}) = \frac{1}{4\pi\epsilon_0}\int \mathbf{g}(\mathbf{r},\mathbf{r}')\rho(\mathbf{r}')dV', \tag{3.22}$$

where $\mathbf{g}(\mathbf{r},\mathbf{r}') = (\mathbf{r}-\mathbf{r}')/||\mathbf{r}-\mathbf{r}'||^3$.

Equation (3.18), with $\sigma > 0$, is a Poisson equation for V with the solution

$$V(\mathbf{r}) = -\frac{1}{4\pi\sigma}\int_G \frac{\nabla'\cdot\mathbf{J}^{\mathrm{p}}(\mathbf{r}')}{||\mathbf{r}-\mathbf{r}'||}dV', \tag{3.23}$$

where the integration is over a region G containing the source current $\mathbf{J}^{\mathrm{p}}$. Due to the identity

$$\nabla'\cdot\left(\frac{\mathbf{J}^{\mathrm{p}}(\mathbf{r}')}{||\mathbf{r}-\mathbf{r}'||}\right) = \frac{1}{||\mathbf{r}-\mathbf{r}'||}\nabla'\cdot\mathbf{J}^{\mathrm{p}}(\mathbf{r}') + \mathbf{J}^{\mathrm{p}}(\mathbf{r}')\cdot\nabla'\left(\frac{1}{||\mathbf{r}-\mathbf{r}'||}\right), \tag{3.24}$$

with $\nabla'(||\mathbf{r}-\mathbf{r}'||^{-1}) = ||\mathbf{r}-\mathbf{r}'||^{-3}(\mathbf{r}-\mathbf{r}')$ and the Gauss theorem, $\int_V \nabla\cdot\mathbf{u}\,dV = \int_S \mathbf{u}\cdot\mathbf{dS}$, we can transform equation (3.23) to the form

$$V(\mathbf{r}) = \frac{1}{4\pi\sigma}\int_G \mathbf{J}^{\mathrm{p}}(\mathbf{r}')\cdot\frac{\mathbf{r}-\mathbf{r}'}{||\mathbf{r}-\mathbf{r}'||^3}dV', \tag{3.25}$$

because the surface intergral $\int_{\partial G}\mathbf{J}^{\mathrm{p}}(\mathbf{r}')||\mathbf{r}-\mathbf{r}'||^{-1}\cdot d\mathbf{S}' = 0$ with $\mathbf{J}^{\mathrm{p}}$ vanishing on the boundary ∂G of G. The advantage of this representation of $V(\mathbf{r})$ over equation (3.23) is that instead of $\nabla\cdot\mathbf{J}^{\mathrm{p}}$, the source current $\mathbf{J}^{\mathrm{p}}$ itself appears in equation (3.25).

Next we transform the Ampère–Laplace law (3.15) to a form similar to equation (3.25). Due to the identity $\nabla'\times(\mathbf{J}(\mathbf{r}')||\mathbf{r}-\mathbf{r}'||^{-1}) = ||\mathbf{r}-\mathbf{r}'||^{-1}\nabla'\times\mathbf{J}(\mathbf{r}') + \nabla'(||\mathbf{r}-\mathbf{r}'||^{-1})\times\mathbf{J}(\mathbf{r}')$ and the Stokes theorem, equation (3.15) implies that

$$\mathbf{B}(\mathbf{r}) = \frac{\mu_0}{4\pi}\int_G \frac{\nabla'\times\mathbf{J}(\mathbf{r}')}{||\mathbf{r}-\mathbf{r}'||}dV'. \tag{3.26}$$

Now, $\mathbf{J} = \mathbf{J}^{\mathrm{p}} - \sigma\nabla V$ by equation (3.13), and so $\nabla\times\mathbf{J} = \nabla\times\mathbf{J}^{\mathrm{p}} - \sigma\nabla\times\nabla V$. Because $\nabla\times\nabla = 0$, we obtain

$$\mathbf{B}(\mathbf{r}) = \frac{\mu_0}{4\pi}\int_G \frac{\nabla'\times\mathbf{J}^{\mathrm{p}}(\mathbf{r}')}{||\mathbf{r}-\mathbf{r}'||}dV', \tag{3.27}$$

and performing the previous transforms in a reverse order, we get

$$\mathbf{B}(\mathbf{r}) = \frac{\mu_0}{4\pi}\int_G \mathbf{J}^{\mathrm{p}}(\mathbf{r}')\times\frac{\mathbf{r}-\mathbf{r}'}{||\mathbf{r}-\mathbf{r}'||^3}dV'. \tag{3.28}$$

This equation, among other things, shows an important result. In (infinite) homogeneous space, the total current $\mathbf{J}$ in the Ampère–Laplace equation (3.15) can be replaced by the primary current $\mathbf{J}^{\mathrm{p}}$, or, in other words, the volume current $\sigma\mathbf{E} = -\sigma\nabla V$ is not explicitly needed to calculate $\mathbf{B}$ in this case.

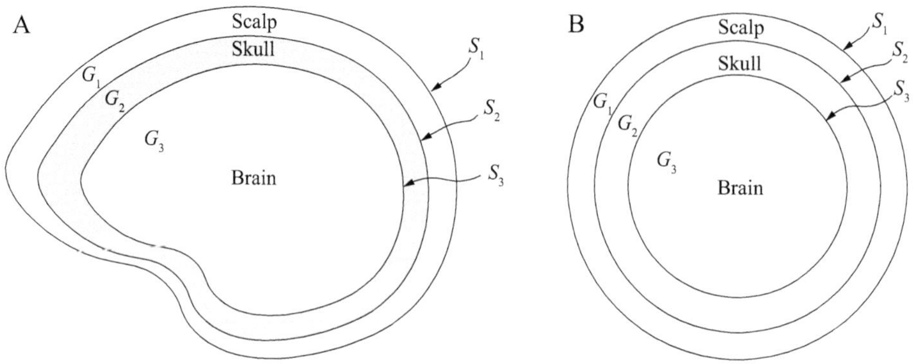

Figure 3.3
Schematic illustrations of (A) nonspherical and (B) spherical three-layer models of the head, including scalp, skull, and the intracranial cavity that contains the brain.

Next, we consider the potential $V(\mathbf{r})$ and the magnetic field $\mathbf{B}(\mathbf{r})$ due to *a current dipole* $\mathbf{J}^{\mathrm{p}}_Q(\mathbf{r}) = \delta(\mathbf{r} - \mathbf{r}_Q)\mathbf{Q}$ at point $\mathbf{r}_Q$ with moment $\mathbf{Q}$. In fact, equations (3.25) and (3.28) immediately yield that those fields in an infinite homogeneous space are

$$V(\mathbf{r}) = \frac{1}{4\pi\sigma}\mathbf{Q}\cdot\frac{\mathbf{r}-\mathbf{r}_Q}{||\mathbf{r}-\mathbf{r}_Q||^3}, \tag{3.29}$$

$$\mathbf{B}(\mathbf{r}) = \frac{\mu_0}{4\pi}\mathbf{Q}\times\frac{\mathbf{r}-\mathbf{r}_Q}{||\mathbf{r}-\mathbf{r}_Q||^3}. \tag{3.30}$$

If the source current is distributed on a line or on a surface, equations (3.25) and (3.28) remain valid if we replace the volume density $\mathbf{J}^{\mathrm{p}}$ by a line or surface density and the volume integral by a line or a surface integral, respectively.

3.4 Fields in an Inhomogeneous Medium

Here, and this is already useful for practical purposes, we assume that the medium comprises a finite conductor G, lying in vacuum, with a *piecewise constant conductivity* σ and with constant $\epsilon = \epsilon_0$ and $\mu = \mu_0$ (see figure 3.3). Furthermore, we assume that G is divided into nested subregions G_i, $i = 1, \ldots, n$, with constant conductivity $\sigma = \sigma_i$ in each G_i. Let the outer surface of G_i be S_i, $i = 1, \ldots, n$. Let S_1 be the boundary of G and S_n be the boundary of the innermost region G_n. Let the source current $\mathbf{J}^{\mathrm{p}}$ lie in G_n. We note that all our later results can easily be transferred to a more general case where all subregions are not necessarily nested and the source current may lie in any subregion.

Next we derive a useful representation for the magnetic field $\mathbf{B}$ and the potential V due to $\mathbf{J}^{\mathrm{p}}$ in terms of $\mathbf{J}^{\mathrm{p}}$ and the values of V on the surfaces S_i, $i = 1, \ldots, n$. We start with the magnetic field. From equations (3.13) and (3.15), we get

$$\mathbf{B}(\mathbf{r}) = \frac{\mu_0}{4\pi} \int_G \left(\mathbf{J}^{\mathrm{p}}(\mathbf{r}') - \sigma(\mathbf{r}')\nabla' V(\mathbf{r}')\right) \times \frac{\mathbf{r}-\mathbf{r}'}{||\mathbf{r}-\mathbf{r}'||^3} dV'$$
$$= \mathbf{B}_0(\mathbf{r}) - \frac{\mu_0}{4\pi}\sum_{i=1}^{n} \sigma_i \int_{G_i} \nabla' V(\mathbf{r}') \times \frac{\mathbf{r}-\mathbf{r}'}{||\mathbf{r}-\mathbf{r}'||^3} dV', \tag{3.31}$$

where

$$\mathbf{B}_0(\mathbf{r}) = \frac{\mu_0}{4\pi} \int_G \mathbf{J}^{\mathrm{p}}(\mathbf{r}') \times \frac{\mathbf{r}-\mathbf{r}'}{||\mathbf{r}-\mathbf{r}'||^3} dV' \tag{3.32}$$

is the magnetic field due to $\mathbf{J}^{\mathrm{p}}(\mathbf{r}')$ in homogeneous space. Using the identity $\nabla \times (V\nabla g) = \nabla V \times \nabla g$ with $g = ||\mathbf{r}-\mathbf{r}'||^{-1}$, $\nabla' g = ||\mathbf{r}-\mathbf{r}'||^{-3}(\mathbf{r}-\mathbf{r}')$, and Stokes's theorem, we get

$$\int_{G_i} \nabla' V(\mathbf{r}') \times \frac{\mathbf{r}-\mathbf{r}'}{||\mathbf{r}-\mathbf{r}'||^3} dV' = \int_{S_i} V(\mathbf{r}')\,\mathbf{n}(\mathbf{r}') \times \frac{\mathbf{r}-\mathbf{r}'}{||\mathbf{r}-\mathbf{r}'||^3} dS', \tag{3.33}$$

where $\mathbf{n}(\mathbf{r}')$ is the outer unit normal of surface S_i. This equation, with equation (3.31), yields the desired representation for the magnetic field [99],

$$\mathbf{B}(\mathbf{r}) = \mathbf{B}_0(\mathbf{r}) - \frac{\mu_0}{4\pi}\sum_{i=1}^{n}(\sigma_i' - \sigma_i'') \int_{S_i} V(\mathbf{r}')\,\mathbf{n}(\mathbf{r}') \times \frac{\mathbf{r}-\mathbf{r}'}{||\mathbf{r}-\mathbf{r}'||^3} dS' \tag{3.34}$$

for all $\mathbf{r}$ not on any surface S_i. Here, σ_i' and σ_i'' are the conductivities on the inner and outer sides of S_i, respectively. Comparing this equation with the magnetic field (3.30) of a current dipole, we see that the contribution of the volume current $-\sigma\nabla V$ to $\mathbf{B}$ is equal to the magnetic field that would arise from the source-current surface distributions $-(\sigma_i' - \sigma_i'')V(\mathbf{r}')\mathbf{n}(\mathbf{r}')$, $\mathbf{r}' \in S_i$, $i = 1, \ldots, n$, in an infinite homogeneous space. These (fictitious) current sources are called *equivalent dipolar surface distributions* or sometimes *secondary currents* and the potentials due to them *double-layer potentials*.

Using Green's theorem and boundary conditions (3.19), one can obtain a representation for V [100] similar to equation (3.34),

$$\sigma(\mathbf{r})V(\mathbf{r}) = \sigma_n V_0(\mathbf{r}) - \sum_{i=1}^{n} \frac{\sigma_i' - \sigma_i''}{4\pi} \int_{S_i} V(\mathbf{r}')\,\mathbf{n}(\mathbf{r}') \cdot \frac{\mathbf{r}-\mathbf{r}'}{||\mathbf{r}-\mathbf{r}'||^3} dS', \tag{3.35}$$

for $\mathbf{r}$ in G but not on any surface S_i, V_0 is the electric potential due to $\mathbf{J}^{\mathrm{p}}(\mathbf{r}')$ in homogeneous space.

Equation (3.35) can now be used to derive a surface integral equation for V, which then can be employed for numerically computing V. To that end, we let $\mathbf{r}$ tend to a point

$\mathbf{s}$ on S_i from the inside of G_i. Let us denote the integral over S_i in equation (3.35) by $F_i(\mathbf{r})$. It is known (e.g., see [179]) that the limit of $F_i(\mathbf{r})$ as $\mathbf{r}$ tends to $\mathbf{s}$ from the inside of G_i is

$$\lim_{\mathbf{r}\to\mathbf{s}} F_i(\mathbf{r}) = -2\pi V(\mathbf{s}) + F_i(\mathbf{s}) . \tag{3.36}$$

With this result and equation (3.35), we get the desired *surface-integral equation* for V, stating that for each $\mathbf{r}$ on S_k, $k=1,\ldots,n$, we have

$$\frac{\sigma_k' + \sigma_k''}{2} V(\mathbf{r}) = \sigma_n V_0(\mathbf{r}) - \sum_{i=1}^{n} \frac{\sigma_i' - \sigma_i''}{4\pi} \int_{S_i} V(\mathbf{r}')\,\mathbf{n}(\mathbf{r}') \cdot \frac{\mathbf{r}-\mathbf{r}'}{||\mathbf{r}-\mathbf{r}'||^3} dS' . \tag{3.37}$$

With this equation and equation (3.34), we can compute V and $\mathbf{B}$ for given G and $\mathbf{J}^{\mathrm{p}}$. We first numerically solve equation (3.37) for V on surfaces S_k using BEM; see, for example, Stenroos and Sarvas [295]. Thereafter, we get V in G from equation (3.35) and $\mathbf{B}$ from equation (3.34) (first obtain $\mathbf{B}$ outside surfaces S_i; then the $\mathbf{B}$ values on S_i can be taken as limits, because $\mathbf{B}$ is continuous across the interfaces).

3.5 Computing the Lead Field

By equation (2.15), the lead field $\mathbf{L}_i(\mathbf{r}')$ of sensor i assigned to the source location $\mathbf{r}'$ is

$$\mathbf{L}_i(\mathbf{r}') = [L_{i1}(\mathbf{r}'),\, L_{i2}(\mathbf{r}'),\, L_{i3}(\mathbf{r}')]^{\mathrm{T}} , \tag{3.38}$$

where $L_{ij}(\mathbf{r}')$ is the signal in sensor i due to unit source dipole $\mathbf{J}^{\mathrm{p}}(\mathbf{r}) = \delta(\mathbf{r}-\mathbf{r}')\,\mathbf{e}_j$ at location $\mathbf{r}'$, and $\mathbf{e}_j$, $j=1,2,3$, are the *xyz*-coordinate unit vectors.

For numerical computing of the lead fields for $\mathbf{B}$ and V, we need a forward field solver (e.g., the finite-element method (FEM [123]) or the boundary-element method [295]), for the potential $V(\mathbf{r})$ in the given head model. Usually, the field solver is built for the potential $V(\mathbf{r})$, satisfying $V(\mathbf{r}) \to 0$ as $||\mathbf{r}|| \to \infty$.

Because the potential is determined only up to an additive constant, in EEG, one usually fixes a location $\mathbf{r}_{\mathrm{ref}}$ on the scalp, the *reference electrode* location, and the sensor signal y_i of the electrode at location $\mathbf{r}_i$ is the potential difference

$$y_i = V(\mathbf{r}_i) - V(\mathbf{r}_{\mathrm{ref}}) . \tag{3.39}$$

Another way to fix the potential (i.e., its zero level) is to demand that $\sum_{i=1}^{N_{\mathrm{ch}}} V(\mathbf{r}_i) = 0$, which is achieved by subtracting the mean $N_{\mathrm{ch}}^{-1} \sum_{i=1}^{N_{\mathrm{ch}}} V(\mathbf{r}_i)$ from each $V(\mathbf{r}_i)$. This is called the average-reference approach [31]. In computing lead-field components $L_{ij}(\mathbf{r}')$, the chosen way of fixing the potential must be taken into account. It should be mentioned that there is a lot of confusion in the literature about the selection of the reference electrode or reference potential; people talk about the *reference electrode problem* [82, 109, 325], although there is no real problem if one understands that the

EEG signal is always a difference between two potentials (i.e., line integral of the electric field; see equation (1.4)) or a linear combination of such differences. When this is taken into account, no problem should remain; however, this issue continues to be debated (see, e.g., Hu et al. [131] and Ilmoniemi [140]).

In general, a magnetic sensor loop measures a surface integral of the magnetic flux density passing through the loop, and so the signal in the magnetic field sensor i is

$$y_i = \int_{S_i} \mathbf{B}(\mathbf{r}) \cdot \mathbf{n}(\mathbf{r})\, \mathbf{dS}\,, \tag{3.40}$$

where S_i is the surface circumvented by loop i and $\mathbf{n}(\mathbf{r})$ is the unit normal of the surface. The surface integral can be computed numerically, and $\mathbf{B}(\mathbf{r})$ in the integral is either given by the Ampère–Laplace law (3.15) with $\mathbf{J} = \mathbf{J}^{\mathrm{p}} - \sigma(\mathbf{r})\nabla V$ if a three-dimensional FEM solver is used, or $\mathbf{B}(\mathbf{r})$ is given by equation (3.34) if a BEM solver is used. Note that the magnetic signal (3.40) is often divided by the area of the pickup loop, yielding the magnetic flux density (typically expressed in femtotesla) or, in the case of a tangential first-order gradiometer, by the area of one loop and the baseline, yielding the magnetic field gradient (expressed in fT/cm).

3.6 Spherical Model

For many purposes, a sufficient approximation for the conductivity geometry of the human head is the *spherically symmetric conductor* model $\sigma(\mathbf{r}) = \sigma(r) > 0$ for $r < r_0$, $\sigma(\mathbf{r}) = 0$ for $r > r_0$, where r_0 is the head radius, $r = ||\mathbf{r}||$, and the center of the conductor is at the origin.

The magnetic field in the spherical model has the following three special properties. Let $\mathbf{B}(\mathbf{r})$ be the magnetic field due to a primary current $\mathbf{J}^{\mathrm{p}}$ in the spherical conductor G. Then it holds:

(a) For any $\mathbf{J}^{\mathrm{p}}$, the *radial component* $B_r(\mathbf{r}) = \mathbf{B}(\mathbf{r}) \cdot \mathbf{e}_r$ with $\mathbf{e}_r = \mathbf{r}/||\mathbf{r}||$ coincides with that of the magnetic field $\mathbf{B}_0(\mathbf{r})$ due to $\mathbf{J}^{\mathrm{p}}$ in an infinite homogeneous medium, that is, by Ampère–Laplace law (3.15),

$$B_r(\mathbf{r}) = \mathbf{B}_0(\mathbf{r}) \cdot \mathbf{e}_r = \frac{\mu_0}{4\pi} \mathbf{e}_r \cdot \int_G \mathbf{J}^{\mathrm{p}}(\mathbf{r}') \times \frac{\mathbf{r} - \mathbf{r}'}{||\mathbf{r} - \mathbf{r}'||^3}\, dV'\,. \tag{3.41}$$

(b) If $\mathbf{J}^{\mathrm{p}}$ is a *radial current dipole* or a *radial current segment*, then $\mathbf{B}(\mathbf{r}) = 0$ outside G.

(c) If $\mathbf{J}^{\mathrm{p}}$ is a *current dipole*, then $\mathbf{B}(\mathbf{r})$ outside G can be presented in a *closed form* in terms of $\mathbf{J}^{\mathrm{p}}$ alone (see the next two sections); thus, $\mathbf{B}(\mathbf{r})$ outside G does not depend on the conductivity profile $\sigma(r)$.

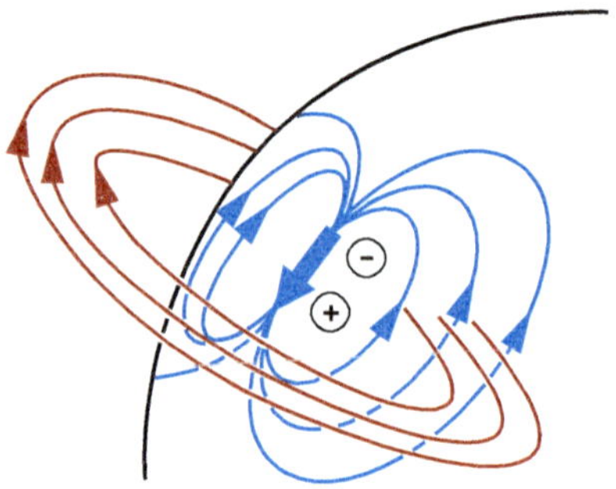

Figure 3.4
Magnetic and electric fields produced by a current dipole. The large arrow depicts a mechanism that transports current from the sink (minus sign indicating that the sink is negatively charged) to the source (plus sign indicating positive charge). The charges at the ends of the current element constitute a charge dipole, which gives rise to the electric field (E-field, blue). The E-field within the conductor gives rise to volume-current flow. All currents together produce the magnetic field (red lines and arrows) that extends outside the conducting volume. The electric and magnetic fields are proportional to the current-dipole moment.

To verify claim (a), we first consider a spherical, piecewise homogeneous conductor G. By equation (3.34), we have

$$\mathbf{B}_r(\mathbf{r}) = \mathbf{B}_0(\mathbf{r}) \cdot \mathbf{e}_r - \frac{\mu_0}{4\pi} \sum_{i=1}^{n} (\sigma_i' - \sigma_i'') \int_{S_i} V(\mathbf{r}')\, \mathbf{n}(\mathbf{r}') \times \frac{\mathbf{r} - \mathbf{r}'}{||\mathbf{r} - \mathbf{r}'||^3} \cdot \mathbf{e}_r \, dS \,, \tag{3.42}$$

where the interfaces S_i are concentric spheres. In the above integrals, the triple product vanishes because $\mathbf{n}(\mathbf{r}') \times (\mathbf{r} - \mathbf{r}') \cdot \mathbf{r}/||\mathbf{r}|| = \mathbf{n}(\mathbf{r}') \times \mathbf{r} \cdot \mathbf{r}/||\mathbf{r}|| = 0$, and equation (3.41) follows. Because a general spherical conducter is a limiting case of piecewise homogeneous ones with increasing number of homogeneous spherical layers, equation (3.41) also holds in the general case.

The claim (b) follows from the closed-form formula for $\mathbf{B}(\mathbf{r})$ outside G; see the next section.

The claim (c) is an important property of the spherical model. The shapes of field $\mathbf{B}$ are displayed graphically in figure 3.4. The closed-form presentation in the spherical coordinates was first derived in 1973 by Grynszpan and Geselowitz [107] by making use of spherical-harmonics series expansions. The first simple derivation for this result was derived in 1985 by Ilmoniemi [142]. In the same paper, he also presented for the first time a closed-form formula in rectangular coordinates. A simpler derivation, yielding a different closed-form formula in rectangular coordinates, was presented by Sarvas [269] in 1987. In the next two sections, we give the derivation and formula of Sarvas in rectangular coordinates and those of Ilmoniemi [142] in spherical coordinates.

In the spherical model, the electric field, and so its potential, is strongly influenced by the conductivity values of the different layers: the scalp, skull, cerebrospinal fluid, and the brain. Classical results are that the electric potential V due to a dipole in a sphere can be presented in closed form for a *homogeneous sphere*, and in an analytic form as a spherical-harmonics expansion for a *layered sphere.* Later we derive and present these formulae.

3.7 Magnetic Field for the Spherical Model in Cartesian Coordinates

We want to find the magnetic field $\mathbf{B}(\mathbf{r})$ outside a spherically symmetric conductor G, due to a source dipole with moment $\mathbf{Q}$ at location $\mathbf{r}_Q$ inside G. Because $\nabla\times\mathbf{B}(\mathbf{r})=\mu_0\mathbf{J}(\mathbf{r})=0$ outside G, $\mathbf{B}$ arises from a magnetic scalar potential ϕ_{m} as $\mathbf{B}(\mathbf{r})=-\nabla\phi_{\mathrm{m}}(\mathbf{r})$ for $\mathbf{r}$ outside G. Due to the spherical structure of the conductor, the radial component $B_r(\mathbf{r})=\mathbf{B}(\mathbf{r})\cdot\mathbf{e}_r$ is equal to that of the magnetic field of the dipole in vacuum, as shown in the previous subsection (claim (a)), and so by equation (3.30),

$$B_r(\mathbf{r})=\frac{\mu_0}{4\pi}\mathbf{Q}\times\frac{\mathbf{r}-\mathbf{r}_Q}{||\mathbf{r}-\mathbf{r}_Q||^3}\cdot\mathbf{e}_r\,. \tag{3.43}$$

To find an expression for ϕ_{m}, we fix $\mathbf{r}$ outside G and form the line integral of $\nabla\phi_{\mathrm{m}}$ along the line $\mathbf{r}+t\,\mathbf{e}_r$, $0\le t\le\infty$, from $\mathbf{r}$ to infinity. Because ϕ_{m} vanishes at infinity, we get

$$\phi_{\mathrm{m}}(\mathbf{r})=-\int_0^\infty\nabla\phi_{\mathrm{m}}(\mathbf{r}+t\,\mathbf{e}_r)\cdot\mathbf{e}_r\,dt=\int_0^\infty B_r(\mathbf{r}+t\,\mathbf{e}_r)\,dt \tag{3.44}$$

$$=\frac{\mu_0}{4\pi}\int_0^\infty\frac{\mathbf{Q}\times(\mathbf{r}+t\,\mathbf{e}_r-\mathbf{r}_Q)}{||\mathbf{r}+t\,\mathbf{e}_r-\mathbf{r}_Q||^3}\cdot\mathbf{e}_r\,dt \tag{3.45}$$

$$=-\frac{\mu_0}{4\pi}\mathbf{Q}\times\mathbf{r}_Q\cdot\mathbf{e}_r\int_0^\infty\frac{dt}{||\mathbf{r}+t\,\mathbf{e}_r-\mathbf{r}_Q||^3}\,, \tag{3.46}$$

because $\mathbf{Q}\times(\mathbf{r}+t\,\mathbf{e}_r)\cdot\mathbf{e}_r=0$. The last integral above can be computed in closed form, and we obtain

$$\phi_{\mathrm{m}}(\mathbf{r})=-\frac{\mu_0}{4\pi}\frac{\mathbf{Q}\times\mathbf{r}_Q\cdot\mathbf{r}}{F}\,, \tag{3.47}$$

where

$$F=R\,(rR+r^2-\mathbf{r}\cdot\mathbf{r}_Q) \tag{3.48}$$

with $\mathbf{R}=\mathbf{r}-\mathbf{r}_Q$, $R=||\mathbf{R}||$, and $r=||\mathbf{r}||$. Note that $\phi_{\mathrm{m}}(\mathbf{r})$ in equation (3.47), and so $\mathbf{B}$ outside G, does not depend on $\sigma=\sigma(r)$. By forming the gradient of $\phi_{\mathrm{m}}(\mathbf{r})$, we obtain the desired *closed-form presentation* for $\mathbf{B}$ outside G, as

$$\mathbf{B}(\mathbf{r})=\frac{\mu_0}{4\pi F^2}\left(F\,\mathbf{Q}\times\mathbf{r}_Q-\mathbf{Q}\times\mathbf{r}_Q\cdot\mathbf{r}\,\nabla F\right), \tag{3.49}$$

where

$$\nabla F=(r^{-1}R^2+R^{-1}\mathbf{R}\cdot\mathbf{r}+2R+2r)\,\mathbf{r}-(R+2r+R^{-1}\mathbf{R}\cdot\mathbf{r})\,\mathbf{r}_Q. \tag{3.50}$$

Notice that above we derived the tangential components of $\mathbf{B}$ from the radial component B_r, showing that all components can be calculated directly from the dipole parameters. Although the volume currents significantly contribute to the tangential components outside G, the latter can be derived from the radial component, which can be computed purely from the primary current. This seems paradoxical at first, but there is nothing more to it than the fact that the different field components are not independent; they are tied together by Maxwell's equations.

3.8 Magnetic Field for the Spherical Model in Spherical Coordinates

We again consider a spherically symmetric conductor G. Since radial primary currents yield no magnetic field outside G, it suffices to consider a *tangential* source-current dipole. Let $\mathbf{e}_x$, $\mathbf{e}_y$, and $\mathbf{e}_z$ be the coordinate unit vectors. Without losing generality, we may assume that the dipole is oriented along the y-axis with moment $\mathbf{Q}=Q\,\mathbf{e}_y$, and located in G on the z-axis at $\mathbf{r}_Q=a\mathbf{e}_z$, $a>0$. The magnetic field is produced both by the primary current dipole and by the volume current. Again, due to the spherical structure of G, the radial component of the magnetic field outside G is equal to that of the magnetic field of the dipole in vacuum. Thus, the radial component of the magnetic field outside G can be obtained from the dipole expression:

$$B_r(\mathbf{r})=\mathbf{B}(\mathbf{r})\cdot\mathbf{e}_r=\frac{\mu_0}{4\pi}\frac{\mathbf{Q}\times(\mathbf{r}-a\mathbf{e}_z)}{||\mathbf{r}-a\mathbf{e}_z||^3}\cdot\mathbf{e}_r=-AF^{-3}\sin\theta\,\cos\varphi \quad \text{with} \tag{3.51}$$

$$A=\frac{\mu_0 Q}{4\pi a^2},\;\; F=(1-2\rho\cos\theta+\rho^2)^{1/2}\;\text{and}\;\rho=r/a, \tag{3.52}$$

where $\mathbf{r}=r(\sin\theta\cos\varphi\,\mathbf{e}_x+\sin\theta\sin\varphi\,\mathbf{e}_y+\cos\theta\,\mathbf{e}_z)$ is in the spherical coordinates (r,θ,φ).

In the quasi-static approximation, $\nabla\times\mathbf{B}=0$ outside the head. As we write the curl $\nabla\times$ in spherical coordinates, this implies that

$$r\sin\theta\,\nabla\times\mathbf{B}\cdot\mathbf{e}_\theta=\partial B_r/\partial\varphi-\partial(r\sin\theta\,B_\varphi)/\partial r=0\,,\text{and} \tag{3.53}$$

$$r\,\nabla\times\mathbf{B}\cdot\mathbf{e}_\varphi=\partial(rB_\theta)/\partial r-\partial B_r/\partial\theta=0\,. \tag{3.54}$$

We can solve for the tangential field components by integrating $\partial B_\varphi/\partial r$ and $\partial(rB_\theta)/\partial r$ from infinity to r and get

$$\sin\theta \int_\infty^r \partial(r'B_\varphi)/\partial r'\, dr' = r\sin\theta\, B_\varphi = \int_\infty^r \partial B_r/\partial\varphi\, dr', \text{ and} \tag{3.55}$$

$$\int_\infty^r \partial(r'B_\theta)/\partial r\, dr' = rB_\theta = \int_\infty^r \partial B_r/\partial\theta\, dr'. \tag{3.56}$$

Inserting B_r from equation (3.51), changing the order of integration and partial differentiation, computing the resulting integrals in closed forms, and taking the partial derivatives, we obtain with equation (3.51) the wanted *closed-form presentations* $\mathbf{B}(r,\theta,\varphi)$ outside G,

$$B_r(r,\theta,\varphi) = -AF^{-3}\sin\theta\,\cos\varphi\,, \tag{3.57}$$

$$B_\theta(r,\theta,\varphi) = \frac{-A\cos\varphi}{\rho F^3\sin^2\theta}\Big[1+\rho\cos\theta\left(\rho\cos\theta - 1 - \cos^2\theta - F^2\right) + F^3\cos\theta\Big], \tag{3.58}$$

$$B_\varphi(r,\theta,\varphi) = \frac{A\sin\varphi}{\rho F\sin^2\theta}(\rho - \cos\theta - F)\,, \tag{3.59}$$

where A and F are as in equation (3.52).

3.9 Triangle Construction and the Vector Potential of the Magnetic Field in the Spherical Model

Ilmoniemi et al. [142] introduced the *triangle construction* in the spherical model; with it, one can present the magnetic field outside the spherical conductor due to a current line in the conductor as the magnetic field of a current loop in vacuum. The current-loop construction is depicted in figure 3.5, and is described as follows. Consider a line, straight or curved, of primary current $\mathbf{J}^{\mathrm{p}}$. It includes, in addition to the current, a sink and a source at its ends that, due to the divergence of $\mathbf{J}^{\mathrm{p}}$, produce the volume current in the conducting medium. In the triangle construction, two *radial* primary-current segments are added to the original current line to join its ends to the center of the sphere so that the three lines form a current loop with sinks and sources coinciding in such a manner that they "annihilate" each other. Since in a spherical structure, radial primary currents (together with the associated volume currents) produce no magnetic field outside the conductor, the resulting loop current $\mathbf{J}^{\mathrm{p}}_{\mathrm{loop}}$ with the current strength I

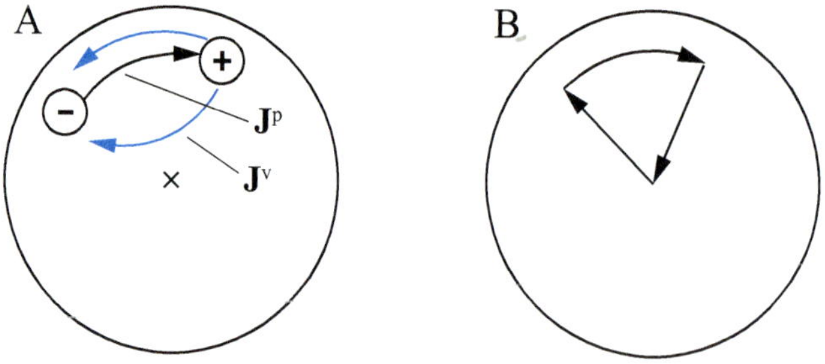

Figure 3.5
Triangle construction in the spherical model. The primary current $\mathbf{J}^{\mathrm{p}}$ (curved arrow from the sink to the source in part A of the figure) and the volume current $\mathbf{J}^{\mathrm{v}}$ together produce an external magnetic field pattern that is exactly identical to that of the current loop in part B of the figure.

produces exactly the same magnetic field outside the conductor as the original primary current line $\mathbf{J}^{\mathrm{p}}$ in the spherical model.

Since a closed current loop is sourceless as a current density, its (generalized) divergence vanishes, that is, $\nabla \cdot \mathbf{J}^{\mathrm{p}}_{\mathrm{loop}} = 0$. It follows that $\mathbf{J}^{\mathrm{p}}_{\mathrm{loop}}$ as a primary current does not produce any (non-constant) potential V, because in the Poisson equation (3.18) for V the source term $\nabla \cdot \mathbf{J}^{\mathrm{p}}_{\mathrm{loop}} = 0$. Therefore, the total current $\mathbf{J}^{\mathrm{p}} = \mathbf{J}^{\mathrm{p}}_{\mathrm{loop}} - \sigma \nabla V = \mathbf{J}^{\mathrm{p}}_{\mathrm{loop}}$, and so by the Ampère–Laplace law (3.15), the magnetic field due to $\mathbf{J}^{\mathrm{p}}_{\mathrm{loop}}$ is

$$\mathbf{B}(\mathbf{r}) = \frac{\mu_0}{4\pi} \int_C \mathbf{J}^{\mathrm{p}}_{\mathrm{loop}}(\mathbf{r}') \times \frac{\mathbf{r} - \mathbf{r}'}{||\mathbf{r} - \mathbf{r}'||^3} dV' \tag{3.60}$$

$$= \frac{\mu_0}{4\pi} I \int_C \mathbf{t}(\mathbf{r}') \times \frac{\mathbf{r} - \mathbf{r}'}{||\mathbf{r} - \mathbf{r}'||^3} dl', \tag{3.61}$$

where I is the current strength in the closed loop C and $\mathbf{t}(\mathbf{r}')$ is the unit tangent vector of C in the direction of the current. So the magnetic field (3.60) outside the conductor is the same as the original magnetic field $\mathbf{B}(\mathbf{r})$ and can be computed by equation (3.61).

This triangle construction has been used to build *dry phantoms* (simplified models that mimic the real head with internal sources) in which triangular loops of wire generate magnetic field patterns that are identical to those generated in wet phantoms where current source elements are immersed in a spherical volume of conducting saline [146, 224, 237–239]. The advantage of dry phantoms over wet ones is that they are easier to make and maintain and that their geometrical form can be very accurately defined.

An important application of the triangle construction in the spherical model is the derivation of a *closed-form presentation* for the *vector potential* $\mathbf{A}(\mathbf{r})$ of the magnetic field $\mathbf{B}(\mathbf{r}) = \nabla \times \mathbf{A}(\mathbf{r})$ *outside* the spherically symmetric conductor, due to a current dipole in the conductor.

The closed-form formula for $\mathbf{A}(\mathbf{r})$ is as follows. Consider a spherically symmetric conductor G centered at the origin. Let the primary current dipole be at location $\mathbf{r}_Q$ in G with moment $\mathbf{Q}$. Then, at a field point $\mathbf{r}$ outside G, the *vector potential* $\mathbf{A}(\mathbf{r})$ of the magnetic field due to that primary-current dipole is

$$\mathbf{A}(\mathbf{r}) = \frac{\mu_0}{4\pi}\left[\left(\frac{1}{||\mathbf{r}-\mathbf{r}_Q||} - \frac{1}{||\mathbf{r}_Q||}\log(\frac{\alpha}{\beta})\right)\boldsymbol{Q}_{\mathrm{t}} + \frac{\mathbf{r}\cdot\boldsymbol{Q}_{\mathrm{t}}}{||\mathbf{r}_Q||}\left(\frac{||\mathbf{r}-\mathbf{r}_Q|| + ||\mathbf{r}_Q||}{\alpha\,||\mathbf{r}-\mathbf{r}_Q||} - \frac{1}{\beta}\right)\mathbf{r}_Q\right], \tag{3.62}$$

where

$$\boldsymbol{Q}_{\mathrm{t}} = \mathbf{Q} - \frac{\mathbf{Q}\cdot\mathbf{r}_Q}{||\mathbf{r}_Q||^2}\mathbf{r}_Q\,, \tag{3.63}$$

$$\alpha = ||\mathbf{r}_Q||\,||\mathbf{r}-\mathbf{r}_Q|| - \mathbf{r}_Q\cdot(\mathbf{r}-\mathbf{r}_Q)\,, \tag{3.64}$$

$$\beta = ||\mathbf{r}||\,||\mathbf{r}_Q|| - \mathbf{r}\cdot\mathbf{r}_Q\,. \tag{3.65}$$

For the proof, see Appendix 3.12.

One application of equation (3.62) is to compute the magnetic flux of $\mathbf{B}$, due to a current dipole in a spherical conductor, through a coil C outside the conductor. The flux Φ is given by

$$\Phi = \int_S \mathbf{B}(\mathbf{r}')\cdot d\mathbf{S}'\,, \tag{3.66}$$

where S is the surface spanned by the coil (loop) C. We can write this surface integral as a line integral by equation $\mathbf{B}(\mathbf{r}') = \nabla \times \mathbf{A}(\mathbf{r}')$, where $\mathbf{A}(\mathbf{r}')$ is a vector potential of $\mathbf{B}(\mathbf{r}')$, and by Stokes's theorem, and we get

$$\Phi = \oint_C \mathbf{A}(\mathbf{r}')\cdot\mathbf{t}(\mathbf{r}')\,dl\,, \tag{3.67}$$

where $\mathbf{t}(\mathbf{r}')$ is the unit tangent vector of loop C, right-hand-oriented with respect to the normal direction $d\mathbf{S}'$ at $\mathbf{r}'$. Here, $\mathbf{A}(\mathbf{r}')$ is given by equation (3.62).

3.10 Electric Potential V in a Layered Sphere and in a Homogeneous Sphere

We first consider a *layered sphere* G lying in vacuum with radii $0<r_1<r_2<r_3$ and comprising the inner sphere $0<||\mathbf{r}||<r_1$, with conductivity $\sigma_1 \geq 0$, and two spherical layers $r_1<||\mathbf{r}||<r_2$ and $r_2<||\mathbf{r}||<r_3$ with conductivities σ_2 and σ_3. This spherical conductor is a head model with the three compartments corresponding to brain, skull, and scalp and, for instance, with the radii $r_1=8.1$ cm, $r_2=8.5$ cm, and $r_3=8.8$ cm and conductivities $\sigma_1=0.33$, $\sigma_2=0.0042$, and $\sigma_3=0.33$ $1/(\Omega$ m). In Appendix 3.13, a layered sphere with more layers is presented.

In the layered sphere, the electric potential due to a current dipole with moment $\mathbf{Q}$ and location $\mathbf{r}_0$ in the inner sphere, that is, $||\mathbf{r}_0||<r_1$, can be presented as a series expansion. We below give $V(\mathbf{r})$ *on the surface of the sphere*, that is, $||\mathbf{r}||=r_3$, in the case of the three-compartment model,

$$V(\mathbf{r})=\frac{1}{4\pi\sigma_1}\sum_{n=1}^{\infty}\gamma_n r_0^{n-1}\Big(\mathbf{Q}\cdot\widehat{\mathbf{r}_0}\,n\,P_n(\widehat{\mathbf{r}}\cdot\widehat{\mathbf{r}_0})+(\mathbf{Q}\times\widehat{\mathbf{r}_0})\cdot(\widehat{\mathbf{r}}\times\widehat{\mathbf{r}_0})\,P_n'(\widehat{\mathbf{r}_0}\cdot\widehat{\mathbf{r}})\Big), \tag{3.68}$$

where $r=||\mathbf{r}||$, $r_0=||\mathbf{r}_0||$, $\widehat{\mathbf{r}}=\mathbf{r}/r$, and $\widehat{\mathbf{r}_0}=\mathbf{r}_0/r_0$; $P_n(t)$ is the Legendre polynomial of order n and $P_n'(t)=\partial P_n(t)/\partial t$ is its derivative. The coefficients γ_n are

$$\gamma_n=\frac{\sigma_1}{\sigma_3}\frac{r_3^n}{\left(\alpha_n\,r_3^{2n+1}-\beta_n\,r_2^{2n+1}\right)}, \tag{3.69}$$

where

$$\alpha_n=\frac{n(n+1)}{(2n+1)^3}\left[\left(1+\frac{n}{n+1}\frac{\sigma_2}{\sigma_3}\right)\left(n+1+n\frac{\sigma_1}{\sigma_2}\right)+\left(n-n\frac{\sigma_2}{\sigma_3}\right)\left(1-\frac{\sigma_1}{\sigma_2}\right)\left(\frac{r_1}{r_2}\right)^{2n+1}\right]$$

and

$$\beta_n=\frac{n(n+1)}{(2n+1)^3}\left[\left(1-\frac{\sigma_2}{\sigma_3}\right)\left(n+1+n\frac{\sigma_1}{\sigma_2}\right)+\left(n+(n+1)\frac{\sigma_2}{\sigma_3}\right)\left(1-\frac{\sigma_1}{\sigma_2}\right)\left(\frac{r_1}{r_2}\right)^{2n+1}\right].$$

Furthermore, $P_n(t)$ and $P_n'(t)$ can easily be computed by the following algorithms:

$P_0(t)=1,\ P_1(t)=t,$ and for $n=1,2,\ldots,$

$$P_{n+1}(t)=\frac{2n+1}{n+1}\,t\,P_n(t)-\frac{n}{n+1}P_{n-1}(t), \tag{3.70}$$

$$P_{n+1}'(t)=P_{n-1}'(t)+(2n+1)P_n(t). \tag{3.71}$$

A sufficient number of terms in the series (3.68), for desired accuracy, depends on how close the dipole is to the surface of the inner sphere, that is, on $s = r_1 - r_0$. The needed number N of terms increases as the distance s decreases. The relative error (error divided by the maximal field value) less than 1 percent is reached with $N = 40$ for $s > 4$ mm for the head model parameters given above.

See Appendix 3.13 for the case with more layers, computing $V(\mathbf{r})$ inside the sphere and the derivation of the series expansion (3.68). For the potential in an *anisotropic* layered sphere, see Zhang [333].

If $\sigma_1 = \sigma_2 = \sigma_3 = \sigma > 0$, we have a *homogeneous sphere* with radius $r_3 > 0$, and $V(\mathbf{r})$ on the surface of the sphere, that is, for $r = r_3$, is given in the following *closed form* [127, 329]:

$$V(\mathbf{r}) = \frac{1}{4\pi\sigma} \mathbf{Q} \cdot \left[\frac{2}{d^3} (\mathbf{r} - \mathbf{r}_0) + \frac{1}{r\,d^2 + d\,\mathbf{r} \cdot (\mathbf{r} - \mathbf{r}_0)} \left(\frac{r + d}{r} \mathbf{r} - \mathbf{r}_0 \right) \right], \tag{3.72}$$

where $r = ||\mathbf{r}||$ and $d = ||\mathbf{r} - \mathbf{r}_0||$.

The potential $V(\mathbf{r})$ inside the sphere, that is, for $r < r_3$, can also be given in closed form but with a more involved formula; see Yao [329].

3.11 Semi-Infinite Homogeneous Conductor

If the primary current and electromagnetic sensors are close to a flat portion of a boundary of a homogeneous volume, the half-space or semi-infinite homogeneous conductor model can be used. In this model, the conductivity is constant $\sigma > 0$ for $z < 0$ and it vanishes for $z > 0$; here, $z = 0$ is the boundary.

Let the primary current $\mathbf{J}^{\mathrm{p}}$ lie in the lower half-space $z < 0$. We first consider the magnetic field $\mathbf{B}$ due to $\mathbf{J}^{\mathrm{p}}$ in this conductor model. We write that $\mathbf{B} = \mathbf{B}_0 + \mathbf{B}_{\mathrm{sec}}$, where $\mathbf{B}_0$ is the primary field due to $\mathbf{J}^{\mathrm{p}}$ in vacuum, and $\mathbf{B}_{\mathrm{sec}}$ is the secondary field, which, by equation (3.34), is

$$\mathbf{B}_{\mathrm{sec}}(\mathbf{r}) = -\frac{\mu_0}{4\pi} \sigma \int_{z=0} V(\mathbf{r}')\, \mathbf{e}_z \times \frac{(\mathbf{r} - \mathbf{r}')}{||\mathbf{r} - \mathbf{r}'||^3}\, dS', \tag{3.73}$$

where $-\sigma V(\mathbf{r}')\,\mathbf{e}_z$ is the equivalent surface current and $\mathbf{e}_z$ is the z-oriented unit vector. Because $\mathbf{e}_z \times (\mathbf{r} - \mathbf{r}')$ is perpendicular to $\mathbf{e}_z$ in the integrand, the entire integral is perpendicular to $\mathbf{e}_z$, and thus the z-component $\mathbf{B}_{\mathrm{sec},z} = 0$. Therefore, the z-component $\mathbf{B}_{\mathrm{sec},z}$ vanishes and $\mathbf{B}_z = \mathbf{B}_{0,z}$.

We want to derive a presentation of $\mathbf{B}(\mathbf{r})$ for $z > 0$. It is sufficient to do that for a current dipole because any $\mathbf{J}^{\mathrm{p}}$ can be approximated by current dipoles. So, let $\mathbf{J}^{\mathrm{p}}$

be a current dipole with moment $\mathbf{Q}$ at $\mathbf{r}_Q$ in the lower half-space. We find a closed-form presentation for $\mathbf{B}$ for $z>0$ in the same way as we found (3.49) for a spherical model. Because $\mathbf{J}^{\mathrm{p}}(\mathbf{r})=0$ for $z\geq 0$, then $\nabla\times\mathbf{B}(\mathbf{r})=0$ for $z>0$, and $\mathbf{B}(\mathbf{r})=-\nabla\phi_{\mathrm{m}}(\mathbf{r})$ for $z>0$, where $\phi_{\mathrm{m}}(\mathbf{r})$ is the magnetic scalar potential for $\mathbf{B}$ in the upper half-space. Then $\mathbf{B}_{0,z}(\mathbf{r})=\mathbf{B}_z(\mathbf{r})=-\partial\phi_{\mathrm{m}}(\mathbf{r})/\partial z$, and for any $\mathbf{r}=(x,y,z)$, $z>0$, we get

$$\phi_{\mathrm{m}}(x,y,z)=-\int_z^{\infty}\frac{\partial_{\mathrm{m}}}{\partial z}(x,y,z+t)\,dl=\int_z^{\infty}\mathbf{B}_{0,z}(x,y,z+t)\,dt.$$

With equation (3.30) for $\mathbf{B}_0$ we can integrate the above integral in a closed form, and get

$$\phi_{\mathrm{m}}=\frac{\mu_0}{4\pi}K^{-1}\mathbf{Q}\times\mathbf{R}\cdot\mathbf{e}_z,\ \text{ with } K=R\,(R+\mathbf{R}\cdot\mathbf{e}_z),\tag{3.74}$$

where $\mathbf{R}=\mathbf{r}-\mathbf{r}_Q$ and $R=||\mathbf{R}||$. We form the gradient $\nabla\phi_{\mathrm{m}}$ and obtain

$$\mathbf{B}=\frac{\mu_0}{4\pi K^2}\left(\mathbf{Q}\times\mathbf{R}\cdot\mathbf{e}_z\nabla K-K\mathbf{e}_z\times\mathbf{Q}\right)\ \text{ for } z>0,\ \text{ where}\tag{3.75}$$

$$\nabla K=\left(2+R^{-1}\mathbf{R}\cdot\mathbf{e}_z\right)\mathbf{R}+R\,\mathbf{e}_z,\tag{3.76}$$

which is the desired closed-form presentation of $\mathbf{B}$ for $z>0$.

Let next $V(\mathbf{r})$ be the potential due to $\mathbf{J}^{\mathrm{p}}$ lying in the lower half-space in the considered conductor model. We want to show that

$$V(\mathbf{r})=\frac{1}{2\pi\sigma}\int\frac{\mathbf{J}^{\mathrm{p}}(\mathbf{r}')\cdot(\mathbf{r}-\mathbf{r}')}{||\mathbf{r}-\mathbf{r}'||^3}\,dV'\tag{3.77}$$

for all $\mathbf{r}=(x,y,0)$ on the interface $z=0$.

It is again sufficient to verify equation (3.77) for a current dipole. To that end, let $\mathbf{J}^{\mathrm{p}}$ be a current dipole with moment $\mathbf{Q}=[q_x,q_y,q_z]^{\mathrm{T}}$ at (x_0,y_0,z_0) with $z_0<0$. By equations (3.18) and (3.19), the potential $V(x,y,z)$ is uniquely determined in the region $z\leq 0$ by the following conditions:

$$\nabla^2 V=\frac{1}{\sigma}\nabla\cdot\mathbf{J}^{\mathrm{p}}\quad\text{for}\quad z<0,\text{ and}\tag{3.78}$$

$$\frac{\partial V}{\partial z}=0\quad\text{for}\quad z=0.\tag{3.79}$$

We construct the desired potential $V(x,y,z)$, $z\leq 0$, with the mirror-source principle. We set a "mirror" dipole to the point $(x_0,y_0,-z_0)$ with the moment $\widetilde{\mathbf{Q}}=[q_x,q_y,-q_z]^{\mathrm{T}}$. By equation (3.29), the potentials due to the original and mirror dipoles in the homogeneous space with conductivity σ are

$$V_1(x,y,z) = \frac{1}{4\pi\sigma} \frac{q_x(x-x_0)+q_y(y-y_0)+q_z(z-z_0)}{||(x-x_0, y-y_0, z-z_0)||^3}, \tag{3.80}$$

$$V_2(x,y,z) = \frac{1}{4\pi\sigma} \frac{q_x(x-x_0)+q_y(y-y_0)-q_z(z+z_0)}{||(x-x_0, y-y_0, z+z_0)||^3} \tag{3.81}$$

for all (x,y,z) in the 3-space, respectively. Let $V(x,y,z)=V_1(x,y,z)+V_2(x,y,z)$. Then, $V_2(x,y,z)=V_1(x,y,-z)$ everywhere, and it follows, as we easily see, that $\partial V(x,y,0)/\partial z = 0$ on the interface $z=0$, and so equation (3.79) is satisfied. On the other hand, $\nabla^2 V_2(x,y,z)=0$ in the half-space $z<0$, because the source dipole of $V_2(x,y,z)$ is in the upper half-space $z>0$. Therefore, in the half-space $z<0$, we obtain,

$$\nabla^2 V(x,y,z) = \nabla^2 V_1(x,y,z) + \nabla^2 V_2(x,y,z) = \frac{1}{\sigma}\nabla\cdot\mathbf{J}^{\mathrm{p}} + 0, \tag{3.82}$$

and so $V(x,y,z)$ also satisfies equation (3.78). Therefore, $V(x,y,z)$ is the desired potential in the half-space $z \le 0$. Because $V(x,y,z)=V_1(x,y,z)+V_2(x,y,z)$ and $V_2(x,y,0)=V_1(x,y,0)$, then $V(x,y,0)=2\,V_1(x,y,0)$ on the interface $z=0$, and so equation (3.77) is valid, if $\mathbf{J}^{\mathrm{p}}$ is a current dipole.

3.12 Appendix: Vector Potential Outside a Spherical Conductor Due to Current Dipole in the Conductor

We want to derive equation (3.62). Let a current dipole with moment $\mathbf{Q}$ be at the location $\mathbf{r}_0$ in a multilayer spherical conductor G with center at the origin. Let $\mathbf{B}(\mathbf{r})$ be the magnetic field at $\mathbf{r}$ outside G due to the dipole. Let $\mathbf{A}(\mathbf{r})$ be the vector potential of $\mathbf{B}(\mathbf{r})$ outside G.

Let $\boldsymbol{Q}_{\mathrm{t}} = \mathbf{Q} - ||\mathbf{r}_0||^{-2}\mathbf{Q}\cdot\mathbf{r}_0\,\mathbf{r}_0$ be the tangential component of $\mathbf{Q}$, and $\mathbf{v} = \boldsymbol{Q}_{\mathrm{t}}/||\boldsymbol{Q}_{\mathrm{t}}||$. $\mathbf{B}(\mathbf{r})$ is equal to the magnetic field due to the tangential dipole with moment $\boldsymbol{Q}_{\mathrm{t}}$ at $\mathbf{r}_0$, because the radial component of $\mathbf{Q}$ does not produce any magnetic field outside G. Next, we approximate this tangential dipole with a current line segment $\mathbf{J}_s^{\mathrm{p}}$ with end points $\mathbf{r}_0 - s\mathbf{v}$ and $\mathbf{r}_0 + s\mathbf{v}$, a (constant) current strength $I_s = ||\boldsymbol{Q}_{\mathrm{t}}||/(2s)$, and current direction $\mathbf{v}$, for a small $s>0$, that is, $\mathbf{J}_s^{\mathrm{p}}(\mathbf{r}') = I_s\mathbf{v}$ at every point $\mathbf{r}'$ on the segment. Let $\mathbf{B}_s(\mathbf{r})$ be the (total) magnetic field due to the primary current segment $\mathbf{J}_s^{\mathrm{p}}$. It follows that $\mathbf{B}(\mathbf{r}) = \lim_{s\to 0} \mathbf{B}_s(\mathbf{r})$. In the following, we derive the vector potential $\mathbf{A}_s(\mathbf{r})$ of $\mathbf{B}_s(\mathbf{r})$ outside G in closed form, and get the desired $\mathbf{A}(\mathbf{r})$ in closed form as $\mathbf{A}(\mathbf{r}) = \lim_{s\to 0} \mathbf{A}_s(\mathbf{r})$.

To form $\mathbf{A}_s(\mathbf{r})$ by the triangle construction [146], we join the end point $\mathbf{r}_0 + s\mathbf{v}$ to the origin with current $\mathbf{J}_{s,1}^{\mathrm{p}}(\mathbf{r}') = -I_s\,||\mathbf{r}_0 + s\mathbf{v}||^{-1}(\mathbf{r}_0 + s\mathbf{v})$ and the origin to the end point $\mathbf{r}_0 - s\mathbf{v}$ to with current $\mathbf{J}_{s,2}^{\mathrm{p}}(\mathbf{r}') = I_s\,||\mathbf{r}_0 - s\mathbf{v}||^{-1}(\mathbf{r}_0 - s\mathbf{v})$. Accordingly, the current segments $\mathbf{J}_s^{\mathrm{p}}$, $\mathbf{J}_{s,1}^{\mathrm{p}}$ and $\mathbf{J}_{s,2}^{\mathrm{p}}$ form a current loop $\mathbf{J}_{\mathrm{loop}}^{\mathrm{p}}(\mathbf{r}')$ (see figure 3.6).

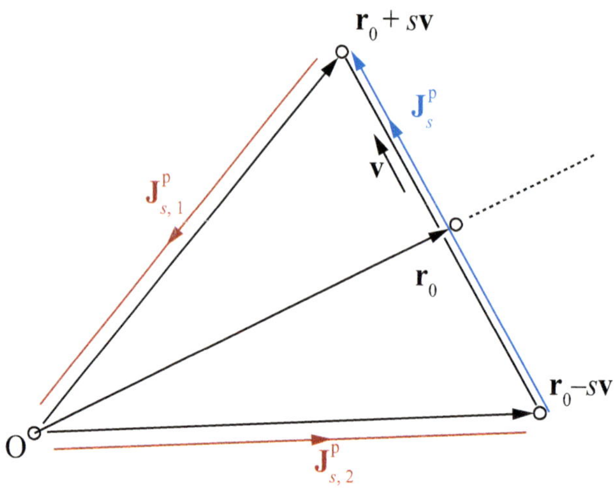

Figure 3.6

Triangular current loop used to derive the vector potential outside a spherically symmetric head model due to a current dipole. The trick is to add two radial current elements (including the associated sources and sinks) to the current element of the dipole. As the sinks and sources cancel pairwise, we get a current loop.

Due to the triangle construction for $\mathbf{r}$ outside G,

$$\begin{aligned} \mathbf{B}_s(\mathbf{r}) &= \frac{\mu_0}{4\pi} \oint \mathbf{J}^{\mathrm{p}}_{\mathrm{loop}}(\mathbf{r}') \times \frac{(\mathbf{r}-\mathbf{r}')}{||\mathbf{r}-\mathbf{r}'||^3}\, dl' \\ &= \nabla \times \left(\frac{\mu_0}{4\pi} \oint \mathbf{J}^{\mathrm{p}}_{\mathrm{loop}}(\mathbf{r}') \frac{1}{||\mathbf{r}-\mathbf{r}'||}\, dl' \right), \end{aligned} \tag{3.83}$$

because $(\mathbf{r}-\mathbf{r}')/||\mathbf{r}-\mathbf{r}'||^3 = -\nabla(1/||\mathbf{r}-\mathbf{r}'||)$ and

$$\nabla \times \left(\mathbf{J}^{\mathrm{p}}_{\mathrm{loop}}(\mathbf{r}') \frac{1}{||\mathbf{r}-\mathbf{r}'||} \right) = \nabla \left(\frac{1}{||\mathbf{r}-\mathbf{r}'||} \right) \times \mathbf{J}^{\mathrm{p}}_{\mathrm{loop}}(\mathbf{r}').$$

Therefore,

$$\mathbf{A}_s(\mathbf{r}) = \frac{\mu_0}{4\pi} \oint \mathbf{J}^{\mathrm{p}}_{\mathrm{loop}}(\mathbf{r}') \frac{1}{||\mathbf{r}-\mathbf{r}'||}\, dl'. \tag{3.84}$$

This integral around the loop is the sum of the line integrals over the three subsegments. Let

$$F(\mathbf{a}, \mathbf{b}, \mathbf{r}) = \int_{\mathbf{a}}^{\mathbf{b}} \frac{1}{||\mathbf{r}-\mathbf{r}'||}\, dl' = \int_0^{||\mathbf{b}-\mathbf{a}||} \frac{1}{||\mathbf{r} - \left(\mathbf{a} + t\frac{(\mathbf{b}-\mathbf{a})}{||\mathbf{b}-\mathbf{a}||}\right)||}\, dt \tag{3.85}$$

be the line integral from $\mathbf{a}$ to $\mathbf{b}$ of $||\mathbf{r}-\mathbf{r}'||^{-1}$, which can be evaluated in closed form as

$$F(\mathbf{a},\mathbf{b},\mathbf{r}) = \log\left(\frac{u + ||\mathbf{b}-\mathbf{r}|| + ||\mathbf{b}-\mathbf{a}||}{u + ||\mathbf{a}-\mathbf{r}||}\right) \text{ with } u = (\mathbf{a}-\mathbf{r})\cdot(\mathbf{b}-\mathbf{a})/||\mathbf{b}-\mathbf{a}||. \tag{3.86}$$

We then get

$$\begin{aligned}\frac{4\pi}{\mu_0}\mathbf{A}_s(\mathbf{r}) =& F(\mathbf{r}_0 - s\mathbf{v}, \mathbf{r}_0 + s\mathbf{v}, \mathbf{r})\, I_s \mathbf{v} - F(0, \mathbf{r}_0 + s\mathbf{v}, \mathbf{r})\, I_s \frac{(\mathbf{r}_0 + s\mathbf{v})}{||\mathbf{r}_0 + s\mathbf{v}||} \\ &+ F(0, \mathbf{r}_0 - s\mathbf{v}, \mathbf{r})\, I_s \frac{(\mathbf{r}_0 - s\mathbf{v})}{||\mathbf{r}_0 - s\mathbf{v}||}.\end{aligned} \tag{3.87}$$

To complete the derivation of equation (3.62), we need to evaluate the limits of the above line integral terms, as $s \to 0$, in closed forms. By using equation (3.86) with $I_s = ||Q_t||/(2s)$, we get

$$F(\mathbf{r}_0 - s\mathbf{v}, \mathbf{r}_0 + s\mathbf{v}, \mathbf{r})\, I_s = \Big(h(s) - h(-s)\Big)\frac{||\mathbf{Q}_t||}{2s}, \tag{3.88}$$

where

$$h(s) = \log(-\mathbf{r}\cdot\mathbf{v} + ||\mathbf{r}_0 - \mathbf{r} + s\mathbf{v}|| + s). \tag{3.89}$$

By differentiation, we see that $h'(0) = 1/||\mathbf{r}_0 - \mathbf{r}||$. Because $\lim_{s\to 0}\Big(h(s) - h(-s)\Big)/s = 2h'(0)$, equation (3.88) yields

$$\lim_{s\to 0} F(\mathbf{r}_0 - s\mathbf{v}, \mathbf{r}_0 + s\mathbf{v}, \mathbf{r})\, I_s \mathbf{v} = \frac{1}{||\mathbf{r}_0 - \mathbf{r}||}\mathbf{Q}_t. \tag{3.90}$$

Next we consider the two latter terms on the right side of equation (3.87), and get

$$F(0, \mathbf{r}_0 - s\mathbf{v}, \mathbf{r})\, I_s \frac{(\mathbf{r}_0 - s\mathbf{v})}{||\mathbf{r}_0 - \mathbf{v}||} - F(0, \mathbf{r}_0 + s\mathbf{v}, \mathbf{r})\, I_s \frac{(\mathbf{r}_0 + s\mathbf{v})}{||\mathbf{r}_0 + s\mathbf{v}||} = \Big(k(-s) - k(s)\Big) I_s, \tag{3.91}$$

where

$$k(s) = F(0, \mathbf{r}_0 + s\mathbf{v}, \mathbf{r}) \frac{(\mathbf{r}_0 + s\mathbf{v})}{||\mathbf{r}_0 + s\mathbf{v}||}, \tag{3.92}$$

By equation (3.86), after differentiation and some algebra, we get

$$k'(0) = \frac{\mathbf{r}\cdot\mathbf{v}}{||\mathbf{r}_0||}\left(\frac{1}{\beta} - \frac{||\mathbf{r}_0|| + ||\mathbf{r}-\mathbf{r}_0||}{\alpha\, ||\mathbf{r}-\mathbf{r}_0||}\right)\mathbf{r}_0 + \frac{1}{||\mathbf{r}_0||}\left(\log\frac{\alpha}{\beta}\right)\mathbf{v}, \tag{3.93}$$

where $\alpha = ||\mathbf{r}_0||\,||\mathbf{r}-\mathbf{r}_0|| - \mathbf{r}_0\cdot(\mathbf{r}-\mathbf{r}_0)$ and $\beta = ||\mathbf{r}||\,||\mathbf{r}_0|| + \mathbf{r}\cdot\mathbf{r}_0$. Because $\lim_{s\to 0}\Big(k(s) - k(-s)\Big)/s = 2\,k'(0)$, we get with equations (3.87), (3.90), and (3.91):

$$\mathbf{A}(\mathbf{r}) = \lim_{s\to 0} \mathbf{A}_s(\mathbf{r}) = \frac{\mu_0}{4\pi}\Big(\frac{1}{||\mathbf{r}_0 - \mathbf{r}||}\,\mathbf{Q}_t - ||\mathbf{Q}_t||\,k'(0)\Big), \tag{3.94}$$

which, with equation (3.93) and $\mathbf{r}_Q = \mathbf{r}_0$, yields the claim (3.62), repeated here:

$$\mathbf{A}(\mathbf{r}) = \frac{\mu_0}{4\pi}\Bigg[\Big(\frac{1}{||\mathbf{r}-\mathbf{r}_Q||} - \frac{1}{||\mathbf{r}_Q||}\log(\frac{\alpha}{\beta})\Big)\,\mathbf{Q}_t + \frac{\mathbf{r}\cdot\mathbf{Q}_t}{||\mathbf{r}_Q||}\Big(\frac{||\mathbf{r}-\mathbf{r}_Q||+||\mathbf{r}_Q||}{\alpha\,||\mathbf{r}-\mathbf{r}_Q||} - \frac{1}{\beta}\Big)\,\mathbf{r}_Q\Bigg]. \tag{3.95}$$

3.13 Appendix: Series Expansion for Potential Due to Current Dipole in Multilayered Sphere

Here we derive a series expansion for potential $V(\mathbf{r})$ due to a current dipole in a multilayered sphere. Let $0 < r_1 < \cdots < r_M$ be the radii of the layered structure with conductivities $\sigma_1, \ldots, \sigma_M$ in the inner sphere $0 < ||\mathbf{r}|| < r_1$ and in layers $r_{i-1} < ||\mathbf{r}|| < r_i$, $i = 2, \ldots, M$. Outside the sphere, $\sigma = 0$. Let the dipole be at location $\mathbf{r}_0$, $0 \le ||\mathbf{r}_0|| < r_1$, with moment $\mathbf{Q}$. Let $r = ||\mathbf{r}||$, $r_0 = ||\mathbf{r}_0||$, $Q = ||\mathbf{Q}||$, $\widehat{\mathbf{r}} = \mathbf{r}/r$, and $\widehat{\mathbf{r}_0} = \mathbf{r}_0/r_0$.

We start with a series expansion of the potential $V_0(\mathbf{r})$ of the dipole in a homogeneous infinite space with conductivity σ_1,

$$\begin{aligned} V_0(\mathbf{r}) &= \frac{1}{4\pi\,\sigma_1}\,\mathbf{Q}\cdot\frac{(\mathbf{r}-\mathbf{r}_0)}{||\mathbf{r}-\mathbf{r}_0||^3} = \frac{1}{4\pi\,\sigma_1}\,\mathbf{Q}\cdot\nabla_{\mathbf{r}_0}\Big(\frac{1}{||\mathbf{r}-\mathbf{r}_0||}\Big) \\ &= \frac{1}{4\pi\,\sigma_1}\sum_{n=1}^{\infty}\frac{1}{r^{n+1}}\,Z_n(\hat{\mathbf{r}}) \end{aligned} \tag{3.96}$$

for $r > r_0$, where $\nabla_{\mathbf{r}_0}$ stands for the gradient with respect to $\mathbf{r}_0$. The functions $Z_n(\hat{\mathbf{r}})$ are given by

$$\begin{aligned} Z_n(\hat{\mathbf{r}}) &= \mathbf{Q}\cdot\nabla_{\mathbf{r}_0}\big(r_0^n\,P_n(\hat{\mathbf{r}}\cdot\hat{\mathbf{r}}_0)\big) = \mathbf{Q}\cdot\nabla_{\mathbf{r}_0}\big(||\mathbf{r}_0||^n\,P_n(\frac{1}{||\mathbf{r}_0||}\hat{\mathbf{r}}\cdot\mathbf{r}_0)\big) \\ &= r_0^{n-1}\Big[\mathbf{Q}\cdot\hat{\mathbf{r}}_0\,n\,P_n(\hat{\mathbf{r}}\cdot\hat{\mathbf{r}}_0) + (\mathbf{Q}\times\hat{\mathbf{r}}_0)\cdot(\hat{\mathbf{r}}\times\hat{\mathbf{r}}_0)\,P_n'(\hat{\mathbf{r}}\cdot\hat{\mathbf{r}}_0)\Big], \end{aligned}$$

where we have used the identity $\mathbf{Q}\cdot\hat{\mathbf{r}} - (\mathbf{Q}\cdot\hat{\mathbf{r}}_0)(\hat{\mathbf{r}}\cdot\hat{\mathbf{r}}_0) = (\mathbf{Q}\times\hat{\mathbf{r}}_0)\cdot(\hat{\mathbf{r}}\times\hat{\mathbf{r}}_0)$. By taking the gradient $\nabla_{\mathbf{r}_0}$ termwise, the expansion (3.96) is obtained from the following well-known series expansion [153] of the potential of the electric unit charge at $\mathbf{r}_0$,

$$\frac{1}{4\pi\,\sigma_1}\,\frac{1}{||\mathbf{r}-\mathbf{r}_0||} = \frac{1}{4\pi\,\sigma_1}\sum_{n=0}^{\infty}\frac{r_0^n}{r^{n+1}}\,P_n(\hat{\mathbf{r}}\cdot\hat{\mathbf{r}}_0) \tag{3.97}$$

for $r > r_0$; if $r < r_0$, then r_0^n/r^{n+1} above is replaced by r^n/r_0^{n+1}.

For the potential $V(\mathbf{r})$, we can now write a series expansion in terms of harmonic functions $r^n Z_n(\hat{\mathbf{r}})$ and $\frac{1}{r^{n+1}} Z_n(\hat{\mathbf{r}})$ as follows:

$$V(\mathbf{r}) = \frac{1}{4\pi\,\sigma_1} \sum_{n=1}^{\infty} \left(a_n(j)\, r^n + b_n(j)\, \frac{1}{r^{n+1}}\right) Z_n(\hat{\mathbf{r}}) \tag{3.98}$$

for $r > r_0$ in the jth layer $r_{j-1} < ||\mathbf{r}|| < r_j$, $j = 1, \ldots, M$. For $r < r_0$, we have

$$V(\mathbf{r}) = V_0(\mathbf{r}) + \frac{1}{4\pi\,\sigma_1} \sum_{n=1}^{\infty} a_n(1)\, r^n\, Z_n(\hat{\mathbf{r}})\,. \tag{3.99}$$

In equation (3.98), the coefficients $b_n(1) = 1$ for all $n = 1, 2, \ldots$. Other coefficients $a_n(j)$, $b_n(j)$ in equation (3.98) can be solved from the following equations:

$$a_n(j)\, r_j^n + b_n(j)\, \frac{1}{r_j^{n+1}} = a_n(j+1)\, r_j^n + b_n(j+1)\, \frac{1}{r_j^{n+1}}\,, \tag{3.100}$$

$$\sigma_j \left(n\, a_n(j)\, r_j^{n-1} - (n+1) b_n(j) \frac{1}{r_j^{n+2}}\right) = \sigma_{j+1} \left(n\, a_n(j+1)\, r_j^{n-1} - (n+1) b_n(j+1) \frac{1}{r_j^{n+2}}\right) \tag{3.101}$$

for $j = 1, \ldots, M-1$, and

$$n\, a_n(M)\, r_M^{n-1} - (n+1) b_n(M) \frac{1}{r_M^{n+2}} = 0. \tag{3.102}$$

Equations (3.100) to (3.102) are the implementations of the boundary conditions for the potential terms $u_n(j, \mathbf{r}) = \left(a_n(j)\, r^n + b_n(j)\, \frac{1}{r^{n+1}}\right) Z_n(\hat{\mathbf{r}})$ in equation (3.98), which state that $u_n(j, \mathbf{r})$ and $\sigma(\mathbf{r})\, \frac{\partial u_n(j,\mathbf{r})}{\partial r}$ must be continuous across the interfaces $||\mathbf{r}|| - r_j$, $j = 1, 2\ldots, M-1$, and $\frac{\partial u_n(j,\mathbf{r})}{\partial r}$ must vanish on the outer boundary $||\mathbf{r}|| = r_M$.

For solving equations (3.100) to (3.102), we fix $n \geq 1$ and briefly write $a_n(j) = a(j)$ and $b_n(j) = b(j)$. The equations can be written in matrix form as

$$\mathbf{A}_j \begin{bmatrix} a(j) \\ b(j) \end{bmatrix} = \mathbf{B}_j \begin{bmatrix} a(j+1) \\ b(j+1) \end{bmatrix}, \quad j = 1, \ldots, M-1, \tag{3.103}$$

where

$$\mathbf{A}_j = \begin{bmatrix} r_j^n & \frac{1}{r_j^{n+1}} \\ n\,\sigma_j r_j^{n-1} & \frac{-(n+1)\,\sigma_j}{r_j^{n+2}} \end{bmatrix}, \quad \mathbf{B}_j = \begin{bmatrix} r_j^n & \frac{1}{r_j^{n+1}} \\ n\,\sigma_{j+1} r_j^{n-1} & \frac{-(n+1)\,\sigma_{j+1}}{r_j^{n+2}} \end{bmatrix}. \tag{3.104}$$

This implies that

$$\begin{bmatrix} a(j+1) \\ b(j+1) \end{bmatrix} = \mathbf{C}_j \begin{bmatrix} a(j) \\ b(j) \end{bmatrix}, \tag{3.105}$$

where

$$\mathbf{C}_j = \mathbf{B}_j^{-1}\mathbf{A}_j = \frac{1}{(2n+1)} \begin{bmatrix} (n+1)+n\frac{\sigma_j}{\sigma_{j+1}} & (n+1)(1-\frac{\sigma_j}{\sigma_{j+1}})\frac{1}{r_j^{2n+1}} \\ n(1-\frac{\sigma_j}{\sigma_{j+1}})r_j^{2n+1} & n+(n+1)\frac{\sigma_j}{\sigma_{j+1}} \end{bmatrix}, \tag{3.106}$$

for $j = 1, \ldots, M-1$, with

$$\det \mathbf{C}_j = \frac{\sigma_j}{\sigma_{j+1}}. \tag{3.107}$$

Equations (3.105) yield

$$\begin{bmatrix} a(j) \\ b(j) \end{bmatrix} = \mathbf{C}_{j-1} \cdots \mathbf{C}_2\mathbf{C}_1 \begin{bmatrix} a(1) \\ b(1) \end{bmatrix}, \tag{3.108}$$

where the matrix products $\mathbf{C}_{j-1} \cdots \mathbf{C}_2\mathbf{C}_1$ are practical to evaluate with the computer. Let

$$\mathbf{C} = \begin{bmatrix} c_{11} & c_{12} \\ c_{21} & c_{22} \end{bmatrix} = \mathbf{C}_{M-1} \cdots \mathbf{C}_2\mathbf{C}_1. \tag{3.109}$$

Then, by equation (3.107),

$$\det \mathbf{C} = \sigma_1/\sigma_M. \tag{3.110}$$

Furthermore, equation (3.102) implies that

$$b(M) = \frac{n}{n+1} r_M^{2n+1} a(M). \tag{3.111}$$

With equations (3.108) to (3.110), we obtain

$$\begin{bmatrix} a(1) \\ b(1) \end{bmatrix} = \mathbf{C}^{-1} \begin{bmatrix} a(M) \\ b(M) \end{bmatrix}$$

$$= a(M)\mathbf{C}^{-1} \begin{bmatrix} 1 \\ \frac{n}{n+1} r_M^{2n+1} \end{bmatrix} = a(M)\frac{\sigma_M}{\sigma_1} \begin{bmatrix} c_{22} - \frac{n}{n+1} r_M^{2n+1} c_{12} \\ -c_{21} + \frac{n}{n+1} r_M^{2n+1} c_{11} \end{bmatrix}, \tag{3.112}$$

which, with $b(1) = 1$, implies

$$a(M) = \frac{\sigma_1}{\sigma_M} \frac{1}{\left(-c_{21} + \frac{n}{n+1} r_M^{2n+1} c_{11}\right)}, \tag{3.113}$$

$$a(1) = \frac{c_{22} - \frac{n}{n+1} r_M^{2n+1} c_{12}}{-c_{21} + \frac{n}{n+1} r_M^{2n+1} c_{11}}. \tag{3.114}$$

Because $b(1) = 1$ and $a(1)$ are given by equation (3.114), we get all coefficients a_j and b_j by equations (3.108), (3.113), and (3.111) for $j = 1, \ldots, M$. By equations (3.98) and (3.99), we can now get the potential $V(\mathbf{r})$ for $||\mathbf{r}|| > r_0$ in the jth layer $r_{j-1} < ||\mathbf{r}|| < r_j$, $j = 1, \ldots, M$, and for $||\mathbf{r}|| < r_0$, respectively. On the surface of the sphere $||\mathbf{r}|| = r_M$, the presentation for $V(\mathbf{r})$ gets the form

$$V(\mathbf{r}) = \frac{1}{4\pi\,\sigma_1} \sum_{n=1}^{\infty} \gamma_n Z_n(\hat{\mathbf{r}}), \tag{3.115}$$

where

$$\gamma_n = \frac{\sigma_1}{\sigma_M} \frac{(2n+1)}{(n+1)} \frac{r_M^n}{\left(\frac{n}{n+1} r_M^{2n+1} \mathbf{C}_n(1,1) - \mathbf{C}_n(2,1)\right)}, \tag{3.116}$$

where $\mathbf{C}_n$ is the matrix $\mathbf{C}$ in equation (3.109) for $n = 1, 2, \ldots$. For a three-layer structure with $M = 3$, we may write $\mathbf{C}_n$ in an explicit form and get the expression (3.68).

4 Review of Linear Algebra and Probability Theory for MEG and EEG Data Analysis

Although the electromagnetic fields, as described by Maxwell's equations, depend nonlinearly on the location coordinates of the currents or charges (the sources), they depend linearly on the source amplitudes (see figure 2.1). This greatly simplifies the analysis of MEG and EEG signals, allowing us to use the powerful arsenal of techniques from linear algebra. In the forward problem (calculating fields from the sources), noise, artifacts, and other distortions translate linearly to the solution; thus, no sophisticated statistics is needed to analyze the forward problem. However, since the inverse problem is generally ill-posed, we will need to use, in addition to linear algebra, methods from estimation theory and to rely on probability theory. To help the reader in the following chapters, we will now review these mathematical topics before entering the data analysis.

4.1 Notation and Terminology

Vectors are denoted by bold lower-case characters, such as $\mathbf{a}$, $\mathbf{u}$, $\boldsymbol{\alpha}$, and matrices by bold capital letters, such as $\mathbf{A}$ and $\mathbf{W}$; and scalars by italic letters, such as a, b, and β. We denote the set of *real numbers* by $\mathbb{R}$ and the *n-dimensional (Euclidean) space* by $\mathbb{R}^n$. The (Euclidean) *norm* of a vector $\mathbf{u} \in \mathbb{R}^n$ is $||\mathbf{u}||$, and the absolute values of scalars a, α are denoted by $|a|$, $|\alpha|$.

Let $\mathbf{A}$ be an $m \times n$ matrix, that is, $\mathbf{A} \in \mathbb{R}^{m \times n}$. The elements of $\mathbf{A}$ are denoted by $\mathbf{A}(i, j)$, or A_{ij}, $1 \leq i \leq m$, $1 \leq j \leq n$. We denote the ith *row* of $\mathbf{A}$ by $\mathbf{A}(i, :)$, $i = 1, \ldots, m$, and the jth *column* by $\mathbf{A}(:, j)$, $j = 1, \ldots, n$; see figure 4.1. The elements, or *components*, of a vector $\mathbf{u} = [u_1, \ldots, u_n]^{\mathrm{T}} \in \mathbb{R}^n$ are denoted by u_i or $\mathbf{u}(i)$, $i = 1, \ldots, n$.

The *transpose* of $\mathbf{A}$ is $\mathbf{A}^{\mathrm{T}}$, and its *trace* is trace($\mathbf{A}$). The *scalar product* of vectors $\mathbf{a} = [a_1, \ldots, a_n]^{\mathrm{T}}$, $\mathbf{b} = [b_1, \ldots, b_n]^{\mathrm{T}} \in \mathbb{R}^n$ is $\mathbf{a}^{\mathrm{T}}\mathbf{b} = \sum_{i=1}^{n} \mathbf{a}(i)\,\mathbf{b}(i)$. The *identity matrix* is denoted by $\mathbf{I}$. If $\mathbf{A}$ is a square matrix and invertible, $\mathbf{A}^{-1}$ stands for its *inverse matrix*.

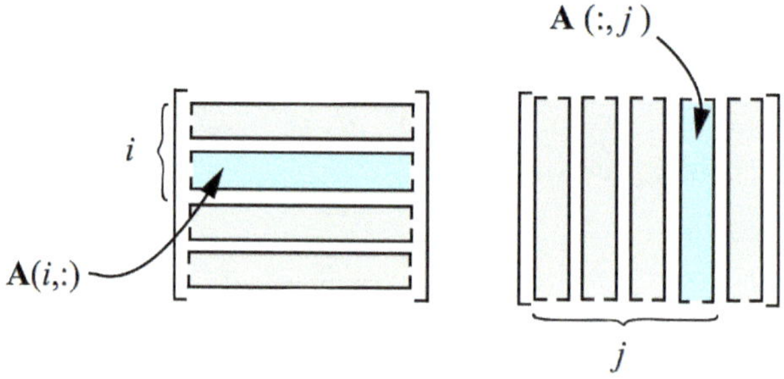

Figure 4.1
The ith row $\mathbf{A}(i,:)$ and the jth column $\mathbf{A}(:,j)$ of a matrix $\mathbf{A}$.

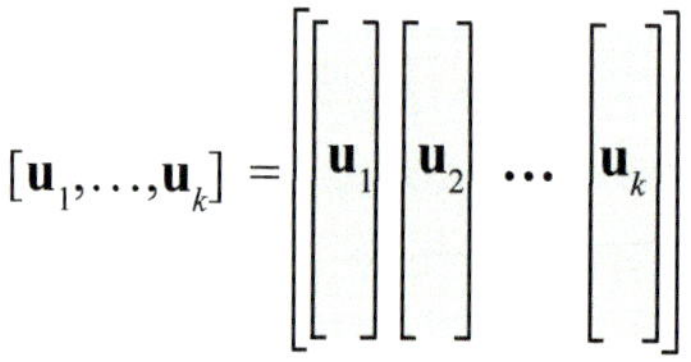

Figure 4.2
A matrix with columns $\mathbf{u}_1,\ldots,\mathbf{u}_k$ is denoted by $[\mathbf{u}_1,\ldots,\mathbf{u}_k]$.

A square matrix $\mathbf{A}$ is *symmetric* if $\mathbf{A}^{\mathrm{T}}=\mathbf{A}$. A symmetric matrix $\mathbf{A}$ is called *positive definite,* if $\mathbf{x}^{\mathrm{T}}\mathbf{A}\mathbf{x}>0$ for all $\mathbf{x}\neq 0$, $\mathbf{x}\in\mathbb{R}^n$, and *positive semi-definite,* if $\mathbf{x}^{\mathrm{T}}\mathbf{A}\mathbf{x}\geq 0$ for all $\mathbf{x}\in\mathbb{R}^n$.

For $1\leq j<k$, we denote the index set $\{j,j+1,\ldots,k\}$ by $j:k$. If $\mathbf{u}_1,\ldots,\mathbf{u}_k\in\mathbb{R}^m$ are column vectors, we let $[\mathbf{u}_1,\ldots,\mathbf{u}_k]$ denote the $m\times k$ matrix with columns $\mathbf{u}_1,\ldots,\mathbf{u}_k$ (see figure 4.2). By $\mathbf{A}(:,j:k)$ we denote the matrix $[\mathbf{A}(:,j),\ldots,\mathbf{A}(:,k)]$, where $j\leqslant k\leqslant n$.

By $\mathbf{diag}_{m\times n}(d_1,\ldots,d_p)$ we denote an *$m\times n$ diagonal matrix* with diagonal elements $d_1,\ldots,d_p$, with $p=\min(m,n)$ (see figure 4.3). If $m=n=p$, we briefly write: $\mathbf{diag}_{m\times m}(d_1,\ldots,d_m)=\mathbf{diag}\,(d_1,\ldots,d_m)$.

We describe three-dimensional (3-D) arrays (matrices) with the following notation. Let $\mathbf{A}$ be a 3-D array with elements $\mathbf{A}(i,j,k)$; $i=1,\ldots,m$, $j=1,\ldots,n$, $k=1,\ldots,p$. For a fixed i, the notation $\mathbf{A}(i,:,:)$ denotes the two-dimensional (2-D) matrix with elements $\mathbf{A}(i,j,k)$; $j=1,\ldots,n$, $k=1,\ldots,p$. For a fixed j or k, the 2-D matrices $\mathbf{A}(:,j,:)$ and $\mathbf{A}(:,:,k)$ are defined in analogous ways. Furthermore, for a fixed i and j, the notation $\mathbf{A}(i,j,:)$ denotes the vector with elements $\mathbf{A}(i,j,k)$, $k=1,\ldots,p$. The vectors $\mathbf{A}(:,j,k)$ and $\mathbf{A}(i,:,k)$ are defined in analogous ways.

Let $f(\mathbf{x})$ be a real-valued function with $\mathbf{x}\in\mathbb{R}^n$ (or $x\in\mathbb{R}$). By the notation

$$\mathbf{x}_0=\arg\max_{\mathbf{x}} f(\mathbf{x})$$

$$\mathbf{diag}_{m\times n}(d_1,\ldots,d_p) = \left.\underbrace{\begin{bmatrix} d_1 & & & & \\ & d_2 & & (0) & \\ & & \ddots & & \\ & (0) & & \ddots & \\ & & & & d_m \end{bmatrix}}_{m}\right\}m \quad \overbrace{}^{n}$$

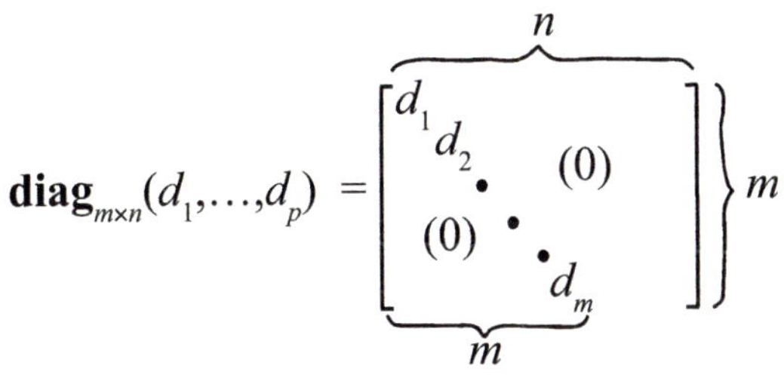

Figure 4.3
An $m \times n$ diagonal matrix, $m \leq n$, with diagonal elements $d_1, \ldots, d_p$, $p = m$.

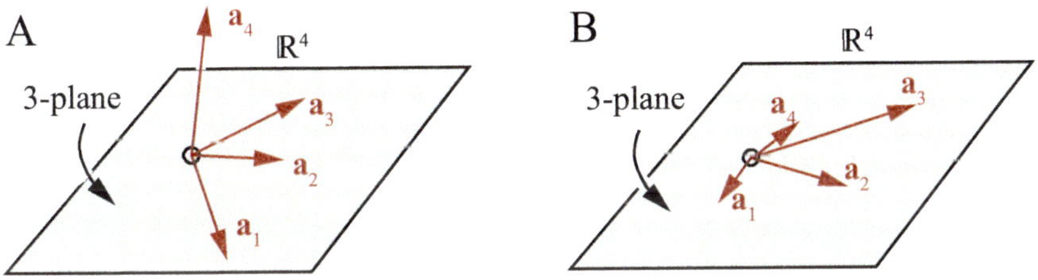

Figure 4.4
(A) If vectors $\mathbf{a}_1, \mathbf{a}_2, \mathbf{a}_3$ span a 3-plane in $\mathbb{R}^3$ and $\mathbf{a}_4$ is not in that plane, then $\mathbf{a}_1, \ldots, \mathbf{a}_4$ are linearly independent. (B) If vectors $\mathbf{a}_1, \ldots, \mathbf{a}_4$ are all in the same 3-plane, then they are linearly dependent.

we mean that the maximum of $f(\mathbf{x})$ is attained at $\mathbf{x} = \mathbf{x}_0$. A similar notation is used with the minimum.

4.2 Review of Linear Algebra for MEG and EEG

Non-zero vectors $\mathbf{a}_1, \ldots, \mathbf{a}_n \in \mathbb{R}^m$ are *linearly independent* if for any $\alpha_1, \ldots, \alpha_n \in \mathbb{R}$, the equation $\alpha_1 \mathbf{a}_1 + \cdots + \alpha_n \mathbf{a}_n = 0$ implies that $\alpha_1 = \cdots = \alpha_n = 0$. If $\mathbf{a}_1, \ldots, \mathbf{a}_n$ are not linearly independent, they are *linearly dependent* (see figure 4.4).

The subspace of $\mathbb{R}^m$ spanned by vectors $\mathbf{a}_1, \ldots, \mathbf{a}_n \in \mathbb{R}^m$ is denoted by span($\mathbf{a}_1, \ldots, \mathbf{a}_n$), that is,

$$\text{span}(\mathbf{a}_1, \ldots, \mathbf{a}_n) = \{\alpha_1 \mathbf{a}_1 + \cdots + \alpha_n \mathbf{a}_n \mid \alpha_1, \ldots, \alpha_n \in \mathbb{R}\} \subset \mathbb{R}^m \quad (4.1)$$

(see figure 4.5). For a matrix $\mathbf{A} \in \mathbb{R}^{m \times n}$, we define

$$\text{span}(\mathbf{A}) = \text{span}(\mathbf{A}(:, 1), \ldots, \mathbf{A}(:, n)). \quad (4.2)$$

Note that every subspace of $\mathbb{R}^m$ is always spanned by some vectors $\mathbf{a}_1, \ldots, \mathbf{a}_n$ with $1 \leq n \leq m$.

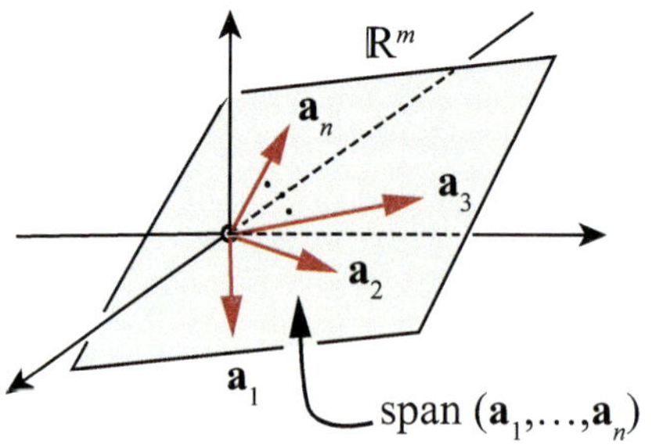

Figure 4.5
Vectors $\mathbf{a}_1,\ldots,\mathbf{a}_n \in \mathbb{R}^m$ span the subspace (or hyperplane) span$(\mathbf{a}_1,\ldots,\mathbf{a}_n)$.

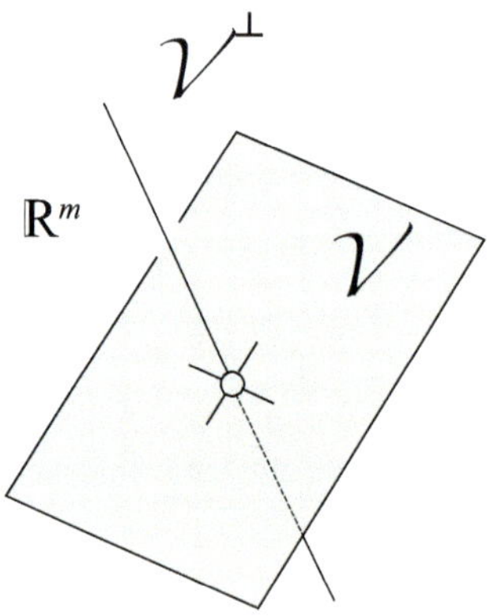

Figure 4.6
Subspace $\mathcal{V} \subset \mathbb{R}^m$ and its orthospace $\mathcal{V}^\perp$.

An orthogonal complement, or briefly *orthospace*, of a subspace $\mathcal{V} \subset \mathbb{R}^m$, denoted by $\mathcal{V}^\perp$, is the subspace

$$\mathcal{V}^\perp = \{\mathbf{x} \in \mathbb{R}^m \mid \mathbf{x}^\mathrm{T}\mathbf{v} = 0 \text{ for all } \mathbf{v} \in \mathcal{V}\}. \tag{4.3}$$

In other words, $\mathcal{V}^\perp$ is the subspace of all vectors $\mathbf{x} \in \mathbb{R}^m$ such that $\mathbf{x}$ is orthogonal to all vectors in $\mathcal{V}$ (see figure 4.6).

We say that vectors $\mathbf{a}_1,\ldots,\mathbf{a}_n$ are (mutually) *orthonormal*, if $||\mathbf{a}_i|| = 1$, $i = 1,\ldots,n$, and ${\mathbf{a}_i}^\mathrm{T}\mathbf{a}_j = 0$ if $i \neq j$. A square matrix $\mathbf{A} \in \mathbb{R}^{m\times m}$ is called *orthogonal* if $\mathbf{A}^\mathrm{T} = \mathbf{A}^{-1}$. The following two properties are both equivalent for $\mathbf{A}$ to be orthogonal: (1) the columns of $\mathbf{A}$ are orthonormal, (2) and the rows of $\mathbf{A}$ are orthonormal (see figure 4.7).

Let vectors $\mathbf{a}_1,\ldots,\mathbf{a}_n \in \mathbb{R}^m$ be linearly independent and belong to a subspace $\mathcal{V} \subset \mathbb{R}^m$. They are defined to form a *basis* of a subspace $\mathcal{V}$ if span$(\mathbf{a}_1,\ldots,\mathbf{a}_n) = \mathcal{V}$. The number n

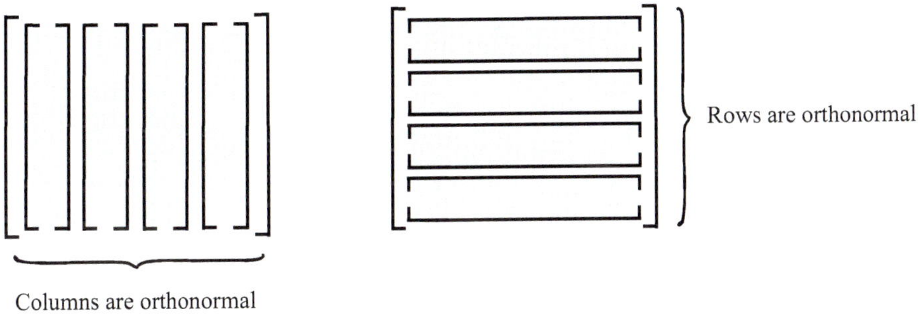

Figure 4.7
The columns of an orthogonal matrix are orthonormal vectors and so are its rows.

is the *dimension* of $\mathcal{V}$. If vectors $\mathbf{a}_1, \dots, \mathbf{a}_n \in \mathbb{R}^m$ are orthonormal and $\operatorname{span}(\mathbf{a}_1, \dots, \mathbf{a}_n) = \mathcal{V}$, we say that $\mathbf{a}_1, \dots, \mathbf{a}_n$ form an *orthonormal basis* of $\mathcal{V}$. Then, every $\mathbf{x} \in \mathcal{V}$ can be expanded in that basis as

$$\mathbf{x} = \sum_{i=1}^{n} (\mathbf{a}_i{}^{\mathrm{T}} \mathbf{x})\, \mathbf{a}_i. \tag{4.4}$$

Let $\mathbf{A} \in \mathbb{R}^{m \times n}$. We also consider $\mathbf{A}$ as a linear mapping $\mathbf{A} : \mathbb{R}^n \longrightarrow \mathbb{R}^m$, which maps each $\mathbf{x} \in \mathbb{R}^n$ to $\mathbf{A}\mathbf{x} \in \mathbb{R}^m$. The *kernel* of $\mathbf{A}$, denoted by $\ker(\mathbf{A})$, is the subspace

$$\ker(\mathbf{A}) = \{\mathbf{x} \mid \mathbf{A}\mathbf{x} = 0\} \subset \mathbb{R}^n. \tag{4.5}$$

The *range* of $\mathbf{A}$, denoted by $\operatorname{ran}(\mathbf{A})$, is the subspace

$$\operatorname{ran}(\mathbf{A}) = \{\mathbf{A}\mathbf{x} \mid \mathbf{x} \in \mathbb{R}^n\} \subset \mathbb{R}^m. \tag{4.6}$$

We can show that

$$\operatorname{ran}(\mathbf{A}) = \operatorname{span}(\mathbf{A}), \tag{4.7}$$

$$\ker(\mathbf{A})^{\perp} = \operatorname{span}(\mathbf{A}^{\mathrm{T}}). \tag{4.8}$$

Namely, by equations (4.1) and (4.2),

$$\begin{aligned} \operatorname{span}(\mathbf{A}) &= \operatorname{span}(\mathbf{A}(:,1), \dots, \mathbf{A}(:,n)) = \left\{ \sum_{i=1}^{n} \alpha_i \mathbf{A}(:,i) \mid \alpha_1, \dots, \alpha_n \in \mathbb{R} \right\} \\ &= \left\{ \sum_{i=1}^{n} \boldsymbol{\alpha}(i) \mathbf{A}(:,i) \mid \boldsymbol{\alpha} \in \mathbb{R}^n \right\} = \left\{ \mathbf{A}\boldsymbol{\alpha} \mid \boldsymbol{\alpha} \in \mathbb{R}^n \right\} = \operatorname{ran}(\mathbf{A}), \end{aligned}$$

which proves equation (4.7). Equation (4.8) can be easily proved by using the *singular value decomposition* (SVD) of the matrix $\mathbf{A}$; we will do it below after we have introduced SVD.

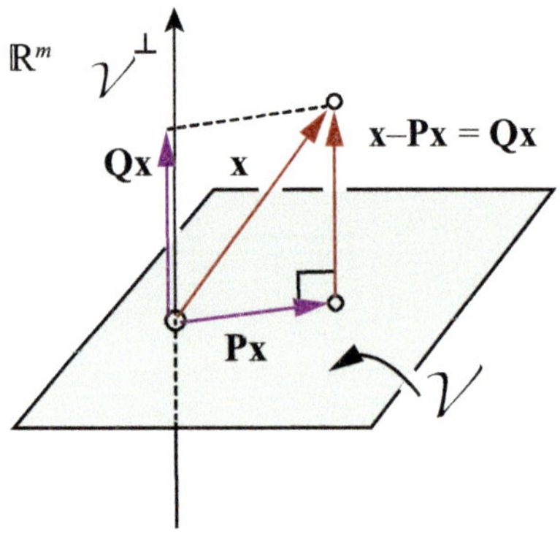

Figure 4.8
An orthogonal projection $\mathbf{P}$ from $\mathbb{R}^m$ onto the subspace $\mathcal{V} = \text{span}(\mathbf{P})$. Any vector $\mathbf{x} \in \mathbb{R}^m$ can be decomposed to orthogonal components $\mathbf{P}\mathbf{x}$ and $\mathbf{Q}\mathbf{x} = \mathbf{x} - \mathbf{P}\mathbf{x}$ where $\mathbf{Q} = \mathbf{I} - \mathbf{P}$ is the orthogonal projection onto $\mathcal{V}^\perp$.

The rank of $\mathbf{A} \in \mathbb{R}^{m\times n}$ is the dimension of span($\mathbf{A}$). Consequently, $1 \le \text{rank}(\mathbf{A}) \le n$. One can show that rank($\mathbf{A}$) is the maximal number of linearly independent columns of $\mathbf{A}$, and also it is the maximal number of linearly independent rows of $\mathbf{A}$. If $m = n = \text{rank}(\mathbf{A})$, then $\mathbf{A}$ is *invertible*, that is, $\mathbf{A}$ has the inverse $\mathbf{A}^{-1}$.

If $\mathbf{A} \in \mathbb{R}^{m\times n}$ and $\mathbf{B} \in \mathbb{R}^{n\times p}$, then we have the following useful formulas,

$$\mathbf{AB} = \sum_{i=1}^{n} \mathbf{A}(:, i)\mathbf{B}(i, :), \tag{4.9}$$

$$(\mathbf{AB})^{\mathrm{T}} = \mathbf{B}^{\mathrm{T}}\mathbf{A}^{\mathrm{T}}, \tag{4.10}$$

which are readily verified by the matrix multiplication rule.

A square matrix $\mathbf{P} \in \mathbb{R}^{m\times m}$ is called an *orthogonal projection* from $\mathbb{R}^m$ onto span($\mathbf{P}$) if $\mathbf{P} = \mathbf{PP}$ and

$$(\mathbf{Px})^{\mathrm{T}}(\mathbf{x} - \mathbf{Px}) = 0 \text{ for all } \mathbf{x} \in \mathbb{R}^m. \tag{4.11}$$

It can be shown that an *orthogonal projection* $\mathbf{P}$ is always a *symmetric* matrix. Notice (figure 4.8) that $\mathbf{x} - \mathbf{P}\mathbf{x} \in \text{span}(\mathbf{P})^\perp$ and, according to the Pythagoras theorem, that

$$||\mathbf{x}||^2 = ||\mathbf{Px}||^2 + ||\mathbf{x} - \mathbf{Px}||^2, \tag{4.12}$$

which implies that

$$||\mathbf{Px}|| \le ||\mathbf{x}|| \text{ for all } \mathbf{x} \in \mathbb{R}^m. \tag{4.13}$$

$$A = UDV^T = \begin{bmatrix} & & \\ & U & \\ & & \end{bmatrix} \begin{bmatrix} d_1 & & & & \\ & \ddots & & & \\ & & d_r & & \\ & & & 0 & \\ & & & & \ddots \\ & & & & & 0 \end{bmatrix} \begin{bmatrix} & V^T & \end{bmatrix}$$

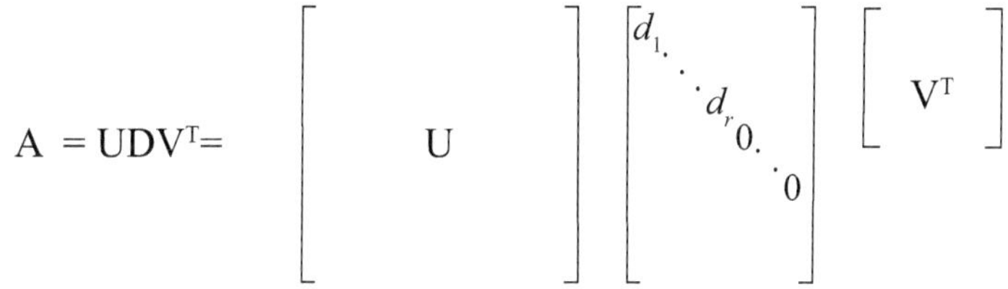

Figure 4.9
Singular value decomposition $A = UDV^T$ of matrix A, where $r = \text{rank}(A)$.

If $u_1, \ldots, u_n$ form an orthonormal basis of subspace $\mathcal{V} \subset \mathbb{R}^m$, and $U = [u_1, \ldots, u_n]$, then the orthogonal projection $P : \mathbb{R}^m \longrightarrow \mathcal{V}$ is given by

$$P = UU^T = \sum_{i=1}^n u_i u_i^T. \tag{4.14}$$

We denote the orthogonal projection P from $\mathbb{R}^m$ onto $\mathcal{V}$ by $\Pi_\mathcal{V}$ to emphasize that it only depends on the subspace $\mathcal{V}$. Furthermore,

$$Q = I - \Pi_\mathcal{V} = \Pi_{\mathcal{V}^\perp} \tag{4.15}$$

is the orthogonal projection from $\mathbb{R}^m$ onto the orthospace $\mathcal{V}^\perp$ of $\mathcal{V}$.

The SVD of matrix $A \in \mathbb{R}^{m \times n}$ is the decomposition

$$A = UDV^T, \tag{4.16}$$

where U and V are $m \times m$ and $n \times n$ orthogonal matrices, respectively, $D = \text{diag}_{m \times n}(d_1, \ldots, d_p)$ is a diagonal matrix with $p = \min(m, n)$, and singular values

$$d_1 \geq \cdots \geq d_r > 0 = d_{r+1} = \cdots = d_p, \tag{4.17}$$

where $r = \text{rank}(A)$ (see figure 4.9).

It follows that the "economy" form of the SVD of A can be written as

$$A = U(:, 1:r) \, \text{diag}(d_1, \ldots, d_r) V(:, 1:r)^T \tag{4.18}$$

$$= \sum_{i=1}^r d_i u_i v_i^T, \tag{4.19}$$

where $u_i = U(:, i)$ and $v_i = V(:, i)$, $i = 1, \ldots, r$, are the left and right singular vectors. Vectors $u_1, \ldots, u_r$ are orthonormal as columns of the orthogonal matrix U; we see that they form an orthonormal basis of $\text{span}(A)$ and so

$$\Pi_{\text{span}(A)} = U(:, 1:r) U(:, 1:r)^T. \tag{4.20}$$

Similarly, $\mathbf{v}_1,\dots,\mathbf{v}_r$ form an orthonormal basis of $\ker(\mathbf{A})^{\perp}$ and $\mathbf{v}_{r+1},\dots,\mathbf{v}_n$ form an orthonormal basis of $\ker(\mathbf{A})$. By equation (4.19),

$$\mathbf{A}^{\mathrm{T}} = \mathbf{V}(:,1:r)\,\mathbf{diag}\,(d_1,\dots,d_r)\mathbf{U}(:,1:r)^{\mathrm{T}}, \tag{4.21}$$

and we see that $\mathbf{v}_1,\dots,\mathbf{v}_r$ also form an orthonormal basis of $\mathrm{span}(\mathbf{A}^{\mathrm{T}})$. This with the above result implies that $\mathrm{span}(\mathbf{A}^{\mathrm{T}}) = \ker(\mathbf{A})^{\perp}$, proving equation (4.8).

For matrix $\mathbf{A} \in \mathbb{R}^{m\times n}$, the *condition number* $\mathrm{cond}(\mathbf{A})$ is defined by

$$\mathrm{cond}(\mathbf{A}) = \frac{d_1}{d_r}, \tag{4.22}$$

where d_1 and d_r are the largest and smallest non-zero singular values of $\mathbf{A}$.

Assume that a square matrix $\mathbf{A} \in \mathbb{R}^{m\times m}$ is *symmetric and positive semi-definite*; for example, $\mathbf{A} = \mathbf{B}\mathbf{B}^{\mathrm{T}}$, or $\mathbf{A} = \mathbf{C}^{\mathrm{T}}\mathbf{C}$, for any $\mathbf{B} \in \mathbb{R}^{m\times n}$, $\mathbf{C} \in \mathbb{R}^{n\times m}$. Then, the SVD of $\mathbf{A}$ is also its *eigenvalue decomposition* (EVD)

$$\mathbf{A} = \mathbf{U}\mathbf{D}\mathbf{U}^{\mathrm{T}}, \tag{4.23}$$

where the diagonal elements $\lambda_i = \mathbf{D}(i,i) \geq 0$ are the *eigenvalues* of $\mathbf{A}$, and $\mathbf{u}_i = \mathbf{U}(:,i)$ are the corresponding *eigenvectors* with $\mathbf{A}\mathbf{u}_i = \lambda_i \mathbf{u}_i$, $i = 1,\dots,m$. The eigenvectors $\mathbf{u}_1,\dots,\mathbf{u}_m$ can be chosen to be orthonormal, so that they form an orthonormal basis of $\mathbb{R}^m$.

With the EVD (4.23), we can form the *square root* $\mathbf{A}^{\frac{1}{2}}$ *of the matrix* $\mathbf{A}$ by

$$\mathbf{A}^{\frac{1}{2}} = \mathbf{U}\,\mathbf{diag}\,(\lambda_1^{\frac{1}{2}},\dots,\lambda_m^{\frac{1}{2}})\,\mathbf{U}^{\mathrm{T}}. \tag{4.24}$$

Clearly, $\mathbf{A}^{\frac{1}{2}}\mathbf{A}^{\frac{1}{2}} = \mathbf{A}$. If the eigenvalues $\lambda_i > 0$ for all $i = 1,\dots,m$, then $\mathbf{A}^{\frac{1}{2}}$ has an inverse, denoted by $\mathbf{A}^{-\frac{1}{2}}$, and given by

$$\mathbf{A}^{-\frac{1}{2}} = \mathbf{U}\,\mathbf{diag}\,(\lambda_1^{-\frac{1}{2}},\dots,\lambda_m^{-\frac{1}{2}})\,\mathbf{U}^{\mathrm{T}}, \tag{4.25}$$

With the EVD (4.23), we can also give the *constrained* maximum and minimum of the *quadratic form* $\mathbf{x}^{\mathrm{T}}\mathbf{A}\mathbf{x}$, $\mathbf{x} \in \mathbb{R}^m$ as

$$\max_{\|\mathbf{x}\|=1} \mathbf{x}^{\mathrm{T}}\mathbf{A}\mathbf{x} = \lambda_1 \quad \text{and} \quad \min_{\|\mathbf{x}\|=1} \mathbf{x}^{\mathrm{T}}\mathbf{A}\mathbf{x} = \lambda_m, \tag{4.26}$$

where λ_1 and λ_m are the largest and smallest eigenvalues of $\mathbf{A}$, and the maximum is reached with $\mathbf{x} = \mathbf{u}_1$ and the minimum with $\mathbf{x} = \mathbf{u}_m$, where $\mathbf{u}_1$ and $\mathbf{u}_m$ are the corresponding eigenvectors, respectively.

The *generalized inverse, or (Moore–Penrose) pseudoinverse,* $\mathrm{pinv}(\mathbf{A})$ of a matrix $\mathbf{A} \in \mathbb{R}^{m\times n}$, often denoted by $\mathbf{A}^{\dagger}$, is given by

$$\mathbf{A}^{\dagger} = \mathrm{pinv}(\mathbf{A}) = \mathbf{V}\,\mathbf{diag}_{n\times m}(d_1^{-1},\dots,d_r^{-1},0,\dots,0)\,\mathbf{U}^{\mathrm{T}}, \tag{4.27}$$

$$\mathbf{A}^{\dagger} = \begin{bmatrix} \\ \mathbf{V} \\ \\ \end{bmatrix} \begin{bmatrix} d_1^{-1} & & & & (0) \\ & \ddots & & & \\ & & d_r^{-1} & & \\ & & & 0 & \\ (0) & & & & \ddots \; 0 \end{bmatrix} \begin{bmatrix} \\ \mathbf{U}^{\mathrm{T}} \\ \\ \end{bmatrix}$$

Figure 4.10
The (Moore–Penrose) pseudoinverse $\mathbf{A}^{\dagger}$ of an $m \times n$ matrix $\mathbf{A}$ whose singular value decomposition (SVD) is $\mathbf{A} = \mathbf{U}\,\mathbf{D}\,\mathbf{V}^{\mathrm{T}}$ with $\mathbf{D} = \mathbf{diag}_{m\times n}(d_1, \ldots, d_r, 0, \ldots, 0)$.

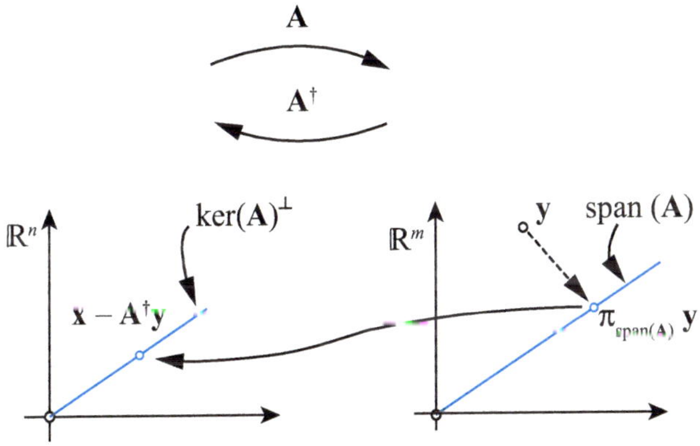

Figure 4.11
Linear mapping $\mathbf{A}: \mathbb{R}^n \to \mathbb{R}^m$ and its pseudoinverse $\mathbf{A}^{\dagger}: \mathbb{R}^m \to \mathbb{R}^n$, which maps span(**A**) one-to-one onto $\ker(\mathbf{A})^{\perp} = \mathrm{span}(\mathbf{A}^{\mathrm{T}})$. Figure shows how $\mathbf{A}^{\dagger}$ maps $\mathbf{y} \in \mathbb{R}^m$ to $\mathbf{A}^{\dagger}\mathbf{y}$. First, $\mathbf{y}$ is orthogonally projected by $\mathbf{\Pi}_{\mathrm{span}(\mathbf{A})}$ to $\mathbf{\Pi}_{\mathrm{span}(\mathbf{A})}\,\mathbf{y}$, and that vector is then mapped by $\mathbf{A}^{\dagger}$ to $\mathbf{A}^{\dagger}\mathbf{\Pi}_{\mathrm{span}(\mathbf{A})}\,\mathbf{y} = \mathbf{A}^{\dagger}\mathbf{y}$.

where $\mathbf{A} = \mathbf{U}\mathbf{D}\mathbf{V}^{\mathrm{T}}$ is the SVD of $\mathbf{A}$ with singular values $d_i = \mathbf{D}(i,i) > 0$, $i = 1, \ldots, r$ and $r = \mathrm{rank}(\mathbf{A})$ (see figure 4.10).

The pseudoinverse $\mathbf{A}^{\dagger}$ is a linear mapping from $\mathbb{R}^m$ to $\mathbb{R}^n$ so that $\mathbf{A}^{\dagger}$ is a one-to-one mapping from span(**A**) onto $\ker(\mathbf{A})^{\perp}$ (see figure 4.11). Using the structure of the SVD, we can show the following results:

$$\mathbf{A}\mathbf{A}^{\dagger} = \Pi_{\mathrm{span}(\mathbf{A})}, \tag{4.28}$$

$$\mathbf{A}^{\dagger}\mathbf{A} = \Pi_{\ker(\mathbf{A})^{\perp}}, \tag{4.29}$$

$$\mathbf{A}^{\dagger} = \mathbf{A}^{\dagger}\Pi_{\mathrm{span}(\mathbf{A})}, \tag{4.30}$$

$$\mathrm{span}(\mathbf{A}^{\dagger}) = \ker(\mathbf{A})^{\perp}, \tag{4.31}$$

$$\mathbf{A}^{\dagger} = (\mathbf{A}^{\mathrm{T}}\mathbf{A})^{-1}\mathbf{A}^{\mathrm{T}} \text{ if } m \geq n = \mathrm{rank}(\mathbf{A}), \tag{4.32}$$

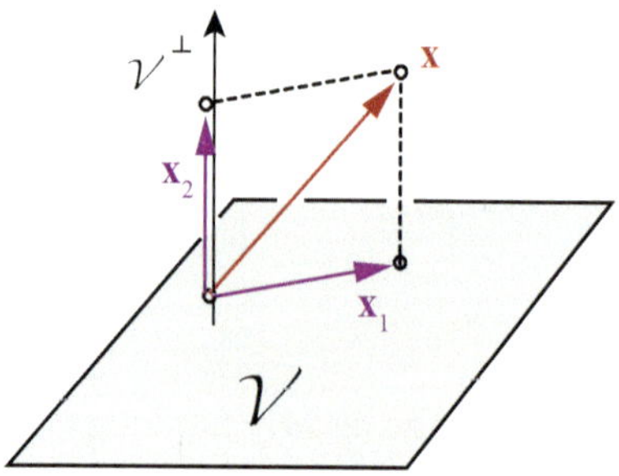

Figure 4.12
Subspace $\mathcal{V} \subset \mathbb{R}^m$ and its orthospace $\mathcal{V}^\perp$ decompose any vector $\mathbf{x} \in \mathbb{R}^m$ to orthogonal components $\mathbf{x}_1 \in \mathcal{V}$ and $\mathbf{x}_2 \in \mathcal{V}^\perp$, where $\mathbf{x}_1 = \mathbf{\Pi}_\mathcal{V}\mathbf{x}$ is the orthogonal projection of $\mathbf{x}$ onto $\mathcal{V}$ and $\mathbf{x}_2 = \mathbf{\Pi}_{\mathcal{V}^\perp}\mathbf{x}$ is the orthogonal projection of $\mathbf{x}$ onto $\mathcal{V}^\perp$.

$$\mathbf{A}^\dagger = \mathbf{A}^\mathrm{T}(\mathbf{A}\mathbf{A}^\mathrm{T})^{-1} \text{ if } n \geq m = \mathrm{rank}(\mathbf{A}), \tag{4.33}$$

$$\mathbf{A}^\dagger = \lim_{\lambda\to 0}(\mathbf{A}^\mathrm{T}\mathbf{A} + \lambda\mathbf{I})^{-1}\mathbf{A}^\mathrm{T} = \lim_{\lambda\to 0}\mathbf{A}^\mathrm{T}(\mathbf{A}\mathbf{A}^\mathrm{T} + \lambda\mathbf{I})^{-1}. \tag{4.34}$$

Let $\mathbf{a}_1, \ldots, \mathbf{a}_n$ be vectors in $\mathbb{R}^m$. By equation (4.28), we get a useful formula (e.g., with MATLAB) to form the orthogonal projection from $\mathbb{R}^m$ onto $\mathrm{span}(\mathbf{a}_1, \ldots, \mathbf{a}_n)$. Namely,

$$\mathbf{\Pi}_{\mathrm{span}(\mathbf{a}_1,\ldots,\mathbf{a}_n)} = \mathbf{A}\mathbf{A}^\dagger \text{ with } \mathbf{A} = [\mathbf{a}_1, \ldots, \mathbf{a}_n], \tag{4.35}$$

and so also

$$\mathbf{\Pi}_{\mathrm{span}(\mathbf{a}_1,\ldots,\mathbf{a}_n)^\perp} = \mathbf{I} - \mathbf{A}\mathbf{A}^\dagger. \tag{4.36}$$

Let $\mathcal{V} = \mathrm{span}(\mathbf{a}_1, \ldots, \mathbf{a}_n) \subset \mathbb{R}^m$. We can decompose any vector $\mathbf{x} \in \mathbb{R}^m$ into two *mutually orthogonal components* $\mathbf{x}_1 \in \mathcal{V}$ and $\mathbf{x}_2 \in \mathcal{V}^\perp$ so that $\mathbf{x} = \mathbf{x}_1 + \mathbf{x}_2$ by setting

$$\mathbf{x}_1 = \mathbf{\Pi}_\mathcal{V}\mathbf{x} \text{ and } \mathbf{x}_2 = \mathbf{x} - \mathbf{\Pi}_\mathcal{V}\mathbf{x} = \mathbf{\Pi}_{\mathcal{V}^\perp}\mathbf{x}, \tag{4.37}$$

where $\mathbf{\Pi}_\mathcal{V}$ is obtained by equation (4.35). Then,

$$\mathbf{x}_1{}^\mathrm{T}\mathbf{x}_2 = (\mathbf{\Pi}_\mathcal{V}\mathbf{x})^\mathrm{T}(\mathbf{x} - \mathbf{\Pi}_\mathcal{V}\mathbf{x}) = 0 \tag{4.38}$$

by equation (4.11), and so $\mathbf{x}_1$ and $\mathbf{x}_2$ are mutually orthogonal (see figure 4.12).

Furthermore, as in equation (4.12), $\mathbf{x}$, $\mathbf{x}_1$, and $\mathbf{x}_2$ satisfy the Pythagoras theorem,

$$||\mathbf{x}||^2 = ||\mathbf{x}_1||^2 + ||\mathbf{x}_2||^2, \tag{4.39}$$

as we can also see by writing $||\mathbf{x}||^2 = (\mathbf{x}_1 + \mathbf{x}_2)^\mathrm{T}(\mathbf{x}_1 + \mathbf{x}_2) = ||\mathbf{x}_1||^2 + 2\mathbf{x}_1{}^\mathrm{T}\mathbf{x}_2 + ||\mathbf{x}_2||^2$ and using the orthogonality condition $\mathbf{x}_1{}^\mathrm{T}\mathbf{x}_2 = 0$.

4.3 Review of Elementary Probability Theory

Let $\mathbf{x}$ be an n-dimensional random vector, or briefly *random n-vector*. Let its probability density be $p(\mathbf{x})$. Its expectation value, or *mean*, $\mathbf{E}(\mathbf{x})$ is defined by

$$\mathbf{E}(\mathbf{x}) = \int \mathbf{x}\, p(\mathbf{x})\, d\mathbf{x} \in \mathbb{R}^n. \tag{4.40}$$

Clearly, $\mathbf{E}$ is a linear operator (on the density functions), meaning that if $\mathbf{x}$ and $\mathbf{y}$ are random n-vectors and $\alpha \in \mathbb{R}$, then $\alpha\mathbf{x}$ and $\mathbf{x}+\mathbf{y}$ are random vectors and

$$\mathbf{E}(\alpha\mathbf{x}) = \alpha\mathbf{E}(\mathbf{x}) \text{ and } \mathbf{E}(\mathbf{x}+\mathbf{y}) = \mathbf{E}(\mathbf{x}) + \mathbf{E}(\mathbf{y}). \tag{4.41}$$

In particular, if $\mathbf{A}$ is an $m \times n$ matrix and $\mathbf{B}$ an $n \times q$ matrix, then $\mathbf{A}\,\mathbf{x}$ and $\mathbf{x}^{\mathrm{T}}\mathbf{B}$ are a random m-vector and a random q-vector and

$$\mathbf{E}(\mathbf{A}\,\mathbf{x}) = \mathbf{A}\,\mathbf{E}(\mathbf{x}), \tag{4.42}$$

$$\mathbf{E}(\mathbf{x}^{\mathrm{T}}\mathbf{B}) = \mathbf{E}(\mathbf{x})^{\mathrm{T}}\mathbf{B}. \tag{4.43}$$

The random n-vectors $\mathbf{x}$ and $\mathbf{y}$ are called *(statistically) independent* if their joint density $p(\mathbf{x},\mathbf{y})$ is a product of the density functions $u(\mathbf{x})$ and $v(\mathbf{y})$ of $\mathbf{x}$ and $\mathbf{y}$, respectively, that is,

$$p(\mathbf{x},\mathbf{y}) = u(\mathbf{x})\, v(\mathbf{y})$$

The *(cross-)correlation matrix* $\operatorname{corr}(\mathbf{x},\mathbf{y})$ of two random n-vectors $\mathbf{x}$ and $\mathbf{y}$ with the joint density $p(\mathbf{x},\mathbf{y})$ is given by

$$\operatorname{corr}(\mathbf{x},\mathbf{y}) = \mathbf{E}(\mathbf{x}\,\mathbf{y}^{\mathrm{T}}) = \int \mathbf{x}\,\mathbf{y}^{\mathrm{T}}\, p(\mathbf{x},\mathbf{y})\, d\mathbf{x}\, d\mathbf{y}. \tag{4.44}$$

We say that $\mathbf{x}$ and $\mathbf{y}$ are *uncorrelated*, if and only if

$$\operatorname{corr}(\mathbf{x}-\mathbf{E}(\mathbf{x}), \mathbf{y}-\mathbf{E}(\mathbf{y})) = \mathbf{E}\Big(\big(\mathbf{x}-\mathbf{E}(\mathbf{x})\big)\big(\mathbf{y}-\mathbf{E}(\mathbf{y})\big)^{\mathrm{T}}\Big) = 0. \tag{4.45}$$

We easily see that $\mathbf{E}\Big(\big(\mathbf{x}-\mathbf{E}(\mathbf{x})\big)\big(\mathbf{y}-\mathbf{E}(\mathbf{y})\big)^{\mathrm{T}}\Big) = \mathbf{E}(\mathbf{x}\mathbf{y}^{\mathrm{T}}) - \mathbf{E}(\mathbf{x})\mathbf{E}(\mathbf{y})^{\mathrm{T}}$, which implies

$$\operatorname{corr}(\mathbf{x},\mathbf{y}) = \mathbf{E}\big((\mathbf{x}-\mathbf{E}(\mathbf{x}))(\mathbf{y}-\mathbf{E}(\mathbf{y}))^{\mathrm{T}}\big) + \mathbf{E}(\mathbf{x})\,\mathbf{E}(\mathbf{y})^{\mathrm{T}}, \tag{4.46}$$

showing that random vectors $\mathbf{x}$ and $\mathbf{y}$ are uncorrelated, if and only if

$$\operatorname{corr}(\mathbf{x},\mathbf{y}) = \mathbf{E}(\mathbf{x})\mathbf{E}(\mathbf{y})^{\mathrm{T}}. \tag{4.47}$$

Furthermore, if $\mathbf{x}$ and $\mathbf{y}$ are uncorrelated, then by equation (4.47) and by its transpose, we get

$$\mathbf{E}(\mathbf{x}\,\mathbf{y}^{\mathrm{T}}) = \mathbf{E}(\mathbf{x})\,\mathbf{E}(\mathbf{y})^{\mathrm{T}}, \tag{4.48}$$

$$\mathbf{E}(\mathbf{y}^{\mathrm{T}}\mathbf{x}) = \mathbf{E}(\mathbf{y})^{\mathrm{T}}\mathbf{E}(\mathbf{x}), \tag{4.49}$$

where in the latter equation, the dimensions of the random vectors $\mathbf{x}$ and $\mathbf{y}$ must be equal.

If $\mathbf{x}$ and $\mathbf{y}$ are independent, then it is easily shown that

$$\mathbf{E}(\mathbf{x}\,\mathbf{y}^{\mathrm{T}}) = \mathbf{E}(\mathbf{x})\,\mathbf{E}(\mathbf{y})^{\mathrm{T}}, \tag{4.50}$$

which by equation (4.47) implies that if $\mathbf{x}$ and $\mathbf{y}$ are independent, then they are also uncorrelated. Note that if $\mathbf{x}$ and $\mathbf{y}$ are uncorrelated, then they do not need to be independent.

The *(auto-)correlation matrix* $\mathrm{corr}(\mathbf{x})$ of a random vector $\mathbf{x}$ is

$$\begin{aligned}\mathrm{corr}(\mathbf{x}) &= \mathrm{corr}(\mathbf{x}, \mathbf{x}) \\ &= \mathbf{E}(\mathbf{x}\,\mathbf{x}^{\mathrm{T}}) = \int \mathbf{x}\,\mathbf{x}^{\mathrm{T}}\, p(\mathbf{x})\, d\mathbf{x}.\end{aligned} \tag{4.51}$$

The *covariance matrix* $\mathrm{cov}(\mathbf{x})$ of $\mathbf{x}$ is defined by

$$\begin{aligned}\mathrm{cov}(\mathbf{x}) &= \mathbf{E}\Big(\big(\mathbf{x} - \mathbf{E}(\mathbf{x})\big)\big(\mathbf{x} - \mathbf{E}(\mathbf{x})\big)^{\mathrm{T}}\Big) \\ &= \int \big(\mathbf{x} - \mathbf{E}(\mathbf{x})\big)\big(\mathbf{x} - \mathbf{E}(\mathbf{x})\big)^{\mathrm{T}} p(\mathbf{x})\, d\mathbf{x}.\end{aligned} \tag{4.52}$$

By equation (4.46), we see that

$$\mathrm{cov}(\mathbf{x}) = \mathrm{corr}(\mathbf{x}) - \mathbf{E}(\mathbf{x})\,\mathbf{E}(\mathbf{x})^{\mathrm{T}}. \tag{4.53}$$

For uncorrelated random vectors $\mathbf{x}$ and $\mathbf{y}$, one gets the *summation rule of covariances*:

$$\mathrm{cov}(\mathbf{x} + \mathbf{y}) = \mathrm{cov}(\mathbf{x}) + \mathrm{cov}(\mathbf{y}). \tag{4.54}$$

If x is a scalar random variable, $\mathrm{cov}(x)$ is the *variance* of x, denoted by $\mathrm{var}(x)$, that is,

$$\mathrm{var}(x) = \mathbf{E}\Big(\big(x - \mathbf{E}(x)\big)^2\Big). \tag{4.55}$$

The *standard deviation* of x is

$$\sigma = \sqrt{\mathrm{var}(x)}. \tag{4.56}$$

We call a random vector (or scalar random variable) $\mathbf{x}$, with $E(\mathbf{x}) = 0$, *white noise* if $\mathrm{cov}(\mathbf{x}) = \sigma^2\mathbf{I}$ for some $\sigma > 0$.

For scalar random variables x and y, the correlation coefficient $\rho(x, y)$ is given by

$$\rho(x, y) = \frac{\mathrm{corr}\big(x - \mathbf{E}(x), y - \mathbf{E}(y)\big)}{\sqrt{\mathrm{var}(x)}\sqrt{\mathrm{var}(y)}} = \frac{\mathbf{E}\Big(\big(x - \mathbf{E}(x)\big)\big(y - \mathbf{E}(y)\big)\Big)}{\sqrt{\mathrm{var}(x)}\sqrt{\mathrm{var}(y)}}. \tag{4.57}$$

If $\mathbf{x} = [x_1, \ldots, x_n]^{\mathrm{T}}$ is a random vector, and $\mathbf{\Gamma} = \mathrm{cov}(\mathbf{x})$ its covariance matrix, then the diagonal elements $\mathbf{\Gamma}(j,j)$ are the variances of the components x_j, that is,

$$\mathbf{\Gamma}(j,j) = \mathrm{var}(x_j),\ j = 1, \ldots, n. \tag{4.58}$$

Let $\mathbf{x}$ be a random n-vector and A an $m \times n$-matrix. With the linearity properties of the mean, we easily get the following useful equations:

$$\mathrm{corr}(\mathrm{A}\,\mathbf{x}) = \mathrm{A}\,\mathrm{corr}(\mathbf{x})\mathrm{A}^{\mathrm{T}} \text{ and } \mathrm{cov}(\mathrm{A}\,\mathbf{x}) = \mathrm{A}\,\mathrm{cov}(\mathbf{x})\mathrm{A}^{\mathrm{T}}. \tag{4.59}$$

Let $\mathbf{x}$ be a random n-vector. Let $\mathrm{X} \in \mathbb{R}^{n \times N}$ be a *sample matrix*, where the columns $\mathrm{X}(:, i)$, $i = 1, \ldots, N$, are random samples of $\mathbf{x}$. Then we can estimate $\mathbf{E}(\mathbf{x})$, $\mathrm{corr}(\mathbf{x})$, and $\mathrm{cov}(\mathbf{x})$ by their *sample values*, obtaining $\overline{\mathbf{x}}$, $\mathrm{corr}(\mathrm{X})$, and $\mathrm{cov}(\mathrm{X})$ as follows:

$$\overline{\mathbf{x}} = \frac{1}{N}\sum_{i=1}^{N} \mathrm{X}(:, i), \tag{4.60}$$

$$\mathrm{corr}(\mathrm{X}) = \frac{1}{N}\sum_{i=1}^{N} \mathrm{X}(:, i)\mathrm{X}(:, i)^{\mathrm{T}} = \frac{1}{N}\mathrm{X}\mathrm{X}^{\mathrm{T}}, \tag{4.61}$$

$$\mathrm{cov}(\mathrm{X}) = \frac{1}{N-1}\sum_{i=1}^{N} (\mathrm{X}(:, i) - \overline{\mathbf{x}})(\mathrm{X}(:, i) - \overline{\mathbf{x}})^{\mathrm{T}} \tag{4.62}$$

$$= \frac{1}{N-1}\mathrm{X_c}\mathrm{X_c}^{\mathrm{T}}, \tag{4.63}$$

where $\mathrm{X_c}$ is the row-centered X given by

$$\mathrm{X_c}(:, i) = \mathrm{X}(:, i) - \overline{\mathbf{x}},\ i = 1, \ldots, N. \tag{4.64}$$

Note that often in equations (4.62) and (4.63) for large N, we replace $(N-1)^{-1}$ by N^{-1}, for convenience.

Sample correlation and covariance matrices $\mathrm{corr}(\mathrm{X})$ and $\mathrm{cov}(\mathrm{X})$ are also called the *normalized* correlation and covariance matrices. The *unnormalized* ones are denoted by $\mathrm{Corr}(\mathrm{X})$ and $\mathrm{Cov}(\mathrm{X})$ and given by

$$\mathrm{Corr}(\mathrm{X}) = \sum_{i=1}^{N} \mathrm{X}(:, i)\mathrm{X}(:, i)^{\mathrm{T}} = \mathrm{X}\mathrm{X}^{\mathrm{T}}, \tag{4.65}$$

$$\mathrm{Cov}(\mathrm{X}) = \sum_{i=1}^{N} (\mathrm{X}(:, i) - \overline{\mathbf{x}})(\mathrm{X}(:, i) - \overline{\mathbf{x}})^{\mathrm{T}} = \mathrm{X_c}\mathrm{X_c}^{\mathrm{T}}, \tag{4.66}$$

where $\mathrm{X_c}$ is as in (4.64).

4.4 Solving Noisy Linear Equations

Consider the equation

$$\mathbf{y} = \mathbf{A}\mathbf{x} + \boldsymbol{\varepsilon}, \tag{4.67}$$

where $\mathbf{y} \in \mathbb{R}^m$, $\mathbf{A} \in \mathbb{R}^{m \times n}$, $\mathbf{x} \in \mathbb{R}^n$, and $\boldsymbol{\varepsilon} \in \mathbb{R}^m$ is random noise. Here, $\mathbf{y} = \mathbf{y}_0 + \boldsymbol{\varepsilon}$ is the known measurement vector, while the noiseless data vector $\mathbf{y}_0 = \mathbf{A}\mathbf{x}$ and noise $\boldsymbol{\varepsilon}$ are unknown. Therefore, finding the exactly correct solution of equation (4.67) is not possible, but we can form an estimate $\widehat{\mathbf{x}}$ of the true $\mathbf{x}$ in equation (4.67) as the solution of the equation

$$\mathbf{A}\widehat{\mathbf{x}} = \mathbf{y}. \tag{4.68}$$

If $\mathbf{A}$ is a square matrix and invertible,

$$\widehat{\mathbf{x}} = \mathbf{A}^{-1}\mathbf{y}. \tag{4.69}$$

If $\mathbf{A}$ is not a square matrix, or $\text{rank}(\mathbf{A}) < m = n$, equation (4.68) may have one solution, several solutions, or no solution. In this case, we must resort to the (Moore–Penrose) *generalized solution*

$$\widehat{\mathbf{x}} = \mathbf{A}^{\dagger}\mathbf{y} \tag{4.70}$$

with $\mathbf{A}^{\dagger} = \text{pinv}(\mathbf{A})$. The generalized solution $\widehat{\mathbf{x}}$ has the following two properties.

First,

$$||\mathbf{A}\widehat{\mathbf{x}} - \mathbf{y}||^2 \leq ||\mathbf{A}\mathbf{x}' - \mathbf{y}||^2 = \sum_{i=1}^{m} \left((\mathbf{A}\mathbf{x}')(i) - \mathbf{y}(i)\right)^2 \tag{4.71}$$

for all $\mathbf{x}' \in \mathbb{R}^n$, that is, $\widehat{\mathbf{x}}$ is the *least-squares* (LS) solution of equation (4.68).

Second,

$$||\widehat{\mathbf{x}}|| \leq ||\widetilde{\mathbf{x}}|| \tag{4.72}$$

for all LS solutions $\widetilde{\mathbf{x}}$ of equation (4.68), that is, $\widehat{\mathbf{x}}$ is the *minimum-norm least-squares* (MNLS), or briefly (MN), solution of (4.68).

If $m > n = \text{rank}(\mathbf{A})$ (overdetermined system) (see figure 4.13), then $\widehat{\mathbf{x}}$ is the only LS solution, and is given by

$$\widehat{\mathbf{x}} = (\mathbf{A}^{\mathrm{T}}\mathbf{A})^{-1}\mathbf{A}^{\mathrm{T}}\mathbf{y}. \tag{4.73}$$

If $m < n$ and $m = \text{rank}(\mathbf{A})$ (underdetermined system) (see figure 4.13), then there are infinitely many LS solutions, and $\widehat{\mathbf{x}}$ is given by

$$\widehat{\mathbf{x}} = \mathbf{A}^{\mathrm{T}}(\mathbf{A}\mathbf{A}^{\mathrm{T}})^{-1}\mathbf{y}. \tag{4.74}$$

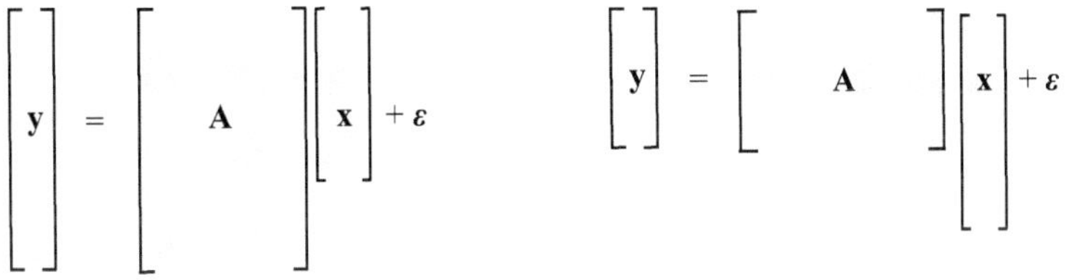

Overdetermined Underdetermined

Figure 4.13
Two noisy matrix equations. *Left-hand figure:* an overdetermined matrix equation with more scalar equations $y_i = \sum_{j=1}^{n} A(i,j)\, x_j + \varepsilon_i,\ i = 1,\dots,m > n,$ than unknowns $x_1,\dots,x_n$. *Right-hand figure:* an underdetermined matrix equation with less scalar equations than unknowns.

Because, by equations (4.33) and (4.34), which state that

$$\mathbf{A}^{\dagger} = \lim_{\lambda\to 0}(\mathbf{A}^{\mathrm{T}}\mathbf{A}+\lambda\mathbf{I})^{-1}\mathbf{A}^{\mathrm{T}} \tag{4.75}$$

$$= \lim_{\lambda\to 0}\mathbf{A}^{\mathrm{T}}(\mathbf{A}\mathbf{A}^{\mathrm{T}}+\lambda\mathbf{I})^{-1}, \tag{4.76}$$

we get, for $\lambda > 0$, an approximation $\widehat{\mathbf{x}}_\lambda$ of $\widehat{\mathbf{x}}$ as

$$\widehat{\mathbf{x}} \simeq \widehat{\mathbf{x}}_\lambda = (\mathbf{A}^{\mathrm{T}}\mathbf{A}+\lambda\mathbf{I})^{-1}\mathbf{A}^{\mathrm{T}}\mathbf{y} \tag{4.77}$$

$$= \mathbf{A}^{\mathrm{T}}(\mathbf{A}\mathbf{A}^{\mathrm{T}}+\lambda\mathbf{I})^{-1}\mathbf{y}, \tag{4.78}$$

which is called the *regularized Tikhonov solution* of equation (4.68) with a *regularization parameter* $\lambda > 0$. The approximation of $\widehat{\mathbf{x}}$ is more accurate the smaller $\lambda > 0$ is. If the condition number cond(A) of A is high, that is, if equation (4.68) is *ill-posed*, then $\widehat{\mathbf{x}}$ is very sensitive to noise in y, while $\widehat{\mathbf{x}}_\lambda$ is less sensitive and often more useful in practice than $\widehat{\mathbf{x}}$. In the next section, among other things, we discuss how a good λ can be chosen.

4.5 Solving Noisy Equations with Estimators

Again, consider the noisy equation

$$\mathbf{y} = \mathbf{A}\mathbf{x} + \boldsymbol{\varepsilon}, \tag{4.79}$$

where $\mathbf{y} \in \mathbb{R}^m$, $\mathbf{A} \in \mathbb{R}^{m\times n}$, $\mathbf{x} \in \mathbb{R}^n$, and $\boldsymbol{\varepsilon}$ is a random noise m-vector with $\mathrm{E}(\boldsymbol{\varepsilon}) = 0$, where E stands for the *expectation*, or mean, operator. Note that $\boldsymbol{\varepsilon}$ also makes y a random vector. As observed in the previous subsection, solving equation (4.79) for x is a task of finding an *estimator* $\widehat{\mathbf{x}}(\mathbf{y})$ for x. The estimator $\widehat{\mathbf{x}}(\mathbf{y})$ is also a random vector as a function of the random vector y.

We call an estimator $\widehat{\mathbf{x}}(\mathbf{y})$ *unbiased* if

$$\mathbf{E}(\widehat{\mathbf{x}}(\mathbf{y})) = \mathbf{x}. \tag{4.80}$$

An estimator $\widehat{\mathbf{x}}(\mathbf{y})$ is called *linear,* if it is a linear function of $\mathbf{y}$, that is, $\widehat{\mathbf{x}}(\mathbf{y}) = \mathbf{My}$ for some matrix $\mathbf{M} \in \mathbb{R}^{n\times m}$.

A popular measure of the goodness of $\widehat{\mathbf{x}}(\mathbf{y})$ is the mean of the square of the distance, or "error," between the estimate $\widehat{\mathbf{x}}(\mathbf{y})$ and the true $\mathbf{x}$, that is, the expression

$$\mathbf{E}(\|\widehat{\mathbf{x}}(\mathbf{y}) - \mathbf{x}\|^2), \tag{4.81}$$

which is called the *mean square error* (MSE) of $\widehat{\mathbf{x}}(\mathbf{y})$. Next we consider different types of estimators.

Let $m \geq n = \text{rank}(\mathbf{A})$. Assume that $\boldsymbol{\varepsilon}$ is *white noise,* that is, $\text{cov}(\boldsymbol{\varepsilon}) = \sigma^2\mathbf{I}$, $\sigma > 0$. Then, the classical *Gauss–Markov theorem* (for instance, see [289]) states that the LS solution of equation (4.79) (also called LS estimator),

$$\widehat{\mathbf{x}}(\mathbf{y}) = (\mathbf{A}^{\mathrm{T}}\mathbf{A})^{-1}\mathbf{A}^{\mathrm{T}}\mathbf{y}, \tag{4.82}$$

is the *best linear unbiased estimator* (BLUE) of $\mathbf{x}$ in equation (4.79). Here, "best" means that $\widehat{\mathbf{x}}(\mathbf{y})$ has the smallest MSE among all unbiased linear estimators.

If $m \geq n = \text{rank}(\mathbf{A})$ but the noise $\boldsymbol{\varepsilon}$ is not white, then the BLUE of $\mathbf{x}$ is

$$\widehat{\mathbf{x}}(\mathbf{y}) = (\mathbf{A}^{\mathrm{T}}\mathbf{C}_{\mathrm{ns}}^{-1}\mathbf{A})^{-1}\mathbf{A}^{\mathrm{T}}\mathbf{C}_{\mathrm{ns}}^{-1}\mathbf{y}, \tag{4.83}$$

where $\mathbf{C}_{\mathrm{ns}} = \text{cov}(\boldsymbol{\varepsilon})$ is the noise covariance matrix.

This claim follows from the Gauss–Markov theorem as follows. Multiply ("whiten") the equation $\mathbf{y} = \mathbf{Ax} + \boldsymbol{\varepsilon}$ by $\mathbf{C}_{\mathrm{ns}}^{-1/2}$ and get

$$\mathbf{z} = \mathbf{Bx} + \boldsymbol{\tau}, \tag{4.84}$$

where $\mathbf{z} = \mathbf{C}_{\mathrm{ns}}^{-1/2}\mathbf{y}$, $\mathbf{B} = \mathbf{C}_{\mathrm{ns}}^{-1/2}\mathbf{A}$, and $\boldsymbol{\tau} = \mathbf{C}_{\mathrm{ns}}^{-1/2}\boldsymbol{\varepsilon}$. Then,

$$\begin{aligned}\text{cov}(\boldsymbol{\tau}) &= \mathbf{E}(\boldsymbol{\tau}\boldsymbol{\tau}^{\mathrm{T}}) = \mathbf{E}(\mathbf{C}_{\mathrm{ns}}^{-1/2}\boldsymbol{\varepsilon}\boldsymbol{\varepsilon}^{\mathrm{T}}\mathbf{C}_{\mathrm{ns}}^{-1/2}) \\ &= \mathbf{C}_{\mathrm{ns}}^{-1/2}\mathbf{E}(\boldsymbol{\varepsilon}\boldsymbol{\varepsilon}^{\mathrm{T}})\mathbf{C}_{\mathrm{ns}}^{-1/2} = \mathbf{C}_{\mathrm{ns}}^{-1/2}\mathbf{C}_{\mathrm{ns}}\mathbf{C}_{\mathrm{ns}}^{-1/2} = \mathbf{I}.\end{aligned} \tag{4.85}$$

Therefore, $\boldsymbol{\tau}$ is white noise, and we can apply the Gauss–Markov theorem to equation (4.84) and get that the BLUE estimator of $\mathbf{x}$ in equation (4.84) is

$$\begin{aligned}\widehat{\mathbf{x}} = \widehat{\mathbf{x}}(\mathbf{z}) &= (\mathbf{B}^{\mathrm{T}}\mathbf{B})^{-1}\mathbf{B}^{\mathrm{T}}\mathbf{z} \\ &= (\mathbf{A}^{\mathrm{T}}\mathbf{C}_{\mathrm{ns}}^{-1/2}\mathbf{C}_{\mathrm{ns}}^{-1/2}\mathbf{A})^{-1}\mathbf{A}^{\mathrm{T}}\mathbf{C}_{\mathrm{ns}}^{-1/2}\mathbf{C}_{\mathrm{ns}}^{-1/2}\mathbf{y} \\ &= (\mathbf{A}^{\mathrm{T}}\mathbf{C}_{\mathrm{ns}}^{-1}\mathbf{A})^{-1}\mathbf{A}^{\mathrm{T}}\mathbf{C}_{\mathrm{ns}}^{-1}\mathbf{y},\end{aligned}$$

as claimed in equation (4.83).

The drawback of estimators $\widehat{\mathbf{x}}(\mathbf{y})$ in equations (4.82) and (4.83) is that if $(\mathbf{A}^T\mathbf{A})^{-1}\mathbf{A}^T$ or $(\mathbf{A}^T\mathbf{C}_{ns}^{-1}\mathbf{A})^{-1}\mathbf{A}^T\mathbf{C}_{ns}^{-1}$ is ill-posed (i.e., their condition numbers are high), then $\widehat{\mathbf{x}}(\mathbf{y})$ is very sensitive to noise $\boldsymbol{\varepsilon}$ in $\mathbf{y}$, which may turn $\widehat{\mathbf{x}}(\mathbf{y})$ useless in practice.

To alleviate this drawback, we must regularize $\widehat{\mathbf{x}}(\mathbf{y})$. For good regularizing methods, we need some prior knowledge of the unknown $\mathbf{x}$, that is, knowledge about $\mathbf{x}$ before it is observed or measured. The prior knowledge must be given in statistical form. Therefore, we also consider $\mathbf{x}$ as a random vector. Here the prior knowledge is assumed to be the prior covariance matrix $\mathbf{C}_{pr} = \mathrm{cov}(\mathbf{x})$ of $\mathbf{x}$. Also, we need to know the noise covariance matrix $\mathbf{C}_{ns} = \mathrm{cov}(\boldsymbol{\varepsilon})$.

We define the MSE of $\widehat{\mathbf{x}}(\mathbf{y})$ for random vectors $\mathbf{x}$ and $\mathbf{y} = \mathbf{A}\mathbf{x} + \boldsymbol{\varepsilon}$ as

$$\mathbf{E}\left(||\widehat{\mathbf{x}}(\mathbf{y}) - \mathbf{x}||^2\right) = \int ||\widehat{\mathbf{x}}(\mathbf{y}) - \mathbf{x}||^2 p(\mathbf{x}, \mathbf{y})\, d\mathbf{x}d\mathbf{y}. \tag{4.86}$$

Note that in equation (4.81) for the MSE, the expectation was taken in the probability distribution of $\mathbf{y}$ alone.

We can assume that $\mathbf{E}(\mathbf{x}) = \mathbf{x}_0 = 0$ (if not, consider $\mathbf{x} - \mathbf{x}_0$ instead of $\mathbf{x}$). Then, the classical result is that the estimator

$$\widehat{\mathbf{x}}(\mathbf{y}) = \mathbf{C}_{pr}\mathbf{A}^T(\mathbf{A}\mathbf{C}_{pr}\mathbf{A}^T + \mathbf{C}_{ns})^{-1}\mathbf{y} \tag{4.87}$$

$$= (\mathbf{A}^T\mathbf{C}_{ns}^{-1}\mathbf{A} + \mathbf{C}_{pr}^{-1})^{-1}\mathbf{A}^T\mathbf{C}_{ns}^{-1}\mathbf{y} \tag{4.88}$$

is the unbiased, linear, and *minimum* MSE (MMSE) *estimator* of $\mathbf{x}$ in the equation $\mathbf{y} = \mathbf{A}\mathbf{x} + \boldsymbol{\varepsilon}$. This MMSE estimator is also called the *Wiener estimator* (in signal processing, the Wiener estimator is approached in a slightly different way).

We note that if the a priori probability distribution of $\mathbf{x}$ and the noise probability distribution of $\boldsymbol{\varepsilon}$ are Gaussian, then $\widehat{\mathbf{x}}(\mathbf{y})$ is also the (Bayesian) posterior mean of $\mathbf{x}$ (and also the MAP = maximal a posteriori probability).

The MMSE estimator could also be called BLUE, because it has the least MSE among all unbiased, linear estimators for the random vector $\mathbf{x}$ with $\mathbf{C}_{pr}$ and $\mathbf{C}_{ns}$. We, however, want to keep the two cases of constant $\mathbf{x}$ and random $\mathbf{x}$ apart and, therefore, use different names.

If $\mathbf{C}_{pr} = \kappa^2\mathbf{I}$ and $\mathbf{C}_{ns} = \sigma^2\mathbf{I}$, then equation (4.87) gets the form

$$\widehat{\mathbf{x}}(\mathbf{y}) = \kappa^2\mathbf{A}^T(\kappa^2\mathbf{A}\mathbf{A}^T + \sigma^2\mathbf{I})^{-1}\mathbf{y},$$

and so

$$\widehat{\mathbf{x}}(\mathbf{y}) = \mathbf{A}^T(\mathbf{A}\mathbf{A}^T + \frac{\sigma^2}{\kappa^2}\mathbf{I})^{-1}\mathbf{y} \tag{4.89}$$

$$= (\mathbf{A}^T\mathbf{A} + \frac{\sigma^2}{\kappa^2}\mathbf{I})^{-1}\mathbf{A}^T\mathbf{y}, \tag{4.90}$$

which is just the Tikhonov-regularized estimator in equations (4.77) and (4.78) with the regularization parameter

$$\lambda = \frac{\sigma^2}{\kappa^2}. \tag{4.91}$$

This λ is the *best regularization parameter in the MMSE sense.* To get that best λ, one should know κ^2 and σ^2. If κ and σ are not available, one can resort to estimating them. Assume that we already have a reasonable estimate for σ^2. If we know for $\mathbf{x} = [x_1, \ldots, x_n]^{\mathrm{T}}$ that each x_i is bounded as $|x_i| \leq L$, then we can choose

$$\kappa^2 = \frac{1}{3} L^2. \tag{4.92}$$

Namely, in this case, a natural choice for the prior distribution is a uniform distribution of $\mathbf{x}$ in the n-dimensional box $|x_i| \leq L$, $i = 1, \ldots, n$, with probability density $p(\mathbf{x}) = (2L)^{-n}$ if $\mathbf{x}$ is in the box, and $p(\mathbf{x}) = 0$ otherwise. Then, $\mathbf{C}_{\mathrm{pr}} = 3^{-1} L^2 \mathbf{I}$, which yields equation (4.92).

If no reasonable estimates for κ^2 and σ^2 are available, we can use heuristic methods to find a suitable λ, like the *L-curve method,* which works as follows.

Set for any $\lambda > 0$,

$$\mathbf{x}_\lambda = \mathbf{A}^{\mathrm{T}} (\mathbf{A}\mathbf{A}^{\mathrm{T}} + \lambda \mathbf{I})^{-1} \mathbf{y}.$$

Draw the graph of $\log(||\mathbf{A}\mathbf{x}_\lambda - \mathbf{y}||)$ as a function of $\log(||\mathbf{x}_\lambda||)$ with varying $\lambda > 0$, for instance, $10^{-8} M \leq \lambda \leq M$, with M being the largest of the diagonal elements of $\mathbf{A}\mathbf{A}^{\mathrm{T}}$.

The obtained curve often has a characteristic L-shape. Choose $\log(||\mathbf{x}_\lambda||)$ so that it corresponds to the "kink" of the L-curve. Find the corresponding λ (for instance, by picking its logarithm from the graph of $\log(\lambda)$ as function of $\log(||\mathbf{x}_\lambda||)$), and that is the Tikhonov λ (see figure 4.14).

4.6 MNLS Solution and Tikhonov Regularization in Other Norms

Again, consider the equation

$$\mathbf{y} = \mathbf{A}\,\mathbf{x} + \boldsymbol{\varepsilon} \tag{4.93}$$

with a noisy data vector as in equation (4.79). We want to find an estimate for the true $\mathbf{x} \in \mathbb{R}^n$, but without any prior knowledge about it or without knowing the noise covariance matrix. We again find the estimate by using a minimum-norm least-squares solution, but here in terms of the *norm determined by a symmetric, positive definite matrix*

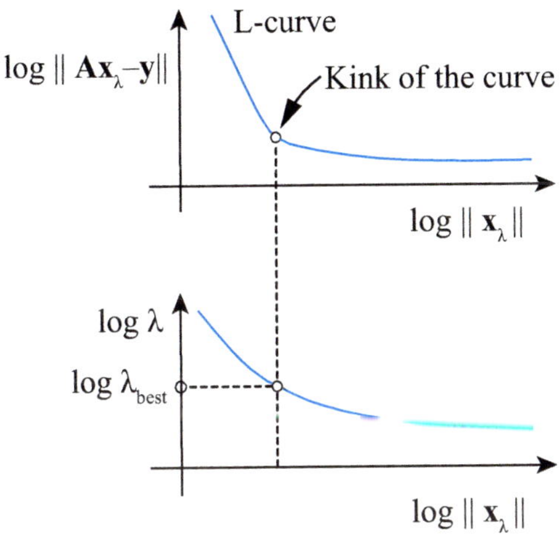

Figure 4.14
Finding the "best" Tikhonov-regularizing parameter λ by the L-curve method.

$\mathbf{B} \in \mathbb{R}^{n \times n}$. This norm for any $\mathbf{x} \in \mathbb{R}^n$ is

$$||\mathbf{x}||_B = (\mathbf{x}^T\mathbf{B}\mathbf{x})^{\frac{1}{2}}, \tag{4.94}$$

The minimum $||\mathbf{x}||_B$ norm least-squares (MNLS) solution of equation (4.93) is given by

$$\widehat{\mathbf{x}}_B = \mathbf{B}^{-1}\mathbf{A}^T\text{pinv}(\mathbf{A}\mathbf{B}^{-1}\mathbf{A}^T)\,\mathbf{y}. \tag{4.95}$$

Solution $\widehat{\mathbf{x}}_B$ is also called the **B**-*weighted MNLS solution* of equation (4.93). A Tikhovov-regularized solution of equation (4.93) with respect to the norm $||\mathbf{x}||_B$ with $\lambda > 0$ is given by

$$\widehat{\mathbf{x}}_{B,\lambda} = \mathbf{B}^{-1}\mathbf{A}^T(\mathbf{A}\mathbf{B}^{-1}\mathbf{A}^T + \lambda\mathbf{I})^{-1}\mathbf{y} \tag{4.96}$$

$$= (\mathbf{A}^T\mathbf{A} + \lambda\mathbf{B})^{-1}\mathbf{A}^T\,\mathbf{y}. \tag{4.97}$$

These equations are derived as follows. Make the substitution $\mathbf{z} = \mathbf{B}^{1/2}\mathbf{x}$ in equation (4.93), and get the equation

$$\mathbf{y} = \mathbf{A}\mathbf{B}^{-1/2}\mathbf{z} + \boldsymbol{\varepsilon}.$$

Apply the MNLS solution (4.70) to this equation, substitute $\widehat{\mathbf{x}} = \mathbf{B}^{-1/2}\widehat{\mathbf{z}}$ to the solution, and get $\widehat{\mathbf{x}} = \mathbf{B}^{-1/2}\text{pinv}(\mathbf{A}\mathbf{B}^{-1/2})\,\mathbf{y}$. You get $\widehat{\mathbf{x}}$ to the form in equation (4.95) by using the identity $\text{pinv}(\mathbf{F}) = \mathbf{F}^T\text{pinv}(\mathbf{F}\mathbf{F}^T)$ for any $\mathbf{F} \in \mathbb{R}^{m \times n}$, which you can derive by the structure (4.27) of the pseudoinverse. The solution $\widehat{\mathbf{x}}_B$ is the minimum-norm solution with respect to the norm $||\mathbf{x}||_B$ because

$$||\mathbf{x}||_B^2 = \mathbf{x}^T\mathbf{B}\mathbf{x} = \mathbf{x}^T\mathbf{B}^{1/2}\mathbf{B}^{1/2}\mathbf{x} = \mathbf{z}^T\mathbf{z} = ||\mathbf{z}||^2.$$

With the same substitutions and with the Tikhonov solutions (4.77) and (4.78), and noticing that

$$\mathbf{B}^{-1/2}(\mathbf{B}^{-1/2}\mathbf{A}^T\mathbf{A}\mathbf{B}^{-1/2} + \lambda\mathbf{I})^{-1}\mathbf{B}^{-1/2} = (\mathbf{A}^T\mathbf{A} + \lambda\mathbf{B})^{-1},$$

you obtain equations (4.96) and (4.97).

By minimizing $F(\mathbf{x}) = ||\mathbf{A}\mathbf{x} - \mathbf{y}||^2 + \lambda||\mathbf{x}||_B^2$ with respect to $\mathbf{x}$ (recall that the minimizing $\mathbf{x}$ is the solution of the equation $\nabla F(\mathbf{x}) = 2(\mathbf{A}^T\mathbf{A}\mathbf{x} + \lambda\mathbf{B}^T\mathbf{x} - \mathbf{A}^T\mathbf{y}) = 0$), we see that $\mathbf{x} = \widehat{\mathbf{x}}_{B,\lambda}$ yields the minimum, that is,

$$\widehat{\mathbf{x}}_{B,\lambda} = \arg\min_{\mathbf{x}} \left(||\mathbf{A}\mathbf{x} - \mathbf{y}||^2 + \lambda||\mathbf{x}||_B^2\right). \tag{4.98}$$

This shows that the solution $\widehat{\mathbf{x}}_{B,\lambda}$ is a trade-off between the sizes of the residual $||\mathbf{A}\mathbf{x} - \mathbf{y}||$ and the norm $||\mathbf{x}||_B$ controlled by $\lambda > 0$.

The term $\lambda||\mathbf{x}||_B^2$ in equation (4.98) is called the *penalty term*. Often the weighting matrix $\mathbf{B}$ is chosen so that small $||\mathbf{x}_B||$ suppresses a specific unwanted property in $\mathbf{x}$. Consequently, large $\lambda > 0$ suppresses that property in the solution $\widehat{\mathbf{x}}_{B,\lambda}$ at the cost of increasing the residual $||\mathbf{A}\mathbf{x} - \mathbf{y}||^2$. A good $\lambda > 0$ must be chosen so that these two properties are in an appropriate balance. It can be searched, for instance, in the range

$$\lambda = c\frac{M_a}{M_b} \text{ with } 10^{-6} \le c \le 10^{-1}, \tag{4.99}$$

where M_a and M_b are the maxima of the diagonal elements of $\mathbf{A}^T\mathbf{A}$ and $\mathbf{B}$, respectively.

One unwanted property of an MNLS solution is its sensitivity to noise if $\mathbf{A}$ is ill-posed, which can be damped by choosing $\mathbf{B}$ so that it makes $\widehat{\mathbf{x}}_{B,\lambda}$ a smoothed version of the MNLS solution $\widehat{\mathbf{x}} = \text{pinv}(\mathbf{A})\mathbf{y}$. Popular choices for a smoothing $\mathbf{B}$ are as follows:

1. $\mathbf{B} = \mathbf{I}$; this yields the usual regularized Tikhonov solution, where $\widehat{\mathbf{x}}_{B,\lambda}$ is smoothed (regularized) by suppressing $||\mathbf{x}||_B^2 = ||\mathbf{x}||^2$ in equation (4.98);
2. $\mathbf{B} = \mathbf{H}^T\mathbf{H}$, where $\mathbf{H}$ is a numerical Hessian operator;
3. $\mathbf{B} = \mathbf{L}^T\mathbf{L}$, where $\mathbf{L}$ is a numerical Laplace operator.

In the two latter cases, $\mathbf{x}$ is thought to represent a real-valued function, whose domain of definition is on a line, on a 2-D surface, or in a 3-D volume. The components $\mathbf{x}(i)$, $i = 1, \ldots, n$, are samples of the function in a discretization grid of that domain. Components $(\mathbf{H}\mathbf{x})(j)$ and $(\mathbf{L}\mathbf{x})(j)$ are numerical approximations of the Hessian or Laplacian of the function at grid points $\mathbf{p}_j$. Note that if the function is defined on a line, then the Hessian is the second derivative of the function.

5 Interpreting MEG and EEG Data

In this chapter, we will present some well-established methods to solve the inverse problem, that is, to determine the primary-current distribution $\mathbf{J}^{\mathrm{p}}(\mathbf{r})$ in the brain on the basis of measurement data and other available information. Estimation theory deals with these kinds of problems. The electromagnetic inverse problem of MEG and EEG has been discussed and solutions have been offered for decades, but we are still in a situation where there is a lot of confusion about how to deal with the inverse problem, even among developers of these techniques, not to speak of the users of MEG and EEG. It would be extremely valuable to minimize the confusion and provide, in addition to the various solutions to the inverse problem, understanding of the meaning and uncertainties of the solutions.

The general logic in data interpretation, which we hope becomes clearer when reading this chapter, is quite simple and should be thoroughly examined by the reader before we elaborate on some sophisticated but straightforward and well-developed approaches (chapters 6–9) about how the inverse problem can be solved.

5.1 Approaches to the Interpretation of MEG and EEG Data

Because of the many different types of MEG and EEG measurement systems and the laboratory-, instrumentation-, and patient-dependent challenges with noise, artifacts, often inaccurately known sensor placement, computational limitations, and the nonuniqueness of the inverse problem, very different approaches have been adopted for the analysis and interpretation of the data.

Originally—and this is still a common practical approach—people just looked at the spontaneous EEG data; paid attention to the characteristic frequencies, spikes, or other features in the waveforms; and learned to correlate these with various disorders or conditions such as sleep stages, epileptic seizures, or coma. Spatial information was obtained by observing which electrodes over the head produce the strongest signals.

This spatial analysis of the signals evolved to include the presentation of interpolated contour or color maps of the recorded potential values and magnetic fields or the topographies of, for example, their Fourier or wavelet components or other derived measures. This approach, which has been widely used in biofeedback therapy, is often called quantitative EEG or qEEG [90]. Sometimes such maps have been analyzed by searching for "microstates," that is, topographical patterns in EEG that would reflect fixed spatial distributions of neuronal sources in the brain [245]. However, such approaches are generally only descriptive and correlative and therefore not discussed further in this book.

EEG and MEG studies can be performed in three different basic modes: (1) the measurement of spontaneous activity during rest or during task performance, (2) the measurement of evoked responses, and (3) closed-loop or feedback-based studies. In mode 1, one can often detect characteristic frequencies in the signal; with Fourier or wavelet analysis, one can obtain measures of activity in different sensors with a high signal-to-noise ratio. In mode 2, the same stimulus is typically presented to the subject multiple times, which allows averaging so that the contribution of random noise can be reduced. In mode 3, where the aim may be the control by EEG signals of an external device such as a computer or wheelchair or just a game, feedback is given to the subject depending on the measured signals. For each of these modes, a multitude of experimental paradigms and data-analysis approaches has been developed. For most of the approaches, we only can refer to books [45, 94, 116, 117, 121, 193, 194, 201, 206, 215, 221, 223, 228, 240, 243, 256, 268, 270, 278, 283, 301] and reviews or other presentations on MEG or EEG [5, 20, 22, 60, 61, 70, 77–79, 110, 111, 119, 120, 148, 149, 151, 168, 186, 191, 217, 222, 251, 253, 271, 272, 288, 296, 323, 332] that the interested reader might find useful.

A large number of software packages are available for the analysis of MEG or EEG data [21]. These include: ARTIST [326], Biosig [318], BrainNetVis [53], Brainstorm [302], CARTOOL [43], Craniux [76], eConnectome [125], EEGIFT [89], EEGLAB [80, 81], ELAN [3], EMEGS [248], FAST [183], FieldTrip [235], HADES [47], OpenMEEG [104], LIMO EEG [246], MathBrain [263] MEGMRIAn [311], MEG-SIM [9], MNE [103], NUTMEG [68], PyEEG [23], Ragu [173], rtMEG [300], sLORETA [244], SOUND [214], SPM [190], TMSEEG [17], and TopoToolbox [309]. In addition, commercial hardware vendors often provide their own analysis tools specific to their instrumentation or approach. Descriptions of the different techniques are thus scattered in many different papers and books, and in software documentation. Some effort has been put into unifying the approaches; see Gross et al. [106] and Keil et al. [164] and references therein.

Here, we do not delve into the specific pros and cons of the different approaches and packages. Suffice it to say as a general comment that while the mentioned toolboxes are based on a tremendous amount of understanding and sophistication in data analysis and software technology, there is a danger of obtaining meaningless results if one uses them without knowing and understanding what the toolboxes do and how they do it. For example, the best dipole-fitting or dipole-searching toolbox may produce unpredictable results if the data are not due to a sufficiently small number of localized sources, if there are artifacts or too much noise, or if the forward model is not accurate. Likewise, the minimum-norm estimate is not the right choice for data interpretation if there is reason to believe that the electrical activity is localized; for a comparison of EEG-based minimum-norm estimation and dipole fitting of MEG data, see Komssi et al. [175].

Instead of discussing the multifarious benefits and limitations of the various methods, our aim here is to lay the basis for some of the basic techniques in interpreting the signals quantitatively and correctly. We will limit ourselves to methods we are confident in presenting in a straightforward and trustworthy way.

5.2 The Inverse Problem

Our inverse problem is simply this: determine the unknown, time-varying primary-current distribution $\mathbf{J}^{\mathrm{p}}(\mathbf{r}, t)$ in the brain based on MEG or EEG data $\mathbf{Y}(t) = [y_1(t), y_2(t), \cdots, y_{N_{\mathrm{ch}}}(t)]^{\mathrm{T}}$, sensor array geometry, head conductivity structure $\sigma(\mathbf{r})$, and a priori information about $\mathbf{J}^{\mathrm{p}}(\mathbf{r}, t)$ and the noise $\eta(t)$.

The solution of the inverse problem of MEG or EEG is generally nonunique, meaning that many solutions (typically an infinite number) are compatible with the data. The relationship between sources and the measured data values being linear, the nonuniqueness manifests itself as the existence of *silent sources*, that is, primary-current distributions that produce no signal in any of the sensors. Our ability to produce useful solutions to the inverse problem depends on what we know about the silent sources or their relationship with the part of primary-current distribution that causes our measured data. Typically, this a priori knowledge about the sources is implicit and very hard to quantify or use in the analysis of the inverse problem. Often, the existence of a priori knowledge is revealed to a researcher only after some inverse solution is produced: the researcher may be confident that the results are wrong, although the calculations have been done correctly; this situation indicates that not all available a priori knowledge about the primary currents was used in computing the solution. Unfortunately, no good recipe for dealing with this state of affairs has been found. In any case, if we have supplementary information about the sources, in principle the solution to the inverse problem can be made more accurate or reliable.

5.3 The Solution Without A Priori Information

The first thing to realize about the nature of the measured signal is that

$$y_i(t) = \int \mathbf{L}_i(\mathbf{r}) \cdot \mathbf{J}^{\mathrm{p}}(\mathbf{r}, t)\, dV \tag{5.1}$$

is a projection (multiplied by the norm of the lead field $\mathbf{L}_i$) of $\mathbf{J}^{\mathrm{p}}(t)$ onto the lead field; it is the inner product $\langle \mathbf{J}^{\mathrm{p}}, \mathbf{L}_i \rangle$ of $\mathbf{J}^{\mathrm{p}}$ and $\mathbf{L}_i$, see section 2.2 and the beginning of chapter 3. We thus obtain, by measuring $y_i(t)$, $i = 1, 2, \cdots, N_{\mathrm{ch}}$, the projection of $\mathbf{J}^{\mathrm{p}}(t)$ on the subspace spanned by the lead fields. Note that it suffices to define the lead fields only in the volume where $\mathbf{J}^{\mathrm{p}}$ is known to reside, such as in the intracranial space or the cortex. If one also knows, in each spatial location, the direction of the primary current (typically, one can assume that it is roughly perpendicular to the cortex), one can take advantage of this knowledge and improve the inverse solution [187]. The projection that we measure is the *minimum-norm solution* (MNE) of the inverse problem. Any other primary-current distribution that can explain the measurement is the MNE solution plus some distribution of *silent currents* that lie in a subspace orthogonal to the subspace spanned by the lead fields. In the MNE solution, the silent components are assigned the value zero. When nothing is known about the silent sources, MNE is the best solution in the sense that it minimizes the expectation value of the squared error of the estimated $\mathbf{J}^{\mathrm{p}}$.

MEG and EEG thus inform us faithfully (apart from noise) about those components of the unknown primary-current distribution that we measure and give no information whatsoever about the silent currents [124]. Only if we have a priori information about the relationship between the measured and silent-current components, can we gain knowledge of the latter.

5.4 Signal Space and Signal-Space Projection (SSP)

The signals in an N_{ch}-sensor EEG/MEG measurement at one time instant constitute the vector $\mathbf{y} = [y_1, \ldots, y_{N_{\mathrm{ch}}}]^{\mathrm{T}}$ in the *signal space* $\subset \mathbb{R}^{N_{\mathrm{ch}}}$. Each signal y_i is the scalar product $\langle \mathbf{J}^{\mathrm{p}}, \mathbf{L}_i \rangle$, and so $y_i \langle \mathbf{L}_i, \mathbf{L}_i \rangle^{-1} \mathbf{L}_i$ is the (orthogonal) component of $\mathbf{J}^{\mathrm{p}}$ on $\mathbf{L}_i$. MEG and EEG thus directly measure components of the (infinite-dimensional) primary-current vector $\mathbf{J}^{\mathrm{p}}$. It may be appropriate to note that MEG and EEG literature often refers to the concepts of *sensor space* and *source space*. The set of signals in sensor space means just the measured signals in the signal space while signals in the source space usually refer to estimates of the source activity such as source waveforms.

The signal-space projection [133, 141, 226, 227, 255, 281, 306, 307, 312] or SSP is simply a projection of the signal vector onto a subspace of the signal space. SSP is a signal-separation method, but since the signal vectors are projections of source-current vectors, it can equally well be considered a source-separation method. Each spatially fixed source configuration produces a signal vector that varies in amplitude but remains fixed in its orientation in the signal space. Therefore, if one knows (e.g., after having performed a separate measurement) the direction in signal space of a signal vector arising from a particular source, the contribution of that source can be completely eliminated by SSP. This is often very useful in eliminating artifacts that span only a small subspace of the signal space; see section 1.7.6.

For each signal vector direction in the signal space, many different source configurations can produce a signal in that direction; the collection of all source configurations with the same signal vectors in the signal space constitutes an *equivalence class* of current configurations [306]. Also, signal vectors produced by distinct sources are generally not orthogonal, meaning that by projecting out one source component, the topographies, that is, signal vectors of other sources also change. This apparent distortion of the topographies may make it difficult to interpret EEG or MEG topoplots (contour plots of topographies) by eye. To remedy this problem, one can reconstruct the original topography by estimating the primary-current distribution from the artifact-cleaned, projected data and then computing the signals in all sensors. It turns out that for this procedure to work satisfactorily, it is not important to have a very accurate volume-conductor model, because the inverse and forward calculations done in sequence compensate for each other's errors [213].

A simple example of the use of SSP is the removal of an artifact that always produces the same EEG or MEG topography but has variable and unknown amplitude. Once the topography of the artifact is known, that dimension is projected away, and one is left with artifact-free signal vectors that have one fewer dimensions than the original signal vectors. If there are many artifacts, one can project away the subspace spanned by the artifact signals. One can also separate sources with SSP [307].

5.5 The Solution If There Is A Priori Information

Sometimes, we may have sufficient information about the source currents to allow us to *constrain* the solution to be unique. The simplest case is the *equivalent-dipole* model, with which one determines the location, orientation, and amplitude of a current dipole that best explains the measurement. A current dipole is a useful representation of primary currents that are confined to a small volume: then, the dipole parameters correspond approximately to the location, orientation, and total dipole moment of the modeled primary-current distribution. Unique solutions can often be obtained for any source models where the number of unknowns is at most the number of independent measurements. If one has reason to believe that N dipoles suffice to represent the primary-current distribution, one can use an N-dipole model, which has $5N$ unknowns for tangential dipoles (for MEG use) or $6N$ unknowns in the general case. However, as N grows, it quickly becomes very difficult to find the combination of dipole parameters that would best correspond to the data unless the time courses are sufficiently independent and are taken into account; see chapters 6 to 8, or various approaches in [209–211, 273, 274].

In more complicated cases, it may be best to use Bayesian type of thinking as guidance (see section 1.4) [324]. This means that one combines any a priori information about the source currents with information obtained from measurements; the result would be the a posteriori probability distribution in the space of all source-current distributions, mathematically given by Bayes's rule. In most cases, however, one finds it very hard to express the experimenter's or clinician's knowledge as a probability distribution—or even if one could, it would usually be computationally very hard to determine the posterior distribution.

As pointed out above, many inverse-problem solutions have been presented in the literature. The usual requirement is to demand that the estimated primary-current distribution, whether discrete as in multiple-dipole models or continuous as in MNE, would be in no contradiction with the data, that is, the estimated current distribution should reproduce the data at the level of accuracy determined by noise. However, very little emphasis is usually put on the requirement that the solution should also be compatible with the a priori knowledge. Often the implicit assumptions, used to constrain the inverse solution, about the nature of the source currents are wrong (not compatible with *knowledge* about the sources), and at other times, the solution does not take prior knowledge into account.

In the following chapters, we will describe several methods to solve the inverse problem or to separate the topographies and time courses of distinct sources. Two points are worth emphasizing from the outset. The first point is that inverse solutions should not be judged by their performance in idealized simulated circumstances or in situations that they are not meant to deal with. Both equivalent-dipole and minimum-norm solutions are optimal in their respective realms of validity, but not in other cases. Dipole modeling is appropriate for locating a small number of spatially confined current sources (for some caveats in dipole modeling, see Kobayashi et al. [172]); MNE is suitable for obtaining optimal estimates of the current distribution when nothing else is known about the sources except the source volume (and possibly primary-current direction); independent component analysis (ICA) cannot be assumed to separate sources correctly unless the sources are (statistically) *independent*. Often the worst thing a scientist can do is to try different methods and choose the result that best fits preconceived expectations or just seems plausible; biased results would then be obtained.

The second point is that one has to pay attention to the requirements of each inverse-problem solution regarding data quality and the accuracy of the conductivity model. For example, although some methods such as beamformer algorithms (chapter 6) or the TRAP-MUSIC algorithm (section 7.5.2) may be otherwise perfect in their domain of validity, they may be worthless if the conductivity model (including estimated tissue conductivity values) is not sufficiently accurate even if all assumptions about the sources are correct. For issues concerning conductivity effects, see references [13, 18, 98, 108].

5.6 Measurement Data

We start by recalling the concepts of a current dipole and lead-field matrices assigned to source dipole locations, as presented in section 2.2, but with slightly different notation. If a source-current distribution $\mathbf{J}^p(\mathbf{r})$ is confined in a *small volume* (say, diameter less than 5 mm), it can be modeled as a *current dipole*,

$$\mathbf{J}^p(\mathbf{r}) = \delta(\mathbf{r}-\mathbf{p})\,\mathbf{q}, \tag{5.2}$$

where $\mathbf{p}$ is the location and $\mathbf{q}$ is the *moment* of the dipole, and $\delta(\mathbf{r}-\mathbf{p})$ is the Dirac delta function with the property that $\int g(\mathbf{r})\delta(\mathbf{r}-\mathbf{p})\,d\mathbf{r} = g(\mathbf{p})$. We also denote this current dipole as the pair $(\mathbf{p}, \mathbf{q})$. A practical model for the current dipole is a small battery in a conductive environment. Such a battery has two spatially separated poles, one positive and the other negative, and a mechanism that moves an electric charge from the negative pole to the positive one (see figure 2.2). Then, the negative pole becomes a sink and the positive pole a source for current in the conducting medium in which the dipole is embedded.

In field computing, a distributed current source *can be approximated* in practice by a collection of current dipoles. Therefore, we need to compute the magnetic field $\mathbf{B}(\mathbf{r})$ or the electric potential $V(\mathbf{r})$ only due to current dipoles. The fields of a distributed source are then obtained by adding together the fields of the approximating dipoles.

Consider a source dipole $(\mathbf{p}, \mathbf{q})$, where $\mathbf{p}$ is the *location* and $\mathbf{q} = s\,\eta$ the *moment*, with the unit vector η being the *orientation* and the scalar s the *amplitude* of the dipole.

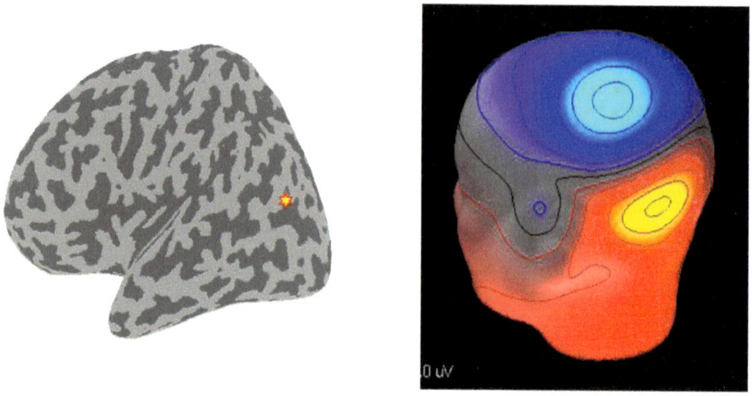

Figure 5.1
Topography of a single (tangential) source dipole: the location of the dipole on the cortex is marked by a star in the left-hand figure, the BEM-computed topography (the potential contour plot) is depicted in the right-hand figure; colors from light blue to yellow correspond to electric potential values from minimum to maximum.

We denote by $\mathbf{l}(\mathbf{p}, \eta)$ the MEG or EEG *measurement* (*sensor*) vector due to the unit dipole $(\mathbf{p}, \eta)$. Then, thanks to linearity, the scalar measurement readings at the sensors (magnetometers or gradiometers for MEG and electrodes for EEG) due to the dipole $\mathbf{q} = s\,\eta$ at $\mathbf{p}$ form the vector

$$\mathbf{y} = [y_1, \ldots, y_m]^{\mathrm{T}} = s\,\mathbf{l}(\mathbf{p}, \eta)\,, \tag{5.3}$$

where m is the number of sensors. The m-vector $\mathbf{l}(\mathbf{p}, \eta)$ is called the *topography* of the dipole at location $\mathbf{p}$ with orientation η and amplitude $s = 1$ (see figure 5.1).

Again, thanks to the linearity with respect to the moment, the topography can be given in the form

$$\mathbf{l}(\mathbf{p}, \eta) = \mathbf{L}(\mathbf{p})\,\eta\,, \tag{5.4}$$

where $\mathbf{L}(\mathbf{p})$ is the (local) $m \times 3$ lead-field matrix at $\mathbf{p}$ given by

$$\mathbf{L} = [\mathbf{l}_x, \mathbf{l}_y, \mathbf{l}_z] \tag{5.5}$$

with the columns

$$\mathbf{l}_x = \mathbf{l}(\mathbf{p}, \mathbf{e}_x),\; \mathbf{l}_y = \mathbf{l}(\mathbf{p}, \mathbf{e}_y),\; \mathbf{l}_z = \mathbf{l}(\mathbf{p}, \mathbf{e}_z)\,, \tag{5.6}$$

where $\mathbf{e}_x$, $\mathbf{e}_y$, and $\mathbf{e}_z$ are the xyz-coordinate unit vectors in 3-space. The matrix $\mathbf{L}(\mathbf{p})$ is numerically computed by a forward field solver, for locations $\mathbf{p}$ in the *region of interest* (ROI) for the given head model.

We say that the source dipoles in the ROI are *fixed-oriented* if their orientations are known functions of location $\eta = \eta(\mathbf{p})$. For instance, if the ROI is the cortical surface, the source dipoles can often be assumed to be perpendicular to that surface. It follows that the topographies of the fixed-oriented dipoles are also known functions of locations and we denote the topography of such a dipole at $\mathbf{p}$ by $\mathbf{l}(\mathbf{p}) = \mathbf{L}(\mathbf{p})\eta(\mathbf{p})$. If the dipoles are not fixed-oriented, we say that they are *freely oriented.*

Assume now that dipoles in the ROI are freely oriented. Let $(\mathbf{p}, \mathbf{q})$ again be a dipole at location $\mathbf{r}$ with moment

$$\mathbf{q} = [q_1, q_2, q_3]^{\mathrm{T}} = q_1\mathbf{e}_x + q_2\mathbf{e}_y + q_3\mathbf{e}_z.$$

Let $\mathbf{y} = \mathbf{l}(\mathbf{p}, \mathbf{q})$ be the measurement vector due to this dipole. Due to linearity and the superposition principle, we can write:

$$\begin{aligned}\mathbf{y} &= \mathbf{l}(\mathbf{p}, \mathbf{q}) = \mathbf{l}(\mathbf{p}, q_1\mathbf{e}_x + q_2\mathbf{e}_y + q_3\mathbf{e}_z) \\ &= \mathbf{l}(\mathbf{p}, q_1\mathbf{e}_x) + \mathbf{l}(\mathbf{p}, q_2\mathbf{e}_y) + \mathbf{l}(\mathbf{p}, q_3\mathbf{e}_z) \\ &= q_1\mathbf{l}(\mathbf{p}, \mathbf{e}_x) + q_2\mathbf{l}(\mathbf{p}, \mathbf{e}_y) + q_3\mathbf{l}(\mathbf{p}, \mathbf{e}_z) \\ &= [\mathbf{l}(\mathbf{p}, \mathbf{e}_x), \mathbf{l}(\mathbf{p}, \mathbf{e}_y), \mathbf{l}(\mathbf{p}, \mathbf{e}_z)] \begin{bmatrix} q_1 \\ q_2 \\ q_3 \end{bmatrix} = \mathbf{L}(\mathbf{p})\,\mathbf{q}.\end{aligned} \tag{5.7}$$

Let the source consist of dipoles $(\mathbf{p}_1, \mathbf{q}_1), \ldots, (\mathbf{p}_n, \mathbf{q}_n)$. By the superposition principle, the measurement vector $\mathbf{y}$ due to this source is

$$\begin{aligned}\mathbf{y} &= \sum_{i=1}^{n} \mathbf{L}(\mathbf{p}_i)\,\mathbf{q}_i \\ &= [\mathbf{L}(\mathbf{p}_1), \ldots, \mathbf{L}(\mathbf{p}_n)] \begin{bmatrix} \mathbf{q}_1 \\ \vdots \\ \mathbf{q}_n \end{bmatrix},\end{aligned} \tag{5.8}$$

or briefly,

$$\mathbf{y} = \mathbf{A}\mathbf{x}, \tag{5.9}$$

where

$$\mathbf{A} = [\mathbf{L}(\mathbf{p}_1), \ldots, \mathbf{L}(\mathbf{p}_n)] \tag{5.10}$$

is the *free-orientation* $m \times 3n$ *mixing (gain, system) matrix* assigned to the source dipole locations $\mathbf{p}_1, \ldots, \mathbf{p}_n$, and

$$\mathbf{x} = \begin{bmatrix} \mathbf{q}_1 \\ \vdots \\ \mathbf{q}_n \end{bmatrix} \tag{5.11}$$

is the long $3n \times 1$ column vector containing the moment vectors.

Assume next that the orientations of dipoles in the ROI are fixed-oriented. Let the source consist of n dipoles with locations $\mathbf{p}_1, \ldots, \mathbf{p}_n$, orientations $\eta_1, \ldots, \eta_n$, topographies $\mathbf{l}(\mathbf{p}_i) = \mathbf{L}(\mathbf{p}_i)\,\eta_i$, $i = 1, \ldots, n$, and amplitudes $x_1, \ldots, x_n$. Then the measurement vector $\mathbf{y}$ due to these source dipoles gets the following form:

$$\mathbf{y} = \sum_{i=1}^{n} x_i \mathbf{l}(\mathbf{p}_i) = [\mathbf{l}(\mathbf{p}_1), \ldots, \mathbf{l}(\mathbf{p}_n)] \begin{bmatrix} x_1 \\ \vdots \\ x_n \end{bmatrix} = \mathbf{A}\,\mathbf{x}, \tag{5.12}$$

where

$$\mathbf{A} = [\mathbf{l}(\mathbf{p}_1), \ldots, \mathbf{l}(\mathbf{p}_n)] \tag{5.13}$$

is the *fixed-orientation* $m \times n$ *mixing (gain, system) matrix.*

5.7 Search for a Single Dipole Source

We consider next the simple inverse problem, where we know that the measurement data are due to a single dipole source. The solution is generally unique and accurate provided that a good forward model is available and the data are not very noisy. Let the source be a single dipole $(\mathbf{p}, \mathbf{q})$. Then, by equation (5.7), the measurement vector $\mathbf{y}$ with additional noise ϵ is given by

$$\mathbf{y} = \mathbf{L}(\mathbf{p})\mathbf{q} + \boldsymbol{\varepsilon}. \tag{5.14}$$

The inverse problem is to find $(\mathbf{p}, \mathbf{q})$ when $\mathbf{y}$ is given. Here we present a very simple and robust method to solve the problem.

We start by approximating the ROI with a sufficiently dense grid of locations $\mathbf{p}_i$, $i = 1, \ldots, n$, and we look for the best approximation of $(\mathbf{p}, \mathbf{q})$ by a dipole that is located in the grid.

Assume first that the dipoles in the ROI are fixed-oriented with topographies $\mathbf{l}_i = \mathbf{l}(\mathbf{p}_i)$, $i = 1, \ldots, n$. The task is to find the dipole with location $\mathbf{p}_k$ and amplitude x_k, so that its measurement vector $x_k \mathbf{l}_k$ best fits to the measured data vector $\mathbf{y} = \mathbf{L}(\mathbf{p})\mathbf{q} + \boldsymbol{\varepsilon}$, that is,

$$||\mathbf{y} - x_k \mathbf{l}_k|| = \min_{1 \le j \le n,\, x \in \mathbb{R}} ||\mathbf{y} - x\mathbf{l}_j|| = \min_{1 \le j \le n} \left(\min_{x \in \mathbb{R}} ||\mathbf{y} - x\mathbf{l}_j|| \right). \tag{5.15}$$

The amplitude $\widehat{x}_j$ that minimizes $||\mathbf{y} - x\mathbf{l}_j||$ is the least-squares (LS) solution of the overdetermined equation

$$\mathbf{l}_j x = \mathbf{y}, \tag{5.16}$$

and so by equation (4.73),

$$\widehat{x}_j = (\mathbf{l}_j{}^{\mathrm{T}} \mathbf{l}_j)^{-1} \mathbf{l}_j{}^{\mathrm{T}} \mathbf{y} = \frac{1}{||\mathbf{l}_j||^2} \mathbf{l}_j{}^{\mathrm{T}} \mathbf{y}. \tag{5.17}$$

The residual is

$$\mathrm{Res}(j) = ||\mathbf{y} - \widehat{x}_j \mathbf{l}_j||. \tag{5.18}$$

It follows that the index k of the location $\mathbf{p}_k$ of the best-fitting dipole is

$$k = \arg \min_{1 \le j \le n} \mathrm{Res}(j), \tag{5.19}$$

and its amplitude is

$$\widehat{x}_k = \frac{1}{||\mathbf{l}_k||^2} \mathbf{l}_k{}^{\mathrm{T}} \mathbf{y}. \tag{5.20}$$

The inverse problem of a single dipole source for fixed-oriented dipoles is solved.

Assume next that the dipoles in the ROI are freely oriented. Let $\mathbf{L}_j = \mathbf{L}(\mathbf{p}_j)$, $j = 1, \ldots, n$. The task is to find a location $\mathbf{p}_k$ and a moment $\mathbf{q}_k$, that is, a dipole $(\mathbf{p}_k, \mathbf{q}_k)$, so that, with $\mathbf{L}_k = \mathbf{L}(\mathbf{p}_k)$,

$$||\mathbf{y} - \mathbf{L}_k \mathbf{q}_k|| = \min_{1 \le j \le n,\, \mathbf{q} \in \mathbb{R}^3} ||\mathbf{y} - \mathbf{L}_j \mathbf{q}|| = \min_{1 \le j \le n} \left(\min_{\mathbf{q} \in \mathbb{R}^3} ||\mathbf{y} - \mathbf{L}_j \mathbf{q}|| \right). \tag{5.21}$$

The moment vector $\widehat{\mathbf{q}}_j$, which minimizes $||\mathbf{y} - \mathbf{L}_j \mathbf{q}||$, is the LS solution of the overdetermined equation

$$\mathbf{L}_j \mathbf{q} = \mathbf{y}, \tag{5.22}$$

and so,

$$\widehat{\mathbf{q}}_j = (\mathbf{L}_j{}^{\mathrm{T}} \mathbf{L}_j)^{-1} \mathbf{L}_j{}^{\mathrm{T}} \mathbf{y}, \tag{5.23}$$

and the residual is

$$\mathrm{Res}(j) = ||\mathbf{y} - \mathbf{L}_j \widehat{\mathbf{q}}_j||, \ j = 1, \ldots, n. \tag{5.24}$$

It follows that the index k of the location $\mathbf{p}_k$ of the best-fitting dipole is

$$k = \arg \min_{1 \le j \le n} \mathrm{Res}(j), \tag{5.25}$$

and so the moment of that dipole is

$$\widehat{\mathbf{q}}_k = (\mathbf{L}_k{}^{\mathrm{T}}\mathbf{L}_k)^{-1}\mathbf{L}_k{}^{\mathrm{T}}\mathbf{y}. \tag{5.26}$$

We note that there is a simpler expression than Res(j) to find the k in equation (5.19). Namely, if we write $\mathbf{a}_j = ||\mathbf{l}_j||^{-1}\,\mathbf{l}_j$ and then $||\mathbf{a}_j|| = 1$, we get

$$\mathrm{Res}(j)^2 = ||\mathbf{y} - \mathbf{a}_j{}^{\mathrm{T}}\mathbf{y}\,\mathbf{a}_j||^2 = ||\mathbf{y}||^2 - 2(\mathbf{a}_j{}^{\mathrm{T}}\mathbf{y})^2 + (\mathbf{a}_j{}^{\mathrm{T}}\mathbf{y})^2 = ||\mathbf{y}||^2 - (\mathbf{a}_j{}^{\mathrm{T}}\mathbf{y})^2. \tag{5.27}$$

This shows that the minimum of Res(j) is reached when $|\mathbf{a}_j{}^{\mathrm{T}}\mathbf{y}|$ reaches its maximum, that is, the topography $\mathbf{l}_j$ correlates maximally with the measured data vector $\mathbf{y}$, and so

$$k = \arg\max_{1\leq j\leq n} \frac{1}{||\mathbf{l}_j||}\,|\mathbf{l}_j{}^{\mathrm{T}}\mathbf{y}|. \tag{5.28}$$

The solution of the inverse problem of a single dipole source is now completed.

The search for a single dipole source is a widely used inverse method in analyzing MEG/EEG data. Naturally, it suits well only cases where only one (equivalent) dipole is expected to have produced the measured data.

Another typical use of the single dipole search is in a visual preliminary review of a time series of N measurements with time points t_i so that $\mathbf{y}_i \in \mathbb{R}^m$ is the measurement vector at time t_i, $i = 1,\ldots,N$, and $\mathbf{Y} = [\mathbf{y}_1,\ldots,\mathbf{y}_N]$ is the measurement data matrix. The review can be carried out as follows.

We want to find interesting features in the data $\mathbf{y}_1,\ldots,\mathbf{y}_N$, which are supposed to be due to single dipoles, each active at different time points.

We first consider EEG data. Let the components $\mathbf{y}_i(j)$, $j = 1,\ldots,m$, of vector $\mathbf{y}_i$ be the potential readings at electrode locations $\mathbf{p}_j$, $j = 1,\ldots,m$, on the scalp. For each time point t_i, we form the standard deviation of the components $\mathbf{y}_i(j)$, $j = 1,\ldots,m$, as

$$\sigma(t_i) = \left(\frac{1}{m}\sum_{j=1}^{m}(\mathbf{y}_i(j) - \mu_i)^2\right)^{1/2}, \text{ with } \mu_i = \frac{1}{m}\sum_{j=1}^{m}\mathbf{y}_i(j), \tag{5.29}$$

where $\sigma(t_i)$ is also called the *global mean field amplitude* (GMFA). Next, we depict the graph of $\sigma(t_i)$ as a function of t_i and look for peaks, or locations of strong deflections, in the graph; say, they are t_{i_k}, $k = 1,\ldots,K$. Each t_{i_k} indicates high neural activity at that instant. A contour plot of $\mathbf{y}_{i_k}(j)$, as a function of $\mathbf{p}_j$, $j = 1,\ldots,m$, is drawn on the scalp. Often, for that contour plot, the head is approximated by a sphere, with the center at an appropriate location in the head, and $\mathbf{p}_j$ radially projected on the surface of the sphere. The resulting contour plot (often called *topoplot*) is drawn on that spherical surface. For a visual inspection, this approximating contour plot is usually satisfactory. If the plot looks dipolar (i.e., due to a single dipole source), a single dipole search is applied to the measurement vector $\mathbf{y}_{i_k}$, yielding a source dipole $(\widehat{\mathbf{p}}_k, \widehat{\mathbf{q}}_k)$. The plot and the relative residual norm $||\mathbf{y}_{i_k} - \mathbf{L}(\widehat{\mathbf{p}}_k)\widehat{\mathbf{q}}_k||/||\mathbf{y}_{i_k}||$, as a goodness-of-fit measure, suggest how credible it is that $\mathbf{y}_{i_k}$ is due to a single dipole source.

For MEG data, the procedure is similar, but instead of potentials at sensor locations $\mathbf{p}_j$, we consider either one component, say radial, or the norm (possibly appropriately weighted) of the measured magnetic field $\mathbf{B}(\mathbf{p}_j)$ for each time point t_i (see figure 5.2).

5.8 The EEG/MEG Inverse Problem and Its Solution by the MNE Method

As noted earlier, the EEG/MEG inverse problem is generally *ill-posed*, meaning that the solution is *not unique* (different source currents $\mathbf{J}^\mathrm{p}$ may cause the same measurement vector), or *measurement noise* or *background activity* in the brain (also called neural noise) may greatly distort the solution.

The *numerical solving* of the EEG/MEG inverse problem by the *minimum-norm estimate* (MNE) method [22, 112–114, 254] works as follows. Discretize the ROI with a sufficiently dense grid of locations $\mathbf{p}_1, \ldots, \mathbf{p}_n$. Let $\mathbf{y}$ be the measurement vector due to the unknown source distribution $\mathbf{J}^\mathrm{p}$. The noiseless $\mathbf{y}$ is then given by

$$\mathbf{y} = \mathrm{A}\,\mathbf{x}, \tag{5.30}$$

where A is the mixing matrix and $\mathbf{x}$ contains the unknowns. If the orientations are not fixed,

$$\mathrm{A} = [\mathbf{L}(\mathbf{p}_1), \ldots, \mathbf{L}(\mathbf{p}_n)] \text{ and } \mathbf{x} = \begin{bmatrix} \mathbf{q}_1 \\ \vdots \\ \mathbf{q}_n \end{bmatrix}, \tag{5.31}$$

by equation (5.8), where $\mathbf{q}_1, \ldots, \mathbf{q}_n$ are unknown dipole moments at locations $\mathbf{p}_1, \ldots, \mathbf{p}_n$, and only those $\mathbf{q}_i$ are non-vanishing that correspond to active source locations. If the orientations are fixed, $\mathrm{A} = [\mathbf{l}(\mathbf{p}_1), \ldots, \mathbf{l}(\mathbf{p}_n)]$ and $\mathbf{x} = [x_1, \ldots, x_n]$ is the vector of the unknown dipolar amplitudes at locations $\mathbf{p}_1, \ldots, \mathbf{p}_n$, and again only those x_i are non-vanishing that correspond to active source locations. The inverse problem is now reduced to solving the matrix equation (5.30) for $\mathbf{x}$.

Usually this equation is *underdetermined*, because the number m of sensors is smaller than the number n of the grid points. Therefore, the solution $\mathbf{x}$ must be estimated by the *minimum norm least-squares* (MNLS) estimate $\widehat{\mathbf{x}}$, also called the *minimum-norm estimate* (MNE):

$$\widehat{\mathbf{x}} = \mathrm{A}^\mathrm{T}(\mathrm{A}\,\mathrm{A}^\mathrm{T})^{-1}\,\mathbf{y}. \tag{5.32}$$

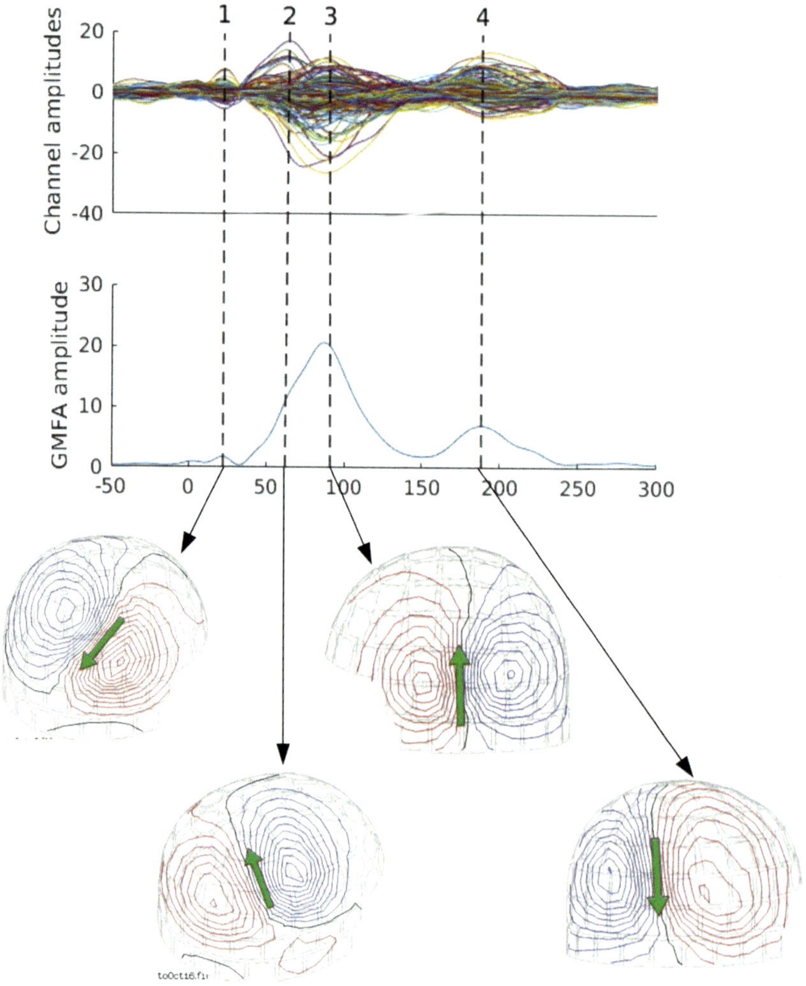

Figure 5.2
A visual review of the MEG data of the radial component of the measured magnetic field. *Uppermost figure*: graphs of the radial magnetometer readings $y_i(j)$ in the jth gradiometer as a function of time t_i, $j=1,\ldots,N$. *Middle figure*: The graph of the GMFA $\sigma(t_i)$ as a function of time t_i. *Lowest figure*: single dipoles fitted to the sensor pattern $y_{i_k}(j)$, $j=1,\ldots,m$, for the peak instants t_{i_k}, $k=1,\ldots,4$. The fitted dipoles (approximately on the cortex) and the corresponding sensor patterns are shown.

Taking into account noise, equation (5.30) becomes

$$\mathbf{y} = \mathbf{A}\,\mathbf{x} + \boldsymbol{\varepsilon}. \tag{5.33}$$

To stabilize the inverse problem and to suppress the influence of noise, the solution must be regularized. A practical choice is to solve the equation by *Tikhonov regularization* (see equations (4.77) and (4.78)), and get the solution

$$\widehat{\mathbf{x}} = \mathbf{A}^{\mathrm{T}}(\mathbf{A}\,\mathbf{A}^{\mathrm{T}} + \lambda\,\mathbf{I})^{-1}\,\mathbf{y}, \tag{5.34}$$

where $\lambda > 0$ is the *regularization parameter.*

With the solution $\widehat{\mathbf{x}}$, the source-current distribution $\mathbf{J}^{\mathrm{p}}$ is approximated by a grid of dipoles $(\mathbf{p}_i, \mathbf{q}_i)$, $i = 1, \ldots, n$, where for freely-oriented dipoles,

$$\mathbf{q}_i = [\mathbf{x}(k), \mathbf{x}(k+1), \mathbf{x}(k+2)]^{\mathrm{T}}, \quad \text{with}\, k = 3(i-1)+1, \tag{5.35}$$

and for fixed-oriented dipoles,

$$\mathbf{q}_i = \mathbf{x}(i)\eta(\mathbf{p}_i); \tag{5.36}$$

see figure 5.3.

The regularization parameter λ in equation (5.34) can be chosen as

$$\lambda = 3\,\frac{\sigma^2}{s^2} \tag{5.37}$$

where s and σ must be estimated. Here, $s > 0$ can be taken as the bound in the prior assumption that $|x_i| \le s$, $i = 1, \ldots, n$, as in equations (4.91) and (4.92). The noise level $\sigma > 0$ can be estimated roughly so that σ^2 is the mean of the diagonal elements of the noise covariance matrix $\mathbf{C}_{\mathrm{ns}}$. Therefore, to obtain σ, we must estimate $\mathbf{C}_{\mathrm{ns}}$. There are various ways to do that, depending on the type of the measured data. For instance, in evoked-response measurements, the signals measured before the stimulus can be considered noise and used to obtain an estimate for $\mathbf{C}_{\mathrm{ns}}$.

If the noise level σ cannot be estimated reasonably well, then λ can be chosen, for instance, by the *L-curve method*; see section 4.5.

After having solved equation (5.33) for $\mathbf{x}$, we can, if the ROI is a part of a surface like the cortex, draw a contour plot of the source strength $|\widehat{\mathbf{x}}(i)|$ and locate the (estimated) sources as the high-value spots or regions of $|\widehat{\mathbf{x}}(i)|$.

5.8.1 Noise-Normalized MNE Methods

The MNE method is biased to mislocate deep sources to more superficial locations. This defect can be partially removed by appropriate weighting of the MNE solution. There are several types of weighting (e.g., [254]), and here we present two of them, [69] and [244], referred to as *noise-normalizing MNE methods.*

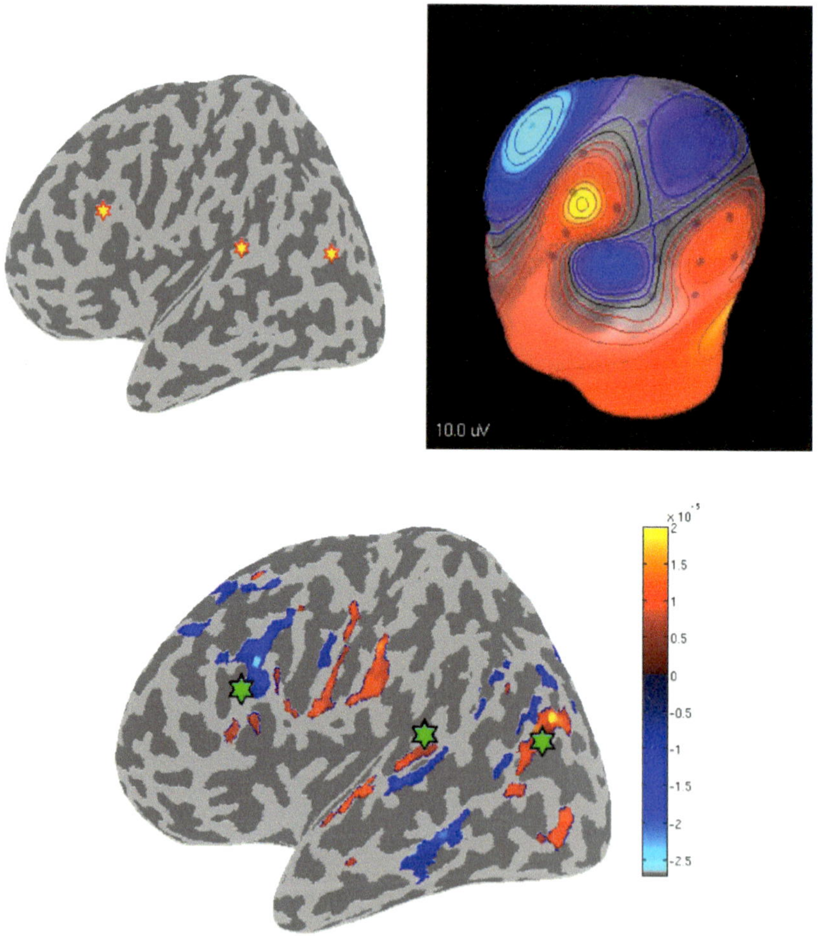

Figure 5.3

Upper figure pair: Three dipoles, marked by stars, are set on the cortex, and their joint potential pattern (EEG) on the scalp, computed by BEM, is presented. *Lower figure*: the MNE of the cortical source-current density with varying signs is depicted (plus sign indicates that the current is normally outward oriented and minus sign the opposite). The estimate is based on computed sixty scalp electrodes recording with 10 percent additive noise. The true source dipole locations are marked by stars. The red and blue areas correspond to strong current density, roughly estimating the true locations. Note the inaccuracy of the estimation.

In both methods, the starting point is the MNE solution of equation (5.33) as a Wiener estimate $\widehat{\mathbf{x}}$ (see (4.87)),

$$\widehat{\mathbf{x}} = \mathbf{W}\,\mathbf{y} = \mathbf{W}(\mathbf{A}\,\mathbf{x} + \boldsymbol{\varepsilon}) = \mathbf{W}\,\mathbf{A}\,\mathbf{x} + \mathbf{W}\,\boldsymbol{\varepsilon}, \tag{5.38}$$

where

$$\mathbf{W} = \mathbf{C}_{\mathrm{pr}}\mathbf{A}^{\mathrm{T}}(\mathbf{A}\,\mathbf{C}_{\mathrm{pr}}\mathbf{A}^{\mathrm{T}} + \mathbf{C}_{\mathrm{ns}})^{-1} \tag{5.39}$$

is the Wiener estimator, $\mathbf{C}_{\mathrm{pr}} = \mathrm{cov}(\mathbf{x})$ is the prior covariance matrix, and $\mathbf{C}_{\mathrm{ns}} = \mathrm{cov}(\boldsymbol{\varepsilon})$ is the noise covariance matrix.

In equation (5.38), we may consider $\mathbf{W}\,\mathbf{A}\,\mathbf{x}$ as the noiseless estimate and $\mathbf{W}\,\boldsymbol{\varepsilon}$ as the estimate's error with the covariance matrix

$$\mathbf{F} = \mathrm{cov}(\mathbf{W}\,\boldsymbol{\varepsilon}) = \mathbf{W}\,\mathrm{cov}(\boldsymbol{\varepsilon})\,\mathbf{W}^{\mathrm{T}} = \mathbf{W}\,\mathbf{C}_{\mathrm{ns}}\mathbf{W}^{\mathrm{T}}. \tag{5.40}$$

The diagonal elements $\mathbf{F}(k,k)$ are the variances of components of the error term $\mathbf{W}\,\boldsymbol{\varepsilon}$, $k = 1, \ldots, M$, and M is the dimension of vector $\mathbf{x}$.

Dale et al. [69] noise-normalized the components $\widehat{\mathbf{x}}(k)$ of $\widehat{\mathbf{x}}$ by $\mathbf{F}(k,k)^{1/2}$, that is, weighted them by $\mathbf{F}(k,k)^{-1/2}$, and so their *noise-normalized MNE estimate* $\widehat{\mathbf{x}}^{NN}$ is given by

$$\widehat{\mathbf{x}}^{NN}(k) = \frac{\widehat{\mathbf{x}}(k)}{\mathbf{F}(k,k)^{1/2}}, \quad k = 1, \ldots, M. \tag{5.41}$$

If $\mathbf{C}_{\mathrm{pr}} = \kappa^2\mathbf{I}$ and $\mathbf{C}_{\mathrm{ns}} = \sigma^2\mathbf{I}$, then

$$\mathbf{W} = \mathbf{A}^{\mathrm{T}}(\mathbf{A}\,\mathbf{A}^{\mathrm{T}} + \frac{\sigma^2}{\kappa^2}\,\mathbf{I})^{-1} \quad \text{and} \quad \mathbf{F} = \mathrm{cov}(\mathbf{W}\,\boldsymbol{\varepsilon}) = \sigma^2\,\mathbf{W}\,\mathbf{W}^{\mathrm{T}}, \tag{5.42}$$

and the noise-normalized $\widehat{\mathbf{x}}^{NN}$ can be written as

$$\widehat{\mathbf{x}}^{NN}(k,k) = \frac{\widehat{\mathbf{x}}(k)}{\widetilde{\mathbf{F}}(k,k)^{1/2}}, \quad \text{with} \tag{5.43}$$

$$\widehat{\mathbf{x}} = \mathbf{A}^{\mathrm{T}}(\mathbf{A}\,\mathbf{A}^{\mathrm{T}} + \frac{\sigma^2}{\kappa^2}\,\mathbf{I})^{-1}\mathbf{y} \quad \text{and} \quad \widetilde{\mathbf{F}} = \mathbf{W}\,\mathbf{W}^{\mathrm{T}}, \tag{5.44}$$

where the constant σ^{-1} has been dropped off in normalizing because it is only a scaling factor when level-curve plots of $\widehat{\mathbf{x}}$ are depicted.

In Pascual-Marqui et al. [244], sLORETA is presented only for the case $\mathbf{C}_{\mathrm{pr}} = \kappa^2\mathbf{I}$ and $\mathbf{C}_{\mathrm{ns}} = \sigma^2\mathbf{I}$, with $\mathbf{W}$ as in equation (5.42). The normalization is carried out with the covariance matrix $\mathbf{G}$ of $\widehat{\mathbf{x}}$. Because

$$\mathrm{cov}(\mathbf{y}) = \mathrm{cov}(\mathbf{A}\,\mathbf{x} + \boldsymbol{\varepsilon}) = \mathbf{A}\,\mathrm{cov}(\mathbf{x})\,\mathbf{A}^{\mathrm{T}} + \mathrm{cov}(\boldsymbol{\varepsilon}) = \kappa^2(\mathbf{A}\,\mathbf{A}^{\mathrm{T}} + \frac{\sigma^2}{\kappa^2}\mathbf{I}), \tag{5.45}$$

$$\mathbf{G} = \mathrm{cov}(\widehat{\mathbf{x}}) = \mathrm{cov}(\mathbf{W}\,\mathbf{y}) = \mathbf{W}\,\mathrm{cov}(\mathbf{y})\,\mathbf{W}^{\mathrm{T}}$$

$$= \kappa^2 \mathbf{W}\,(\mathbf{A}\,\mathbf{A}^{\mathrm{T}} + \frac{\sigma^2}{\kappa^2}\,\mathbf{I})\,\mathbf{W}^{\mathrm{T}} = \kappa^2 \mathbf{W}\,\mathbf{A}, \tag{5.46}$$

where $\mathbf{G}(k,k)$, $k=1,\ldots,M$, are the variances of the components $\widehat{\mathbf{x}}(k)$. The components $\widehat{\mathbf{x}}(k)$ are normalized by numbers $\widetilde{\mathbf{G}}(k,k)^{1/2}$ with $\widetilde{\mathbf{G}} = \mathbf{W}\,\mathbf{A}$, yielding the *sLORETA* MNE solution $\widehat{\mathbf{x}}^{sL}$ by

$$\widehat{\mathbf{x}}^{sL} = \frac{\widehat{\mathbf{x}}(k)}{\widetilde{\mathbf{G}}(k,k)^{1/2}}, \quad k=1,\ldots,M, \quad \text{where} \tag{5.47}$$

$$\widehat{\mathbf{x}} = \mathbf{W}\,\mathbf{y}, \quad \mathbf{W} = \mathbf{A}^{\mathrm{T}}(\mathbf{A}\,\mathbf{A}^{\mathrm{T}} + \frac{\sigma^2}{\kappa^2}\,\mathbf{I})^{-1} \quad \text{and} \quad \widetilde{\mathbf{G}} = \mathbf{W}\,\mathbf{A}, \tag{5.48}$$

where the constant κ^{-1} has been dropped off as a scaling factor.

Note that the solutions $\widehat{\mathbf{x}}^{NN}$ and $\widehat{\mathbf{x}}^{sL}$ do not estimate the true physical source-current vector $\mathbf{x}$ because of the normalization, and they are only meant to be used to localize the spots and regions of neural activity, that is, of the high values of $||\mathbf{x}||$.

Numerical simulations [189] show that for $\mathbf{C}_{\mathrm{pr}} = \kappa^2 \mathbf{I}$ and $\mathbf{C}_{\mathrm{ns}} = \sigma^2 \mathbf{I}$ under realistic noise conditions, the normalized sLORETA and Dale et al.'s MNE methods perform about equally well, and both methods alleviate the depth bias.

5.8.2 Minimum-Norm Estimates with Other Norms

Uutela et al. [313] showed that if source currents consist of discrete patches of primary current, one can find them by minimizing the L_1 norm $\int ||\mathbf{J}^{\mathrm{p}}(\mathbf{r})||dV$ instead of the L_2 norm $(\int ||\mathbf{J}^{\mathrm{p}}(\mathbf{r})||^2 dV)^{1/2}$ while requiring that $\mathbf{J}^{\mathrm{p}}$ would produce the measured magnetic field pattern. They called this solution the *minimum-current estimate*. The appropriateness of the choice of norm (and one is not limited to L_1 or L_2 norms) depends on what one knows or can assume about the primary-current distribution a priori. If nothing is known except for the volume or surface where $\mathbf{J}^{\mathrm{p}}$ is confined to, L_2 norm is the best in the sense of minimizing the expectation value of the squared error; the L_1 may be appropriate if the source distribution is known to consist of discrete hotspots of activity. The L_1 solution was compared with dipole-modeling results by Stenbacka et al. [293]. Gramfort et al. [102, 105] introduced the mixed-norm approach, which combines aspects of the different norms, depending on the data.

6 Beamformers

Beamforming, also called adaptive spatial filtering, is a *spatio-temporal* data-analysis method [181, 314], also applicable to EEG or MEG data [129, 315]. It can be used with a *time series* of EEG/MEG measurement data due to a *limited number* of current sources, usually modeled as current dipoles. The method requires a *forward field solver* for a given *head model*. The goal of beamforming is to recover the unknown number of source dipoles, as well as their locations, orientations, and time courses.

When tackling the ill-posedness of the inverse problem, beamformers take advantage of the assumption that the measured signals are due to separate dipolar sources. Beamformers make use of both spatial and time-domain features of the data.

Recall that we call the dipoles in the region of interest (ROI) *fixed-oriented* if the orientations η are stationary and known functions of locations $\eta = \eta(\mathbf{r})$. If the dipoles are not fixed-oriented, then they are called *freely oriented.*

Beamformers can be divided into scalar and vector types. The *scalar beamformers* are, in the first place, designed for measurement data with fixed-oriented dipoles; if the dipoles in the ROI are freely oriented, one can, however, assign *optimal fixed orientations* to them based on the measurement data, after which scalar beamformers can be applied. If the dipoles are freely oriented, and one does not want to use optimal orientations, one can directly apply vector beamformers to the measurement data.

Conventionally, beamformers are introduced in statistical terms as minimum variance spatial filters [282, 315] (we explain the term later). Our approach here is different. We introduce them in terms of vector algebra and use statistics only in describing how noise affects the outcome. Because the main ideas and structures of beamformers lie in basic vector algebra, we believe that this approach may make beamformers easier to understand and use than the statistical minimum-variance approach.

We present several types of beamformers with different localizers for single-run and iterative multiple-step beamforming. We give only tentative suggestions on how the different beamformers, in comparison with each other, perform in practice, because the performance of a specific beamformer largely depends on measurement data, noise, underlying sources, and the head model and accuracy of tissue-conductivity values used in the forward field solver. Sometimes, it may be instructive to analyze the data with various beamformer types and compare the results. It should be noted that if the assumption of dipolar sources is wrong or if the head model (including tissue-compartment shapes and the conductivities) is inaccurate, the results given by beamformers may not be trusted.

6.1 Measurement Data Matrix for Beamformers

We start by briefly recalling the concepts and notations used to describe measurement data, current dipoles, lead fields, and mixing matrices.

The EEG or MEG *measurement*, or *sensor, vector* is an m-vector $\mathbf{y} = [y_1, \ldots, y_m]^{\mathrm{T}}$, where y_i is the reading at the ith sensor (such as a magnetometer or gradiometer for MEG or an electrode derivation for EEG); m is the number of sensors.

A source dipole (in the brain) is a pair $(\mathbf{p}, \mathbf{q})$, where $\mathbf{p}$ is the *location*, $\mathbf{q} = s\,\eta$ is the *moment*, η is the *orientation*, with $\|\eta\| = 1$, and the scalar s is the *amplitude* of the dipole.

Let $\mathbf{y}$ be the sensor vector due to a single dipole source $(\mathbf{p}, s\eta)$. It has the form

$$\mathbf{y} = s\,\mathbf{l}(\mathbf{p}, \eta), \tag{6.1}$$

where $\mathbf{l}(\mathbf{p}, \eta)$ is the sensor vector, or *topography*, due to a dipole at location $\mathbf{p}$ with orientation η and amplitude $s = 1$. The topography can also be given in the form

$$\mathbf{l}(\mathbf{p}, \eta) = \mathbf{L}(\mathbf{p})\,\eta, \tag{6.2}$$

where $\mathbf{L}(\mathbf{p})$ is the $m \times 3$ *lead-field matrix* at $\mathbf{p}$ given by

$$\mathbf{L}(\mathbf{p}) = [\mathbf{l}(\mathbf{p}, \mathbf{e}_x), \mathbf{l}(\mathbf{p}, \mathbf{e}_y), \mathbf{l}(\mathbf{p}, \mathbf{e}_z)], \tag{6.3}$$

where $\mathbf{e}_x$, $\mathbf{e}_y$, and $\mathbf{e}_z$ are the xyz-coordinate unit vectors in the ordinary 3-space. The matrix $\mathbf{L}(\mathbf{p})$ is numerically computed by a forward field solver for each location $\mathbf{p}$ in the ROI for the given head model.

Let now the measurement data be due to a *finite number* n of source dipoles with locations $\mathbf{p}_1, \ldots, \mathbf{p}_n$ in the ROI, with orientations $\eta_1, \ldots, \eta_n$ and amplitudes $s_1(t), \ldots, s_n(t)$, where t is time. The number n is assumed to be smaller than or equal to the number m of the sensors. The sensor vector $\mathbf{y}_j(t)$ at time t due to a single active dipole $(\mathbf{p}_j, s_j(t)\,\eta)$,

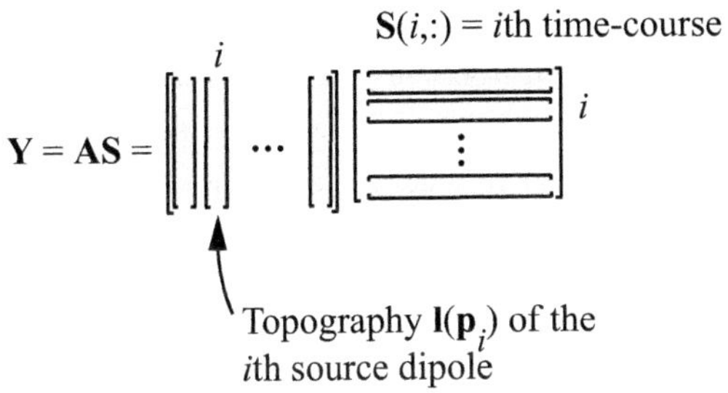

Figure 6.1
The (noiseless) data matrix $\mathbf{Y} = \mathbf{A}\mathbf{S}$, the mixing matrix $\mathbf{A}$ with topographies $\mathbf{l}(\mathbf{p}_i) = \mathbf{l}(\mathbf{p}_i, \eta_i)$, $i = 1, \ldots, n$, as its columns, and the time-course matrix $\mathbf{S}$.

acting alone, gets the form

$$\mathbf{y}_j(t) = s_j(t)\, \mathbf{l}(\mathbf{p}_j, \eta_j), \; j = 1, \ldots, n. \tag{6.4}$$

When all n source dipoles are active at the same time t, the joint sensor vector $\mathbf{y}(t)$ is the sum of $\mathbf{y}_j(t)$ as

$$\mathbf{y}(t) = \sum_{j=1}^{n} s_j(t)\, \mathbf{l}(\mathbf{p}_j, \eta_j). \tag{6.5}$$

Let the measurements be done at N time points $t_1 < t_2 < \cdots < t_N$ and the data be collected in an $m \times N$ *data matrix* $\mathbf{Y}$ with columns $\mathbf{y}(t_1), \ldots, \mathbf{y}(t_N)$, that is,

$$\mathbf{Y} = [\mathbf{y}(t_1), \ldots, \mathbf{y}(t_N)]. \tag{6.6}$$

Let the amplitudes $s_j(t_k)$ be collected in the $n \times N$ *time-course matrix* $\mathbf{S}$ with elements

$$\mathbf{S}(j, k) = s_j(t_k), \; j = 1, \ldots, n, \; k = 1, \ldots, N. \tag{6.7}$$

We can now write $\mathbf{Y}$ in the matrix form

$$\begin{aligned} \mathbf{Y} &= \sum_{j=1}^{n} \mathbf{l}(\mathbf{p}_j, \eta_j)\, \mathbf{S}(j, :) \\ &= [\mathbf{l}(\mathbf{p}_1, \eta_1), \ldots, \mathbf{l}(\mathbf{p}_n, \eta_n)]\, \mathbf{S} = \mathbf{A}\,\mathbf{S}, \end{aligned} \tag{6.8}$$

where $\mathbf{A} = [\mathbf{l}(\mathbf{p}_1, \eta_1), \ldots, \mathbf{l}(\mathbf{p}_n, \eta_n)]$ is the *mixing matrix* (see figure 6.1). We assume that the topographies $\mathbf{l}(\mathbf{p}_1, \eta_1), \ldots, \mathbf{l}(\mathbf{p}_n, \eta_n)$ are *linearly independent* vectors and that $m \geq n$ which implies that rank $(\mathbf{A}) = n$.

In the presence of additive noise $\boldsymbol{\varepsilon}$, the data matrix $\mathbf{Y}$ gets the form

$$\mathbf{Y} = \mathbf{A}\,\mathbf{S} + \boldsymbol{\varepsilon}, \tag{6.9}$$

where $\boldsymbol{\varepsilon}$ is an $m \times N$ noise matrix. We assume that $\boldsymbol{\varepsilon}$ is time-centered, that is, it has zero (temporal) mean and is not correlated with $\mathbf{S}$, that is, $N^{-1} \sum_{k=1}^{N} \boldsymbol{\varepsilon}(i,k) \simeq 0$, for all $i = 1,\ldots,m$, and $\mathbf{S}\boldsymbol{\varepsilon}^{\mathrm{T}} \simeq 0$.

The input to beamformer algorithms, as we will see later, is the (unnormalized sample) *covariance matrix* $\mathbf{C} = \mathbf{Y}_{\mathrm{c}}\,\mathbf{Y}_{\mathrm{c}}{}^{\mathrm{T}}$ of the data matrix $\mathbf{Y}$, where $\mathbf{Y}_{\mathrm{c}}$ is the time-centered $\mathbf{Y}$, that is,

$$\mathbf{Y}_{\mathrm{c}}(i,j) = \mathbf{Y}(i,j) - \frac{1}{N} \sum_{k=1}^{N} \mathbf{Y}(i,k), \tag{6.10}$$

for $i = 1,\ldots,m$, $j = 1,\ldots,N$. We note that beamformers would work equally well if the input were the (unnormalized) data *correlation matrix* $\mathbf{C}_{\mathrm{corr}} = \mathbf{Y}\,\mathbf{Y}^{\mathrm{T}}$, but here we follow the conventional way and use the covariance matrix.

In this section from now on, we assume that $\mathbf{Y}$ is always time-centered, and so the (unnormalized) data covariance matrix is

$$\mathbf{C} = \mathbf{Y}\,\mathbf{Y}^{\mathrm{T}}.$$

This assumption implies that in the data equation $\mathbf{Y} = \mathbf{A}\mathbf{S} + \boldsymbol{\varepsilon}$, $\mathbf{S}$ is also time-centered.

Later, we need the inverse $\mathbf{C}^{-1}$, or pseudoinverse $\mathbf{C}^{\dagger}$ of the data covariance matrix $\mathbf{C}$. If $\mathbf{C}$ is underranked, that is, $\mathrm{rank}(\mathbf{C}) < m$, or $\mathbf{C}$ is ill-posed, the performance of the beamformer may be degraded, and, therefore, $\mathbf{C}$ usually must be regularized. One way to regularize $\mathbf{C}$ is to replace it by $\mathbf{C} + \gamma\, c\,\mathbf{I}$, where c is the maximum of the diagonal elements of $\mathbf{C}$ and γ is appropriately chosen (e.g., $\gamma = 10^{-4}$).

6.1.1 Signal-to-Noise Ratio (SNR)

In the noisy data $\mathbf{Y} = \mathbf{A}\,\mathbf{S} + \boldsymbol{\varepsilon}$, the quality of the data is often expressed as the *signal-to-noise ratio* (SNR), which is usually defined as follows. Consider the time-centered noiseless data $\mathbf{Y}_0 = \mathbf{A}\,\mathbf{S}$, that is, $\sum_{j=1}^{N} \mathbf{Y}_0(i,j) = 0$ for all $i = 1,\ldots,m$ with $\mathbf{Y}_0$ being an $m \times N$ matrix. The variance over time of the noiseless signal in channel i is then

$$\mathrm{var}\big(\mathbf{Y}_0(i,:)\big) = \frac{1}{N} \sum_{j=1}^{N} \mathbf{Y}_0(i,j)^2, \tag{6.11}$$

and the mean of variances over channels is

$$\mathrm{var}_{\mathrm{mean}}(\mathbf{Y}_0) = \frac{1}{mN} \sum_{i=1}^{m} \Big(\sum_{j=1}^{N} \mathbf{Y}_0(i,j)^2 \Big) \tag{6.12}$$

$$= \frac{1}{mN} \operatorname{trace}(\mathbf{Y}_0 \mathbf{Y}_0^{\mathrm{T}}). \tag{6.13}$$

Similarly, we get for the noise $\boldsymbol{\varepsilon}$,

$$\mathrm{var}_{\mathrm{mean}}(\boldsymbol{\varepsilon}) = \frac{1}{mN} \sum_{i=1}^{m} \sum_{j=1}^{N} \boldsymbol{\varepsilon}(i,j)^2 \tag{6.14}$$

$$= \frac{1}{mN} \operatorname{trace}(\boldsymbol{\varepsilon}\, \boldsymbol{\varepsilon}^{\mathrm{T}}). \tag{6.15}$$

Now, the SNR of the data $\mathbf{Y} = \mathbf{A}\,\mathbf{S} + \boldsymbol{\varepsilon}$, with $\mathbf{Y}_0 = \mathbf{A}\,\mathbf{S}$, is defined by

$$\mathrm{SNR} = \frac{\left(\mathrm{var}_{\mathrm{mean}}(\mathbf{Y}_0)\right)^{1/2}}{\left(\mathrm{var}_{\mathrm{mean}}(\boldsymbol{\varepsilon})\right)^{1/2}} \tag{6.16}$$

$$= \left(\frac{\operatorname{trace}(\mathbf{Y}_0 \mathbf{Y}_0^{\mathrm{T}})}{\operatorname{trace}(\boldsymbol{\varepsilon}\, \boldsymbol{\varepsilon}^{\mathrm{T}})}\right)^{1/2} = \left(\frac{\operatorname{trace}(\mathbf{C}_0)}{\operatorname{trace}(\mathbf{C}_\varepsilon)}\right)^{1/2}, \tag{6.17}$$

where $\mathbf{C}_0 = \mathbf{Y}_0 \mathbf{Y}_0^{\mathrm{T}}$ and $\mathbf{C}_\varepsilon = \boldsymbol{\varepsilon}\, \boldsymbol{\varepsilon}^{\mathrm{T}}$ are the noiseless data and noise covariance matrices, respectively.

We note that sometimes (but not in this book), SNR is defined as the ratio $\operatorname{trace}(\mathbf{C}_0)/\operatorname{trace}(\mathbf{C}_\varepsilon)$ (without the square root), and then it is (often) called power SNR. Therefore, one must always pay attention to which definition of SNR is used.

6.2 Scalar Beamformer

We first assume that the dipoles in the ROI are fixed-oriented, and so, for every $\mathbf{p}$ in the ROI, the orientation is a known function of location, $\eta = \eta(\mathbf{p})$. We introduce the *scalar beamformer*, which in the literature is also called the *linearly constrained minimum-variance (LCMV) spatial filter* [282, 315]. Because the topographies of dipoles are determined by their locations, we briefly write $\mathbf{l}(\mathbf{p}) = \mathbf{l}(\mathbf{p}, \eta(\mathbf{p}))$ for the topography assigned to location $\mathbf{p}$.

Consider the measurement data matrix $\mathbf{Y} \in \mathbb{R}^m$, $m \geq n$, due to source dipoles at locations $\mathbf{p}_1, \ldots, \mathbf{p}_n$,

$$\mathbf{Y} = \mathbf{A}\,\mathbf{S} + \boldsymbol{\varepsilon}, \tag{6.18}$$

where $\mathbf{A} = [\mathbf{l}(\mathbf{p}_1), \ldots, \mathbf{l}(\mathbf{p}_n)]$. In analyzing the measurement data with a scalar beamformer, the aim is to *recover* the neuronal sources, that is, the source dipole locations $\mathbf{p}_1, \ldots, \mathbf{p}_n$, their number n, and the time course matrix $\mathbf{S}$.

We assign zero time courses to the locations of the dipoles in the ROI that are not source dipoles. We also denote the time course of a dipole at location $\mathbf{p}$ by $\mathbf{S}_p$. By this notation, $\mathbf{S}_p = \mathbf{S}(j,:)$, if $\mathbf{p}$ coincides with a true source location, that is, $\mathbf{p} \in \{\mathbf{p}_1, \ldots, \mathbf{p}_n\}$, and $\mathbf{S}_p = 0$, if $\mathbf{p} \notin \{\mathbf{p}_1, \ldots, \mathbf{p}_n\}$.

6.3 Scalar Beamformer Filter Vector with Noiseless Data and Uncorrelated Time Courses

We assume in this section that the data $\mathbf{Y}$ in equation (6.18) are noiseless, that is, $\boldsymbol{\varepsilon} = 0$. The basic idea of beamforming with a scalar beamformer is as follows. We try to find a *filter (or weight) vector* $\mathbf{w}_p \in \mathbb{R}^m$ for every $\mathbf{p}$ in the ROI, so that

$$\mathbf{w}_p{}^{\mathrm{T}} \mathbf{Y} = \mathbf{S}_p. \tag{6.19}$$

In fact, for noiseless data, there always is such $\mathbf{w}_p$. Namely, if $\mathbf{p} \notin \{\mathbf{p}_1, \ldots, \mathbf{p}_n\}$, then clearly we can choose $\mathbf{w}_p = 0$. If $\mathbf{p} = \mathbf{p}_k$ for some k, $1 \leq k \leq n$, we find $\mathbf{w}_{p_k}$ as follows. We first notice that for $\mathbf{A}^{\dagger} = (\mathbf{A}^{\mathrm{T}}\mathbf{A})^{-1}\mathbf{A}^{\mathrm{T}}$,

$$\mathbf{A}^{\dagger}\mathbf{A} = \mathbf{I},$$

because $\mathbf{A} \in \mathbb{R}^{m \times n}$ and $m \geq n = \mathrm{rank}(\mathbf{A})$. Therefore,

$$\mathbf{A}^{\dagger}\mathbf{Y} = \mathbf{A}^{\dagger}\mathbf{A}\mathbf{S} = \mathbf{S}. \tag{6.20}$$

We now choose $\mathbf{w}_{p_k}$ so that

$$\mathbf{w}_{p_k}{}^{\mathrm{T}} = \mathbf{A}^{\dagger}(k,:) = \mathbf{e}_k{}^{\mathrm{T}}\mathbf{A}^{\dagger}, \tag{6.21}$$

where $\mathbf{e}_k$ is the kth coordinate unit vector of $\mathbb{R}^m$, that is, $\mathbf{e}_k(j) = 1$ if $j = k$ and $\mathbf{e}_k(j) = 0$ if $j \neq k$. We get

$$\mathbf{w}_{p_k}{}^{\mathrm{T}}\mathbf{Y} = \mathbf{e}_k{}^{\mathrm{T}}\mathbf{A}^{\dagger}\mathbf{A}\mathbf{S} = \mathbf{e}_k{}^{\mathrm{T}}\mathbf{S} = \mathbf{S}(k,:),$$

which proves equation (6.19) for $\mathbf{w}_p = \mathbf{w}_{p_k}$.

So, if $\mathbf{A}$ is known, $\mathbf{w}_p$ is easy to form. A remarkable property of the beamformer is that for every $\mathbf{p}$, we can form the filter vector $\mathbf{w}_p$ only knowing the data $\mathbf{Y}$, provided that, in addition, the source time courses are *uncorrelated*. Namely, we have the following key result.

Assume that the given data $\mathbf{Y}$ are *noiseless* and the time courses $\mathbf{S}(1,:), \ldots, \mathbf{S}(n,:)$ are *uncorrelated*, that is,

$$\mathbf{S}(i,:)\,\mathbf{S}(j,:)^{\mathrm{T}} = 0 \text{ for } i \neq j,$$

or, in other words, they are mutually *orthogonal*. For every $\mathbf{p}$ in the ROI, we form a filter vector $\mathbf{w}_p$ as the *minimizer* of the *quadratic form* $\mathbf{w}^{\mathrm{T}}\mathbf{C}\mathbf{w}$ with the *constraint* $\mathbf{l}(\mathbf{p})^{\mathrm{T}}\mathbf{w} = 1$, that is,

$$\mathbf{w}_p = \underset{\mathbf{l}(\mathbf{p})^{\mathrm{T}}\mathbf{w}=1}{\arg\min}\ \mathbf{w}^{\mathrm{T}}\,\mathbf{C}\,\mathbf{w}, \tag{6.22}$$

where $\mathbf{C} = \mathbf{Y}\,\mathbf{Y}^{\mathrm{T}}$. Then, for noiseless data $\mathbf{Y}$, it follows that

$$\mathbf{w}_p{}^{\mathrm{T}}\mathbf{Y} = \mathbf{S}_p, \tag{6.23}$$

and

$$\min_{\mathbf{l}(\mathbf{p})^{\mathrm{T}}\,\mathbf{w}=1} \mathbf{w}^{\mathrm{T}}\,\mathbf{C}\,\mathbf{w} = \mathbf{w}_p{}^{\mathrm{T}}\mathbf{C}\,\mathbf{w}_p = ||\mathbf{w}_p{}^{\mathrm{T}}\mathbf{Y}||^2 = ||\mathbf{S}_p||^2. \tag{6.24}$$

Equations (6.23) and (6.24) and a closed-form expressions for $\mathbf{w}_p$ are obtained from more general results in the next section. These equations give us the possibility to "beam" our attention to a single location $\mathbf{p}$ and find out if there is an active source at $\mathbf{p}$ and what the time course $\mathbf{S}_p$ is – thus the name "beamformer." The minimum in equation (6.24) is (conventionally) called the *beamformer output (or source) power* at $\mathbf{p}$, and we denote it by

$$\mu(\mathbf{p}) = \min_{\mathbf{l}(\mathbf{p})^{\mathrm{T}}\,\mathbf{w}=1} \mathbf{w}^{\mathrm{T}}\,\mathbf{C}\,\mathbf{w}. \tag{6.25}$$

6.4 Filter Vector with Correlated Time Courses and Noiseless or Noisy Data

We first assume that the data $\mathbf{Y}$ are noiseless, that is, $\mathbf{Y} = \mathbf{AS}$, and let the time courses $\mathbf{S}_1, \ldots, \mathbf{S}_n$ be correlated (non-orthogonal). We again define the filter (or weight) vector $\mathbf{w}_p$ for every $\mathbf{p}$ in the ROI as in equation (6.22) by

$$\mathbf{w}_p = \underset{\mathbf{l}(\mathbf{p})^{\mathrm{T}}\,\mathbf{w}=1}{\arg\min}\ \mathbf{w}^{\mathrm{T}}\,\mathbf{C}\,\mathbf{w}, \tag{6.26}$$

but now, due to the correlatedness of the time courses, the key filter result (6.23) is not valid any longer. However, the following more general result holds:

$$\mathbf{w}_p{}^{\mathrm{T}}\mathbf{Y} = \widetilde{\mathbf{S}}_p, \tag{6.27}$$

where $\widetilde{\mathbf{S}}_p = 0$ if $\mathbf{p} \notin \{\mathbf{p}_1, \ldots, \mathbf{p}_n\}$, and if $\mathbf{p} \in \{\mathbf{p}_1, \ldots, \mathbf{p}_n\}$ with $\mathbf{p} = \mathbf{p}_k$, then $\widetilde{\mathbf{S}}_k$ is the orthogonal projection of $\mathbf{S}_k$ to the orthospace of the subspace spanned by the other time courses $\mathbf{S}_j$, $j \neq k$, or given in explicit terms,

$$\widetilde{\mathbf{S}}_k{}^{\mathrm{T}} = \mathbf{P}_k\mathbf{S}_k{}^{\mathrm{T}}, \tag{6.28}$$

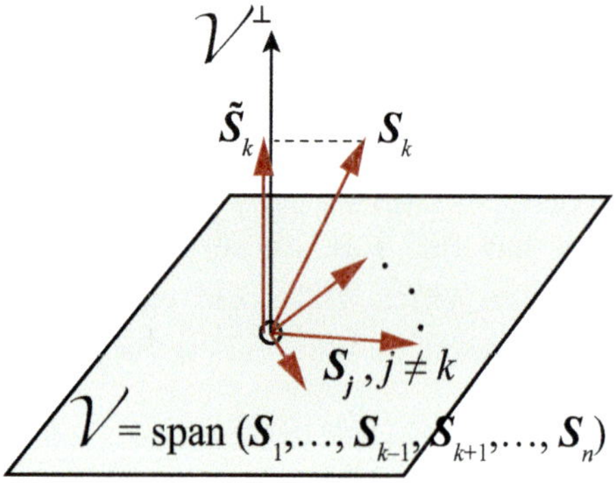

Figure 6.2
The orthogonal projection $\tilde{\mathbf{S}}_k$ of the time course $\mathbf{S}_k$ onto the orthospace $\mathcal{V}^\perp$ of the subspace $\mathcal{V} = \mathrm{span}(\mathbf{S}_1, \ldots, \mathbf{S}_{k-1}, \mathbf{S}_{k+1}, \ldots, \mathbf{S}_n)$.

where $\mathbf{P}_k$ is the orthogonal projection

$$\mathbf{P}_k = \mathbf{I} - \mathbf{G}_k \mathbf{G}_k^\dagger \tag{6.29}$$

with $\mathbf{G}_k = [\mathbf{S}_1{}^\mathrm{T}, \ldots, \mathbf{S}_{k-1}{}^\mathrm{T}, \mathbf{S}_{k+1}{}^\mathrm{T}, \ldots, \mathbf{S}_n{}^\mathrm{T}]$ (see figure 6.2). The proof of equation (6.27) is given in Appendix 6.20.

The beamformer output power $\mu(\mathbf{p})$ is now, as in equation (6.25), given by

$$\mu(\mathbf{p}) = \min_{\mathbf{l}(\mathbf{p})^\mathrm{T}\mathbf{w}=1} \mathbf{w}^\mathrm{T}\mathbf{C}\,\mathbf{w} = ||\mathbf{w}_p{}^\mathrm{T}\mathbf{Y}||^2 = ||\tilde{\mathbf{S}}_p||^2. \tag{6.30}$$

Notice that if $\mathbf{p} = \mathbf{p}_k$, then $\tilde{\mathbf{S}}_k{}^\mathrm{T} = \mathbf{P}_k \mathbf{S}_k{}^\mathrm{T}$, and because $\mathbf{P}_k$ is an orthogonal projection,

$$\mu(\mathbf{p}_k) = ||\tilde{\mathbf{S}}_k||^2 \le ||\mathbf{S}_k||^2. \tag{6.31}$$

In the literature, the difference $||\mathbf{S}_k||^2 - ||\tilde{\mathbf{S}}_k||^2 \ge 0$ is called the *signal cancellation* (of the output power). Note that if the time courses $\mathbf{S}_1, \ldots, \mathbf{S}_n$ are uncorrelated (mutually orthogonal), then $\tilde{\mathbf{S}}_k = \mathbf{S}_k$, and there is no signal cancellation, and equations (6.27) and (6.30) imply equations (6.23) and (6.24), respectively.

Finally, we assume that the data $\mathbf{Y}$ are noisy, that is, $\mathbf{Y} = \mathbf{A}\mathbf{S} + \boldsymbol{\varepsilon}$, and the time courses $\mathbf{S}_1, \ldots, \mathbf{S}_n$ are not necessarily uncorrelated (mutually orthogonal). For every $\mathbf{p}$ in the ROI, we again let the weight (or filter) vector $\mathbf{w}_p$ be

$$\mathbf{w}_p = \arg\min_{\mathbf{l}(\mathbf{p})^\mathrm{T}\mathbf{w}=1} \mathbf{w}^\mathrm{T}\mathbf{C}\mathbf{w}. \tag{6.32}$$

Because of the noise, $\mathbf{C}=\mathbf{Y}\mathbf{Y}^{\mathrm{T}}$ gets the form

$$\begin{aligned}\mathbf{C} &= (\mathbf{AS}+\boldsymbol{\varepsilon})(\mathbf{AS}+\boldsymbol{\varepsilon})^{\mathrm{T}} = (\mathbf{AS}+\boldsymbol{\varepsilon})(\mathbf{S}^{\mathrm{T}}\mathbf{A}^{\mathrm{T}}+\boldsymbol{\varepsilon}^{\mathrm{T}}) \\ &= \mathbf{ASS}^{\mathrm{T}}\mathbf{A}^{\mathrm{T}}+\mathbf{AS}\boldsymbol{\varepsilon}^{\mathrm{T}}+\boldsymbol{\varepsilon}\,\mathbf{S}^{\mathrm{T}}\mathbf{A}^{\mathrm{T}}+\boldsymbol{\varepsilon}\,\boldsymbol{\varepsilon}^{\mathrm{T}} = \mathbf{ASS}^{\mathrm{T}}\mathbf{A}^{\mathrm{T}}+\boldsymbol{\varepsilon}\,\boldsymbol{\varepsilon}^{\mathrm{T}},\end{aligned} \tag{6.33}$$

where we have used the fact that $\boldsymbol{\varepsilon}$ is not correlated with $\mathbf{S}$, that is, $\mathbf{S}\boldsymbol{\varepsilon}^{\mathrm{T}}=0$. We may assume that $\mathrm{rank}(\boldsymbol{\varepsilon})=m$, which makes $\boldsymbol{\varepsilon}\,\boldsymbol{\varepsilon}^{\mathrm{T}}$ positive-definite and $\mathbf{C}=\mathbf{ASS}^{\mathrm{T}}\mathbf{A}^{\mathrm{T}}+\boldsymbol{\varepsilon}\,\boldsymbol{\varepsilon}^{\mathrm{T}}$ invertible. This allows us to present $\mathbf{w}_p$ by the following well-known closed formula:

$$\mathbf{w}_p = \frac{\mathbf{C}^{-1}\mathbf{l}(\mathbf{p})}{\mathbf{l}(\mathbf{p})^{\mathrm{T}}\mathbf{C}^{-1}\mathbf{l}(\mathbf{p})}. \tag{6.34}$$

For the proof, see Appendix 6.21.

We define the (noisy) output power $\mu(\mathbf{p})$ again as the minimum in equation (6.32),

$$\begin{aligned}\mu(\mathbf{p}) &= \min_{\mathbf{l}(\mathbf{p})^{\mathrm{T}}\mathbf{w}=1} \mathbf{w}^{\mathrm{T}}\mathbf{C}\,\mathbf{w} = {\mathbf{w}_p}^{\mathrm{T}}\mathbf{C}\,\mathbf{w}_p \\ &= \frac{1}{\mathbf{l}(\mathbf{p})^{\mathrm{T}}\mathbf{C}^{-1}\mathbf{l}(\mathbf{p})},\end{aligned} \tag{6.35}$$

where the closed form of $\mu(\mathbf{p})$ follows readily by substituting the expression of $\mathbf{w}_p$ in equation (6.34) to ${\mathbf{w}_p}^{\mathrm{T}}\mathbf{C}\,\mathbf{w}_p$.

The filtering outcome ${\mathbf{w}_p}^{\mathrm{T}}\mathbf{Y}$ is now affected by the noise, and equation (6.27) is not exactly valid. However, with a small amount of noise, we have the following approximative equations (see Appendix 6.22):

$${\mathbf{w}_p}^{\mathrm{T}}\mathbf{Y} \simeq \widetilde{\mathbf{S}}_p \text{ and } \mu(\mathbf{p}) \simeq ||\widetilde{\mathbf{S}}_p||^2 \text{ if } \mathbf{p}\in\{\mathbf{p}_1,\ldots,\mathbf{p}_n\}, \text{ and} \tag{6.36}$$

$$\mu(\mathbf{p}) \simeq 0, \text{ if } \mathbf{p}\notin\{\mathbf{p}_1,\ldots,\mathbf{p}_n\}, \tag{6.37}$$

where $\widetilde{\mathbf{S}}_p$ is as in equation (6.27).

Assume next that $\mathbf{C}$ is not invertible, as, for instance, in the case when the data are noiseless and the number of sensors m is larger than the number n of sources. Then, we can also present $\mathbf{w}_p$ and $\mu(\mathbf{p})$ in the following closed forms; for the derivation of the equations, see Appendix 6.21. We form the orthogonal projection $\mathbf{P}=\mathbf{C}\,\mathbf{C}^{\dagger}$ onto $\mathrm{span}(\mathbf{C})$. If $\mathbf{l}(\mathbf{p})\notin\mathrm{span}(\mathbf{C})$, that is, $||\mathbf{P}\mathbf{l}(\mathbf{p})||<||\mathbf{l}(\mathbf{p})||$, then

$$\mathbf{w}_p = \frac{\mathbf{l}(\mathbf{p})-\mathbf{P}\,\mathbf{l}(\mathbf{p})}{||\mathbf{l}(\mathbf{p})-\mathbf{P}\,\mathbf{l}(\mathbf{p})||^2}, \text{ and} \tag{6.38}$$

$$\mu(\mathbf{p})=0. \tag{6.39}$$

If $\mathbf{l}(\mathbf{p}) \in \text{span}(\mathbf{C})$, that is, $||\mathbf{Pl}(\mathbf{p})|| = ||\mathbf{l}(\mathbf{p})||$, then

$$\mathbf{w}_p = \frac{\mathbf{C}^{\dagger}\mathbf{l}(\mathbf{p})}{\mathbf{l}(\mathbf{p})^{\mathrm{T}}\mathbf{C}^{\dagger}\mathbf{l}(\mathbf{p})}, \text{ and} \tag{6.40}$$

$$\mu(\mathbf{p}) = \frac{1}{\mathbf{l}(\mathbf{p})^{\mathrm{T}}\mathbf{C}^{\dagger}\mathbf{l}(\mathbf{p})}. \tag{6.41}$$

6.5 Linear Transform of the Data Equation

In this section, we show that the beamforming results remain valid for *linearly transformed data equations*. We later use transformed data equations, for instance, in whitening the data equation; see below. The linear transforming is carried out by multiplying both sides of the data equation $\mathbf{Y} = \mathbf{AS} + \boldsymbol{\varepsilon}$ by an invertible matrix $\mathbf{B} \in \mathbb{R}^{m \times m}$. That yields a transformed data equation

$$\mathbf{BY} = \mathbf{BAS} + \mathbf{B}\boldsymbol{\varepsilon}$$

with the new data matrix being $\mathbf{BY}$, the covariance matrix $\mathbf{C}_B = \mathbf{BY}(\mathbf{BY})^{\mathrm{T}} = \mathbf{BCB}^{\mathrm{T}}$, the mixing matrix $\mathbf{BA}$, topographies $\mathbf{B}\,\mathbf{l}(\mathbf{p})$ and noise $\mathbf{B}\boldsymbol{\varepsilon}$. Note that the time-course matrix $\mathbf{S}$ and the source dipole locations $\mathbf{p}_1, \ldots, \mathbf{p}_n$ remain unchanged in the transformation. Also the output power remains unchanged. Namely, if $\mu_B(\mathbf{p})$ denotes the power function assigned to the transformed data equation, then

$$\begin{aligned}\mu_B(\mathbf{p}) &= \frac{1}{(\mathbf{B}\,\mathbf{l}(\mathbf{p}))^{\mathrm{T}}\mathbf{C}_B^{-1}\mathbf{B}\,\mathbf{l}(\mathbf{p})} = \frac{1}{\mathbf{l}(\mathbf{p})^{\mathrm{T}}\mathbf{B}^{\mathrm{T}}\left(\mathbf{BCB}^{\mathrm{T}}\right)^{-1}\mathbf{B}\,\mathbf{l}(\mathbf{p})} \\ &= \frac{1}{\mathbf{l}(\mathbf{p})^{\mathrm{T}}\mathbf{B}^{\mathrm{T}}(\mathbf{B}^{\mathrm{T}})^{-1}\mathbf{C}^{-1}\mathbf{B}^{-1}\mathbf{B}\,\mathbf{l}(\mathbf{p})} = \frac{1}{\mathbf{l}(\mathbf{p})\mathbf{C}^{-1}\mathbf{l}(\mathbf{p})} = \mu(\mathbf{p}).\end{aligned} \tag{6.42}$$

It is straightforward to check that all previous beamforming results remain valid for the transformed data equation with the transformed and invariant quantities. Because of equation (6.34), the weight vector $\mathbf{w}_{B,p}$, assigned to the data matrix $\mathbf{BY}$, gets a simple form,

$$\mathbf{w}_{B,p} = \frac{\mathbf{C}_B^{-1}\mathbf{B}\,\mathbf{l}(\mathbf{p})}{(\mathbf{B}\,\mathbf{l}(\mathbf{p}))^{\mathrm{T}}\mathbf{C}_B^{-1}\mathbf{B}\,\mathbf{l}(\mathbf{p})} = \frac{(\mathbf{B}^{\mathrm{T}})^{-1}\mathbf{C}^{-1}\mathbf{B}^{-1}\mathbf{B}\,\mathbf{l}(\mathbf{p})}{\mathbf{l}(\mathbf{p})\mathbf{C}^{-1}\mathbf{l}(\mathbf{p})} = (\mathbf{B}^{\mathrm{T}})^{-1}\mathbf{w}_p, \tag{6.43}$$

where $\mathbf{w}_p$ is as in equation (6.34). It follows that

$$\mathbf{w}_{B,p}{}^{\mathrm{T}}\mathbf{BY} = \mathbf{w}_p{}^{\mathrm{T}}\mathbf{Y}, \tag{6.44}$$

which shows that filtering with the weight vector remains invariant in the transformation.

For a data equation $\mathbf{Y} = \mathbf{AS} + \boldsymbol{\varepsilon}$ with colored noise $\boldsymbol{\varepsilon}$, *whitening* refers to transforming the equation by the *whitening matrix* $\mathbf{C}_{\varepsilon}^{-1/2}$ where $\mathbf{C}_{\varepsilon} = \boldsymbol{\varepsilon}\,\boldsymbol{\varepsilon}^{\mathrm{T}}$ is the noise covariance matrix. The transformed equation is

$$\mathbf{C}_{\varepsilon}^{-1/2}\mathbf{Y} = \mathbf{C}_{\varepsilon}^{-1/2}\mathbf{A}\,\mathbf{S} + \boldsymbol{\nu}, \tag{6.45}$$

where $\boldsymbol{\nu} = \mathbf{C}_{\varepsilon}^{-1/2}\boldsymbol{\varepsilon}$ is white noise, because

$$\boldsymbol{\nu}\,\boldsymbol{\nu}^{\mathrm{T}} = \mathbf{C}_{\varepsilon}^{-1/2}\boldsymbol{\varepsilon}\,\boldsymbol{\varepsilon}^{\mathrm{T}}\mathbf{C}_{\varepsilon}^{-1/2} = \mathbf{C}_{\varepsilon}^{-1/2}\mathbf{C}_{\varepsilon}\mathbf{C}_{\varepsilon}^{-1/2} = \mathbf{I};$$

thus the term "whitening."

6.6 Search for Source Dipole Locations with the Output Power $\mu(\mathbf{p})$

With noisy data $\mathbf{Y} = \mathbf{AS} + \boldsymbol{\varepsilon}$, the output power function $\mu(\mathbf{p})$ is a positive and continuous function of $\mathbf{p}$, as equation (6.35) shows. Using the approximations (6.36) and (6.37) and the equation $\mathbf{S}_p = \widehat{\mathbf{S}}_p = 0$ for $\mathbf{p}$ that is not among the active source locations (i.e., $\mathbf{p} \notin \{\mathbf{p}_1, \ldots, \mathbf{p}_n\}$), we can conclude that with a small amount of noise, the source dipole locations $\mathbf{p}_j$, with $||\widetilde{\mathbf{S}}_{p_j}|| > 0$, (approximately) are (the largest) local maximum points of $\mu(\mathbf{p})$, and furthermore, if $\widehat{\mathbf{p}}_j$ is a local maximum point close to a source location $\mathbf{p}_j$ with $||\widetilde{\mathbf{S}}_j|| > 0$, then

$$\mu(\widehat{\mathbf{p}}_j) \simeq ||\widetilde{\mathbf{S}}_j||^2. \tag{6.46}$$

In practice, these results are usually valid for SNRs greater than about 1/3; for smaller SNRs, they may fail. The results also depend on the number of time points so that in general, they improve with the increasing number.

So, $\mu(\mathbf{p})$ can be used as a *localizer* function in search for source locations. In practice, we discretize the ROI by a sufficiently dense point grid and compute $\mu(\mathbf{p})$ at the grid points. Thereafter, we look for the local maxima of $\mu(\mathbf{p})$, which approximately show the locations of source dipoles with $||\widetilde{\mathbf{S}}_j|| > 0$. If the ROI is a surface, like the surface of the cortex, we can draw a contour plot of $\mu(\mathbf{p})$, where the source locations appear as the highest "hill-tops" (see figure 6.3).

If a local maximum point $\widehat{\mathbf{p}}_j$ is close to a source point $\mathbf{p}_j$, then due to equation (6.37), the maximum value $\mu(\widehat{\mathbf{p}}_j)$ is approximately $||\widetilde{\mathbf{S}}_j||^2$, and therefore, we call $||\widetilde{\mathbf{S}}_j||^2$ the *visibility* of the dipole at $\mathbf{p}_j$. By equation (6.31), the visibility $||\widetilde{\mathbf{S}}_j||^2$ is always smaller than or equal to $||\mathbf{S}_j||^2$. The ratio $||\widetilde{\mathbf{S}}_j||^2/||\mathbf{S}_j||^2$ can be very small, due to the correlation of time courses $\mathbf{S}_1, \ldots, \mathbf{S}_n$, and so the visibility of the dipole at $\mathbf{p}_j$ can be very poor, even when the activity $||\mathbf{S}_j||$ is strong.

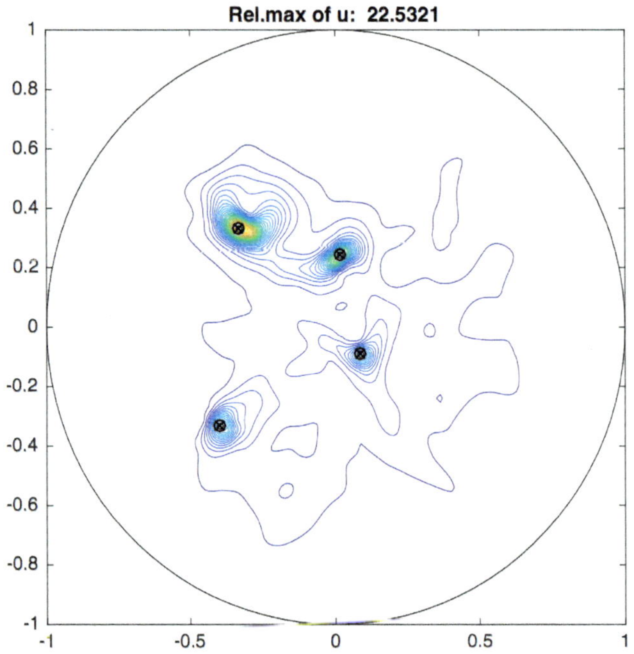

Figure 6.3
Beamformer search for source locations simulated in a three-layer spherical head model. The ROI is the innermost spherical upper-hemisphere surface and the source orientations are fixed to the radial direction. Four dipoles are set to locations marked by black circles. The EEG data have been computed at sixty electrode positions on the upper hemisphere of the sphere. A small amount of colored noise is added to the data. In the figure, the contour plot of the resulting localizer is presented. The "hill-tops" of the plot agree well with the true locations of the source dipoles.

The extreme case is that some time courses, say $\mathbf{S}_1 \ldots \mathbf{S}_k$, are *synchronous*, that is, linearly dependent, meaning that there are $\alpha_j \neq 0$ so that $\alpha_1 \mathbf{S}_1 + \cdots + \alpha_k \mathbf{S}_k = 0$. Then, $||\widetilde{\mathbf{S}}_j|| = 0$ for $j = 1, \ldots, k$, because

$$\mathbf{S}_j{}^{\mathrm{T}} = -\frac{1}{\alpha_j} \sum_{i=1,\, i \neq j}^{k} \alpha_i \mathbf{S}_i{}^{\mathrm{T}} \tag{6.47}$$

is in span$(\mathbf{S}_1{}^{\mathrm{T}}, \ldots, \mathbf{S}_{j-1}{}^{\mathrm{T}}, \mathbf{S}_{j+1}{}^{\mathrm{T}}, \ldots, \mathbf{S}_k{}^{\mathrm{T}})$, and thus $\widetilde{\mathbf{S}}_j{}^{\mathrm{T}} = \mathbf{P}_j \mathbf{S}_k{}^{\mathrm{T}} = 0$, where $\mathbf{P}_j$ is the orthogonal projection onto the orthospace of span$(\mathbf{S}_1{}^{\mathrm{T}}, \ldots, \mathbf{S}_{j-1}{}^{\mathrm{T}}, \mathbf{S}_{j+1}{}^{\mathrm{T}}, \ldots, \mathbf{S}_k{}^{\mathrm{T}})$.

So, the dipoles with synchronous time courses are poorly visible. However, due to noise, they are not completely invisible. Namely, if a dipole is active, that is, $\mathbf{S}_p \neq 0$, but

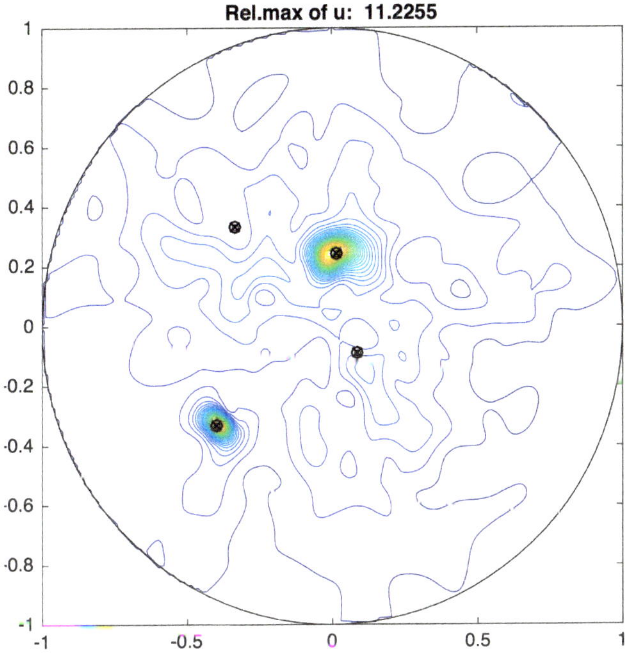

Figure 6.4
The setup is the same as in figure 6.3 except that two of the four dipoles are synchronous. The contour plot correctly shows the locations of the two non-synchronous dipoles while the two synchronous ones remain invisible.

$||\widetilde{\mathbf{S}}_p|| = 0$, then the noise gives it slight visibility and its location may appear as a faint local maximum of the localizer $\mu(\mathbf{p})$. It still can be lost in a conventional beamformer search (see figure 6.4), but as we show later, in an iterative beamformer search, it will likely become clearly visible and can be found.

6.7 Improved Beamformer Localizers for Searching Source Dipoles

When the output power $\mu(\mathbf{p})$, as in equation (6.35), is used as a source localizer, it is affected by noise and its local maximum points do not show the source locations accurately. In particular, this happens if $||\mathbf{l}(\mathbf{p})||$ varies a lot as $\mathbf{p}$ varies in the ROI, typically if the ROI contains points with varying depths. These problems in localizing accuracy can be largely cured by weighting $\mu(\mathbf{p})$ appropriately. The following improved localizers, obtained by weighting, are the most popular:

Array Gain (AG) localizer [282]

$$\tau(\mathbf{p}) = \frac{\mathbf{l}(\mathbf{p})^{\mathrm{T}}\mathbf{l}(\mathbf{p})}{\mathbf{l}(\mathbf{p})^{\mathrm{T}}\mathbf{C}^{-1}\mathbf{l}(\mathbf{p})}, \tag{6.48}$$

(Van Veen's neural) *Activity Index* (AI) localizer [315]

$$\tau(\mathbf{p}) = \frac{\mathbf{l}(\mathbf{p})^{\mathrm{T}}\mathbf{C}_{\varepsilon}^{-1}\mathbf{l}(\mathbf{p})}{\mathbf{l}(\mathbf{p})^{\mathrm{T}}\mathbf{C}^{-1}\mathbf{l}(\mathbf{p})}, \tag{6.49}$$

Pseudo-Z (PZ) localizer (see Robinson and Vrba [260], pages 302 – 305),

$$\tau(\mathbf{p}) = \frac{\mathbf{l}(\mathbf{p})^{\mathrm{T}}\mathbf{C}^{-1}\mathbf{l}(\mathbf{p})}{\mathbf{l}(\mathbf{p})^{\mathrm{T}}\mathbf{C}^{-1}\mathbf{C}_{\varepsilon}\mathbf{C}^{-1}\mathbf{l}(\mathbf{p})}, \tag{6.50}$$

where $\mathbf{C} = \mathbf{Y}\mathbf{Y}^{\mathrm{T}}$ is the data covariance matrix and $\mathbf{C}_{\varepsilon} = \boldsymbol{\varepsilon}\boldsymbol{\varepsilon}^{\mathrm{T}}$ is the noise covariance marix. In the AG and AI localizers, the weights are

$$\mathbf{l}(\mathbf{p})^{\mathrm{T}}\mathbf{l}(\mathbf{p}) \text{ and } \mathbf{l}(\mathbf{p})^{\mathrm{T}}\mathbf{C}_{\varepsilon}^{-1}\mathbf{l}(\mathbf{p}), \tag{6.51}$$

respectively. In the PZ localizer, the weight is d^{-1} for

$$d = \frac{\mathbf{l}(\mathbf{p})^{\mathrm{T}}\mathbf{C}^{-1}\mathbf{C}_{\varepsilon}\mathbf{C}^{-1}\mathbf{l}(\mathbf{p})}{\left(\mathbf{l}(\mathbf{p})^{\mathrm{T}}\mathbf{C}^{-1}\mathbf{l}(\mathbf{p})\right)^{2}}. \tag{6.52}$$

In above localizers, $\mathbf{l}(\mathbf{p})$ can be replaced by the unit vector $||\mathbf{l}(\mathbf{p})||^{-1}\mathbf{l}(\mathbf{p})$ without changing the localizer value. So, the effect of a varying $||\mathbf{l}(\mathbf{p})||$ is removed from the localizers.

The AG localizer is directly formed by normalizing the size $||\mathbf{l}(\mathbf{p})||$. The AI localizer is a whitened version of the AG one, as the following shows.

Consider the data equation $\mathbf{Y} = \mathbf{A}\,\mathbf{S} + \boldsymbol{\varepsilon}$ with $\mathbf{A} = [\mathbf{l}(\mathbf{p}_1), \ldots, \mathbf{l}(\mathbf{p}_n)]$. If we multiply it by the whitening matrix $\mathbf{B} = \mathbf{C}_{\varepsilon}^{-1/2}$, we get the whitened data equation

$$\begin{aligned} \mathbf{B}\,\mathbf{Y} &= \mathbf{B}\,\mathbf{A}\,\mathbf{S} + \mathbf{B}\,\boldsymbol{\varepsilon} \\ &= [\mathbf{B}\,\mathbf{l}(\mathbf{p}_1), \ldots, \mathbf{B}\,\mathbf{l}(\mathbf{p}_n)]\,\mathbf{S} + \mathbf{B}\boldsymbol{\varepsilon} \end{aligned} \tag{6.53}$$

with the data covariance matrix $\mathbf{B}\,\mathbf{Y}(\mathbf{B}\,\mathbf{Y})^{\mathrm{T}} = \mathbf{B}\,\mathbf{C}\,\mathbf{B}$ and topographies $\mathbf{B}\,\mathbf{l}(\mathbf{p})$. Therefore, the AG localizer for this transformed data equation is

$$\begin{aligned} \tau_B(\mathbf{p}) &= \frac{(\mathbf{B}\,\mathbf{l}(\mathbf{p}))^{\mathrm{T}}\mathbf{B}\,\mathbf{l}(\mathbf{p})}{(\mathbf{B}\,\mathbf{l}(\mathbf{p}))^{\mathrm{T}}(\mathbf{B}\,\mathbf{C}\,\mathbf{B})^{-1}\mathbf{B}\,\mathbf{l}(\mathbf{p})} = \frac{\mathbf{l}(\mathbf{p})^{\mathrm{T}}\mathbf{B}^{2}\mathbf{l}(\mathbf{p})}{\mathbf{l}(\mathbf{p})^{\mathrm{T}}\mathbf{C}^{-1}\mathbf{B}^{-1}\mathbf{l}(\mathbf{p})} \\ &= \frac{\mathbf{l}(\mathbf{p})^{\mathrm{T}}\mathbf{C}_{\varepsilon}^{-1}\mathbf{l}(\mathbf{p})}{\mathbf{l}(\mathbf{p})^{\mathrm{T}}\mathbf{C}^{-1}\mathbf{B}^{-1}\mathbf{l}(\mathbf{p})}, \end{aligned}$$

which agrees with the AI localizer of the original data equation.

The idea behind the PZ localizer can be explained in terms of the signal-to-noise ratio as follows. Consider the beamformer estimate $\widetilde{\mathbf{S}}_p$ of a time course $\mathbf{S}_p$ given by

$$\widetilde{\mathbf{S}}_p(\mathbf{p}) = \mathbf{w}_p{}^{\mathrm{T}}\mathbf{Y} = \mathbf{w}_p{}^{\mathrm{T}}\mathbf{AS} + \mathbf{w}_p{}^{\mathrm{T}}\boldsymbol{\varepsilon} \text{ with}$$

$$\mathbf{w}_p = \frac{\mathbf{C}^{-1}\mathbf{l}(\mathbf{p})}{\mathbf{l}(\mathbf{p})^{\mathrm{T}}\mathbf{C}^{-1}\mathbf{l}(\mathbf{p})}.$$

Here, the term $\mathbf{w}_p{}^{\mathrm{T}}\boldsymbol{\varepsilon}$ is the estimation error with "error power"

$$||\mathbf{w}_p{}^{\mathrm{T}}\boldsymbol{\varepsilon}||^2 = \mathbf{w}_p{}^{\mathrm{T}}\mathbf{C}_\varepsilon \mathbf{w}_p = \frac{\mathbf{l}(\mathbf{p})^{\mathrm{T}}\mathbf{C}^{-1}\mathbf{C}_\varepsilon\mathbf{C}^{-1}\mathbf{l}(\mathbf{p})}{(\mathbf{l}(\mathbf{p})^{\mathrm{T}}\mathbf{C}^{-1}\mathbf{l}(\mathbf{p}))^2},$$

and $\mathbf{w}_p{}^{\mathrm{T}}\mathbf{A}\,\mathbf{S}$ is the signal with the "signal power" $||\mathbf{w}_p{}^{\mathrm{T}}\mathbf{A}\,\mathbf{S}||^2$, which approximately is

$$||\mathbf{w}_p{}^{\mathrm{T}}\mathbf{Y}||^2 = \mathbf{w}_p{}^{\mathrm{T}}\mathbf{C}\,\mathbf{w}_p = \frac{1}{\mathbf{l}(\mathbf{p})^{\mathrm{T}}\mathbf{C}^{-1}\mathbf{l}(\mathbf{p})}.$$

It follows that the (approximative) signal-to-noise power ratio is

$$\frac{||\mathbf{w}_p{}^{\mathrm{T}}\mathbf{Y}||^2}{||\mathbf{w}_p{}^{\mathrm{T}}\boldsymbol{\varepsilon}||^2} = \frac{\mathbf{l}(\mathbf{p})^{\mathrm{T}}\mathbf{C}^{-1}\mathbf{l}(\mathbf{p})}{\mathbf{l}(\mathbf{p})^{\mathrm{T}}\mathbf{C}^{-1}\mathbf{C}_\varepsilon\mathbf{C}^{-1}\mathbf{l}(\mathbf{p})},$$

which agrees with the PZ localizer (6.50).

Note that all the above localizers are of the form

$$\tau(\mathbf{p}) = \frac{\mathbf{l}(\mathbf{p})^{\mathrm{T}}\mathbf{F}\,\mathbf{l}(\mathbf{p})}{\mathbf{l}(\mathbf{p})^{\mathrm{T}}\mathbf{G}\,\mathbf{l}(\mathbf{p})}, \tag{6.54}$$

where

$$\mathbf{F} = \mathbf{I}, \quad \mathbf{G} = \mathbf{C}^{-1} \quad \text{for AG,} \tag{6.55}$$

$$\mathbf{F} = \mathbf{C}_\varepsilon^{-1}, \quad \mathbf{G} = \mathbf{C}^{-1} \quad \text{for AI,} \tag{6.56}$$

$$\mathbf{F} = \mathbf{C}^{-1}, \quad \mathbf{G} = \mathbf{C}^{-1}\mathbf{C}_\varepsilon\mathbf{C}^{-1} \quad \text{for PZ.} \tag{6.57}$$

All the above localizers $\tau(\mathbf{p})$ (at least with a small amount of noise) have the property that the source dipole locations approximately are local maximum points of $\tau(\mathbf{p})$; see Appendix 6.23. In particular, if $\mathbf{p}_j$ is a source location, then by equation (6.36),

$$\tau(\mathbf{p}_j) \simeq c(\mathbf{p}_j)||\widetilde{\mathbf{S}}_{\mathbf{p}_j}||^2, \tag{6.58}$$

where $c(\mathbf{p})$ is the weight in equations (6.48) to (6.50), and

$$\tau(\mathbf{p}) = c(\mathbf{p})\mu(\mathbf{p}). \tag{6.59}$$

It follows that the visibility of a source dipole at $\mathbf{p}_j$, when scanning in the ROI with $\tau(\mathbf{p})$, is proportional to $||\widetilde{\mathbf{S}}_{\mathbf{p}_j}||^2$, such as in scanning with $\mu(\mathbf{p})$.

A localizer is said to have *no location bias*, if it exactly finds the location of a source *consisting of a single dipole*. It can be shown that the AI and PZ localizers have no location bias with white or colored noise, while the AG localizer has no location bias only with white noise; see Appendix 6.24.

Although the AI and PZ localizers, in principle, have better locating accuracy than the AG one, it is often difficult to make use of that advantage because one should know $\mathbf{C}_{\boldsymbol{\varepsilon}}$, which is not easily available. This is especially the case in evoked potential EEG and MEG measurements, where noise $\boldsymbol{\varepsilon}$ is due not only to measurement noise but also to so-called brain noise (or neuronal noise), which consists of all that uninteresting neuronal activity during the measurement not modeled in the noiseless data equation $\mathbf{Y}_0 = \mathbf{AS}$.

6.7.1 Regularizing Data and Noise Covariance Matrices

The inverses of the data and noise covariance matrices $\mathbf{C}$ and $\mathbf{C}_\varepsilon$ are used in localizers (6.48) to (6.50). If these matrices are underranked (rank $< m$) or ill-posed (e.g., due to filtering in the time domain or setting the potential zero level in EEG), they must be regularized before inverting.

This can be done, for example, by replacing $\mathbf{C}$ by $\mathbf{C} + \gamma\,\mathbf{I}$ with $\gamma = p^2 m^{-1}\mathrm{trace}(\mathbf{C})$ for an appropriate $p > 0$. This is equivalent to adding (statistically independent) white noise $\boldsymbol{\varepsilon}_{\mathrm{wh}}$, with covariance matrix $\gamma\,\mathbf{I}$, to the data equation $\mathbf{Y} = \mathbf{A\,S} + \boldsymbol{\varepsilon}$.

Namely, in the original data equation, $\mathbf{C} = \mathbf{C}_0 + \mathbf{C}_\varepsilon$, where $\mathbf{C}_0 = \mathbf{A\,S}(\mathbf{A\,S})^{\mathrm{T}}$ is the noiseless data covariance matrix. Due to the addition rule of covariance matrices, in the transformed data equation $\mathbf{Y} = \mathbf{A\,S} + \boldsymbol{\varepsilon} + \boldsymbol{\varepsilon}_{\mathrm{wh}}$, the data covariance matrix is $\mathbf{C}_0 + \mathbf{C}_\varepsilon + \gamma\,\mathbf{I} = \mathbf{C} + \gamma\,\mathbf{I}$, and the new noise covariance matrix is $\mathbf{C}_\varepsilon + \gamma\,\mathbf{I}$.

If we, in the signal-to-noise ratio (SNR), consider data $\mathbf{Y} = \mathbf{A\,S} + \boldsymbol{\varepsilon}$ as "signal" and the added white noise $\boldsymbol{\varepsilon}_{\mathrm{wh}}$ as "noise," then in the transformed data, SNR = $\left(\mathrm{trace}(\mathbf{C})/\mathrm{trace}(\gamma\,\mathbf{I})\right)^{-1/2} = 1/p$. If we, as usual, consider 100/SNR as the noise percentage in the data, we see that replacing $\mathbf{C}$ by $\mathbf{C} + \gamma\,\mathbf{I}$ is equivalent to adding $100\,p$ percent white noise to the measured data $\mathbf{Y}$.

The parameter $p > 0$ controls the amount of regularization. The optimal value of p depends on the data, but often $p = 0.05$ will do, or an appropriate p is found by experimenting.

6.8 Finding Estimates for Time Courses

After we have found estimates $\widehat{\mathbf{p}}_1, \ldots, \widehat{\mathbf{p}}_k$ for source locations with a localizer $\tau(\mathbf{p})$, it may be difficult to know if all source dipole locations $\mathbf{p}_1, \ldots, \mathbf{p}_n$ have been found, and whether there are some false locations among the found ones. Namely, there could be some source dipoles with almost synchronous or very weak time courses, and they may have been missed in the search, because they show very poorly in the scanning with a localizer. We discuss later how this situation can be dealt with using iterative beamformers.

If we may, however, assume that the found locations include good estimates for all true source dipole locations plus some false locations, we can find reasonable estimates for the time courses of the source dipoles and also identify the false locations in the following way.

Given that the time courses of the source dipoles are uncorrelated, we can use equations (6.36) and (6.23), and with

$$\mathbf{w}_i = \frac{\mathbf{C}^{-1}\mathbf{l}(\widehat{\mathbf{p}}_i)}{\mathbf{l}(\widehat{\mathbf{p}}_i)^{\mathrm{T}}\mathbf{C}^{-1}\mathbf{l}(\widehat{\mathbf{p}}_i)} \tag{6.60}$$

get

$$\widehat{\mathbf{S}}_i = \mathbf{w}_i{}^{\mathrm{T}}\mathbf{Y}, \ i = 1, \ldots, k, \tag{6.61}$$

where $\widehat{\mathbf{S}}_i$ is an estimate for the time course of the source dipole close to $\widehat{\mathbf{p}}_i$ or $\widehat{\mathbf{S}}_i \simeq 0$, if $\widehat{\mathbf{p}}_i$ is a false location, and we are done.

If we do not know whether the time courses are correlated or not, we can proceed as follows. We compute the topographies $\mathbf{l}(\widehat{\mathbf{p}}_i), \ldots, \mathbf{l}(\widehat{\mathbf{p}}_k)$ by the forward solver. We can now estimate,

$$\mathbf{Y} \simeq \widehat{\mathbf{A}}\mathbf{S}, \tag{6.62}$$

where $\widehat{\mathbf{A}} = [\mathbf{l}(\widehat{\mathbf{p}}_1), \ldots, \mathbf{l}(\widehat{\mathbf{p}}_k)]$ and $\mathbf{S}_i = \mathbf{S}(i, :)$, $i = 1, \ldots, k$, is the time course at $\widehat{\mathbf{p}}_i$. Next we solve equation (6.62) for $\mathbf{S}$ in a regularized way and get an estimate $\widehat{\mathbf{S}}$ for $\mathbf{S}$,

$$\widehat{\mathbf{S}} = (\widehat{\mathbf{A}}^{\mathrm{T}}\widehat{\mathbf{A}} + \lambda\mathbf{I})^{-1}\widehat{\mathbf{A}}^{\mathrm{T}}\mathbf{Y}, \tag{6.63}$$

where $\lambda > 0$ is a regularization parameter, say, $\lambda = 10^{-3}(d_1 + \cdots + d_k)/k$, where d_i are the diagonal elements of $\widehat{\mathbf{A}}^{\mathrm{T}}\widehat{\mathbf{A}}$. If now $||\widehat{\mathbf{S}}_i|| > 0$, then $\widehat{\mathbf{p}}_i$ estimates the location of some true source dipole and $\widehat{\mathbf{S}}_i$ estimates its time course. If $||\widehat{\mathbf{S}}_i|| \simeq 0$, then $\widehat{\mathbf{p}}_i$ is either a false location or close to a source dipole with a time course so weak that it can be ignored as noise (see figure 6.5). Notice that in (6.63), the possible correlation of the time courses does not play any role.

Finally, we note that the case in which we know that the time courses are uncorrelated can also be analyzed with the latter method, and the results of both methods can be compared. If they do not agree (to a reasonable extent), then we probably have not found all active source dipoles.

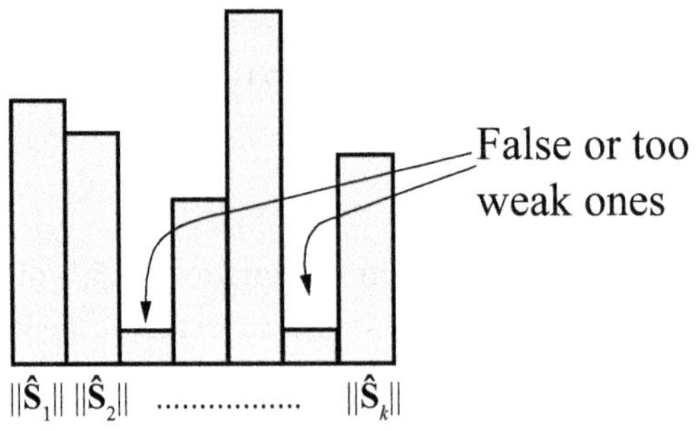

Figure 6.5
The norms $||\widehat{\mathbf{S}}_j||$ of all found time courses can indicate wether the estimate corresponds to a real source or not.

6.9 Scalar Beamformer with Optimal Orientations

If the dipoles in the ROI are freely oriented, one must use optimal orientations of the dipoles when employing scalar beamformers.

For every location $\mathbf{p}$, the *optimal orientation* $\eta_{\text{opt}}(\mathbf{p})$ for the AG, AI, and PZ beamformers is defined by

$$\eta_{\text{opt}}(\mathbf{p}) = \arg\max_{||\eta||=1} \frac{\eta^{\text{T}}\mathbf{L}(\mathbf{p})^{\text{T}}\mathbf{F}\,\mathbf{L}(\mathbf{p})\,\eta}{\eta^{\text{T}}\mathbf{L}(\mathbf{p})^{\text{T}}\mathbf{G}\,\mathbf{L}(\mathbf{p})\,\eta}, \tag{6.64}$$

where $\mathbf{F}$ and $\mathbf{G}$ depend on the beamformer type as in equations (6.55) to (6.57). Note that $\mathbf{F}$ and $\mathbf{G}$ involve the data covariance $\mathbf{C} = \mathbf{Y}\mathbf{Y}^{\text{T}}$ and, for AI and PZ, also the noise covariance $\mathbf{C}_\varepsilon = \boldsymbol{\varepsilon}\boldsymbol{\varepsilon}^{\text{T}}$. With $\eta_{\text{opt}}(\mathbf{p})$, we also assign the (optimal) topography

$$\mathbf{l}_{\text{opt}}(\mathbf{p}) = \mathbf{L}(\mathbf{p})\eta_{\text{opt}}(\mathbf{p}) \tag{6.65}$$

to every $\mathbf{p}$ in the ROI.

One can show that if $(\mathbf{p}_j, \eta_j)$ is a source dipole, then

$$\eta_{\text{opt}}(\mathbf{p}_j) \simeq \pm\eta_j \text{ and } \mathbf{l}_{\text{opt}}(\mathbf{p}_j) \simeq \pm\mathbf{l}(\mathbf{p}_j), \tag{6.66}$$

and the approximations become more accurate with reduced noise levels; see the end of Appendix 6.23.

Finally, we define the localizer $\tau_{\text{opt}}(\mathbf{p})$ of a scalar beamformer with optimal orientation by replacing $\mathbf{l}(\mathbf{p})$ by $\mathbf{l}_{\text{opt}}(\mathbf{p})$ in the definition (6.54) of $\tau(\mathbf{p})$ and get

$$\tau_{\text{opt}}(\mathbf{p}) = \frac{\mathbf{l}_{\text{opt}}(\mathbf{p})^{\text{T}}\mathbf{F}\,\mathbf{l}_{\text{opt}}(\mathbf{p})}{\mathbf{l}_{\text{opt}}(\mathbf{p})^{\text{T}}\mathbf{G}\,\mathbf{l}_{\text{opt}}(\mathbf{p})} = \max_{||\eta||=1} \frac{\eta^{\text{T}}\mathbf{L}(\mathbf{p})^{\text{T}}\mathbf{F}\,\mathbf{L}(\mathbf{p})\eta}{\eta^{\text{T}}\mathbf{L}(\mathbf{p})^{\text{T}}\mathbf{G}\,\mathbf{L}(\mathbf{p})\eta}. \tag{6.67}$$

It follows that the source locations are approximately the local maximum points of $\tau_{opt}(\mathbf{p})$, so if a local maximum point $\widehat{\mathbf{p}}_j$ is close to the true location $\mathbf{p}_j$ of a source dipole $(\mathbf{p}_j, \eta_j)$, then

$$\eta_{opt}(\widehat{\mathbf{p}}_j) \simeq \pm\eta_j \text{ and } \tau_{opt}(\widehat{\mathbf{p}}_j) \simeq c(\mathbf{p})||\widetilde{\mathbf{S}}_j||^2. \tag{6.68}$$

Here, $c(\mathbf{p})$ is the weight in the definition of equation (6.59) of $\tau(\mathbf{p})$. It follows that $\tau_{opt}(\mathbf{p})$ can be used as a localizer, that is, the source locations are local maximum points of $\tau_{opt}(\mathbf{p})$, and the visibility of a source dipole at $\mathbf{p}_j$ (when scanning with $\tau_{opt}(\mathbf{p})$) is proportional to $||\widetilde{\mathbf{S}}_j||^2$.

The AI and PZ localizers $\tau_{opt}(\mathbf{p})$ have no location bias, while the AG localizer $\tau_{opt}(\mathbf{p})$ has it for colored noise, but not for white noise; see the end of Appendix 6.24.

Both $\eta_{opt}(\mathbf{p})$ and $\tau_{opt}(\mathbf{p})$ have the following closed-form representations:

$$\eta_{opt}(\mathbf{p}) = \frac{1}{||\mathbf{Eu}||}\mathbf{Eu}, \tag{6.69}$$

$$\tau_{opt}(\mathbf{p}) = d, \tag{6.70}$$

where

$$\mathbf{E} = \left(\mathbf{L}(\mathbf{p})^T \mathbf{G} \mathbf{L}(\mathbf{p})\right)^{-1/2}, \tag{6.71}$$

and d is the largest eigenvalue of the matrix

$$\mathbf{K} = \mathbf{E}\,\mathbf{L}(\mathbf{p})^T \mathbf{F}\,\mathbf{L}(\mathbf{p})\,\mathbf{E}, \tag{6.72}$$

and $\mathbf{u}$ is the corresponding eigenvector. Here, $\mathbf{F}$ and $\mathbf{G}$ depend on the beamformer types AG, AI, and PZ as in equations (6.55) to (6.57).

Equations (6.69) and (6.70) are derived as follows. First note that

$$\eta^T \mathbf{L}(\mathbf{p})^T \mathbf{G}\,\mathbf{L}(\mathbf{p})\eta = ||\left(\mathbf{L}(\mathbf{p})^T \mathbf{G}\,\mathbf{L}(\mathbf{p})\right)^{1/2} \eta\,||^2.$$

Making in equation (6.64) the change of variable

$$\mathbf{z} = \left(\mathbf{L}(\mathbf{p})^T \mathbf{G}\,\mathbf{L}(\mathbf{p})\right)^{1/2} \eta = \mathbf{E}^{-1}\eta, \tag{6.73}$$

with $\mathbf{E}$ as in equation (6.71), turns the maximizing task in equation (6.64) to that of maximizing the quadratic form $\mathbf{z}^T\mathbf{K}\mathbf{z}$ with constraint $||\mathbf{z}|| = 1$, and $\mathbf{K}$ as in equation (6.72); see equation (4.26). Therefore, the maximum in equation (6.64) will be the largest eigenvalue d of $\mathbf{K}$. Let $\mathbf{u}$ be the eigenvector of $\mathbf{K}$ corresponding to d. Inverting the change of variable in equation (6.73) yields $\eta = \mathbf{E}\,\mathbf{z}$, and thus $||\mathbf{Eu}||^{-1}\mathbf{Eu}$ is the unit vector by which the maximum is reached in equations (6.64), that is, $\eta_{opt}(\mathbf{p}) = ||\mathbf{Eu}||^{-1}\mathbf{Eu}$. This completes the derivation of equations (6.69) and (6.70).

We note that the closed forms of $\eta_{\mathrm{opt}}(\mathbf{p})$ and $\tau_{\mathrm{opt}}(\mathbf{p})$ can also be obtained by the generalized eigenvalue task as follows. Consider the generalized eigenvalue equation

$$\mathbf{A}\,\mathbf{u} = \lambda\,\mathbf{B}\,\mathbf{u}, \tag{6.74}$$

where $\mathbf{A} = \mathbf{L}(\mathbf{p})^{\mathrm{T}}\mathbf{F}\,\mathbf{L}(\mathbf{p})$, $\mathbf{B} = \mathbf{L}(\mathbf{p})^{\mathrm{T}}\mathbf{G}\,\mathbf{L}(\mathbf{p})$, λ is the generalized eigenvalue, and $\mathbf{u}$ is the corresponding eigenvector. Let d be the largest eigenvalue and $\mathbf{u}$ the corresponding eigenvector of equation (6.74). Then, $\tau_{\mathrm{opt}}(\mathbf{p}) = d$ and $\eta_{\mathrm{opt}}(\mathbf{p}) = ||\mathbf{u}||^{-1}\mathbf{u}$, as one can show. With MATLAB, you find the desired eigenvalues and eigenvectors with the command

$$[\mathbf{V}, \mathbf{D}] = \mathrm{eig}(\mathbf{A}, \mathbf{B}), \tag{6.75}$$

where $\mathbf{D}(i, i)$ are the eigenvalues and $\mathbf{V}(:, i)$ the corresponding eigenvectors, $i = 1, 2, 3$, and you only need to pick the largest $\mathbf{D}(i, i)$ and the corresponding eigenvector. When building a MATLAB code for a beamformer with optimal orientations, the use of the generalized eigenvalues for $\eta_{\mathrm{opt}}(\mathbf{p})$ and $\tau_{\mathrm{opt}}(\mathbf{p})$ is technically simpler and leads to a somewhat faster code than the use of equations (6.69) and (6.70), although those formulas with their derivation are more transparent than the use of the generalized eigenvalues.

For given measurement data $\mathbf{Y}$, the search for source dipoles with $\tau_{\mathrm{opt}}(\mathbf{p})$ works similarly as with $\tau(\mathbf{p})$. After having found local maximum points $\widehat{\mathbf{p}}_1, \ldots, \widehat{\mathbf{p}}_k$ of the localizer $\tau_{\mathrm{opt}}(\mathbf{p})$, we compute the optimal orientations $\widehat{\eta}_j = \eta_{\mathrm{opt}}(\widehat{\mathbf{p}}_j)$ by equations (6.64) and (6.69) and assign the topographies $\mathbf{l}(\widehat{\mathbf{p}}_j) = \mathbf{L}(\widehat{\mathbf{p}}_j)\widehat{\eta}_j$ to the locations $\widehat{\mathbf{p}}_j$, $j = 1, \ldots, k$. If we can assume that the true source locations are (approximately) among the candidates $\widehat{\mathbf{p}}_1, \ldots, \widehat{\mathbf{p}}_k$, we can identify them and compute estimates for the corresponding time courses as in section 6.8.

6.10 Time-Dependent Orientations

So far, we have assumed that at source dipole location $\mathbf{p}$, the orientation η_p of the dipole is time-independent with a time course $\mathbf{S}_p \in \mathbb{R}^{1\times N}$. One can relax that assumption and let the orientations be *time-dependent* $\eta_{\mathrm{p}}(t)$. Dipoles with such orientations are called *rotating* [209]. A rotating dipole at $\mathbf{p}$ can be described as a set of three dipoles, set to the same location $\mathbf{p}$, with x-, y-, and z-directed orientations and individual time courses $\mathbf{S}_p(j, :)$, $j = 1, 2, 3$ with $\mathbf{S}_p \in \mathbb{R}^{3\times N}$. With this description, the time-dependent orientation $\eta_p(t_j)$ of the rotating dipole at instant t_j is

$$\eta_p(t_j) = \mathbf{S}_p(:, t_j)/||\mathbf{S}_p(:, t_j)|| \tag{6.76}$$

and its time-dependent amplitude is $||\mathbf{S}_p(:, t_j)||$, that is, the (single) time course is $||\mathbf{S}_p(:, t_j)||$, $j = 1, \ldots, N$. On the other hand, a dipole at $\mathbf{p}$ with time-independent orientation η_p and time course $\mathbf{s} \in \mathbb{R}^{1\times N}$, can be described as a (non-)rotating dipole with $\mathbf{S}_p = \eta_p\,\mathbf{s}$.

Assume now that there is a rotating dipole at $\mathbf{p}$ with x-, y-, and z-oriented dipoles with time courses $\mathbf{S}_p(j,:)$, $j=1,2,3$. Then the rotation dipole spans the subspace span$(\mathbf{S}_p)$, which is one-, two- or three-dimensional depending on whether rank $(\mathbf{S}_p)$ is 1, 2, or 3, respectively. If the dimension is one, then $\mathbf{S}_p$ is of the form $\mathbf{S}_p=\mathbf{a}\,\mathbf{s}$, $\mathbf{a}\in\mathbb{R}^{3\times 1}$, and $\mathbf{s}\in\mathbb{R}^{1\times N}$, and we actually have a dipole with time-independent orientation $\eta_p=\mathbf{a}/||\mathbf{a}||$ and a single time course $||\mathbf{a}||\,\mathbf{s}$.

An important additional property of a scalar beamformer with optimal orientations is that it can handle source dipoles with time-dependent orientations, too. Namely, rotating-dipole locations are also local maximum points of its localizer, as we can see in the following.

Let all source dipoles, either with rotating or nonrotating orientations, have $3\times N$ time courses $\mathbf{S}_{p_j}$, $j=1,\ldots,n$. Then, $\mathbf{Y}=\mathbf{A}\,\mathbf{S}+\boldsymbol{\varepsilon}$ with $\mathbf{A}=[\mathbf{L}(\mathbf{p}_1),\ldots,\mathbf{L}(\mathbf{p}_n)]$ and $\mathbf{S}^{\mathrm{T}}=[\mathbf{S}_1{}^{\mathrm{T}},\ldots,\mathbf{S}_n{}^{\mathrm{T}}]^{\mathrm{T}}$. Consider the localizer $\tau_{\mathrm{opt}}(\mathbf{p})$ in equation (6.67),

$$\tau_{\mathrm{opt}}(\mathbf{p})=\frac{\eta_{\mathrm{opt}}(\mathbf{p})^{\mathrm{T}}\mathbf{L}(\mathbf{p})^{\mathrm{T}}\mathbf{F}\,\mathbf{L}(\mathbf{p})\,\eta_{\mathrm{opt}}(\mathbf{p})}{\eta_{\mathrm{opt}}(\mathbf{p})^{\mathrm{T}}\mathbf{L}(\mathbf{p})^{\mathrm{T}}\mathbf{G}\,\mathbf{L}(\mathbf{p})\,\eta_{\mathrm{opt}}(\mathbf{p})}=\max_{||\eta||=1}\frac{\eta^{\mathrm{T}}\mathbf{L}(\mathbf{p})^{\mathrm{T}}\mathbf{F}\,\mathbf{L}(\mathbf{p})\,\eta}{\eta^{\mathrm{T}}\mathbf{L}(\mathbf{p})^{\mathrm{T}}\mathbf{G}\,\mathbf{L}(\mathbf{p})\,\eta}. \tag{6.77}$$

As in Appendix 6.23, one can show that the localizer reaches a local maximum (approximately) when $\mathbf{p}$ is one of the source locations $\mathbf{p}_j$ and $\mathbf{L}(\mathbf{p})\,\eta_{\mathrm{opt}}\in\mathrm{span}(\mathbf{L}(\mathbf{p}_j)\mathbf{S}_{p_j})$.

After the source locations $\mathbf{p}_1,\ldots,\mathbf{p}_n$ have been found, the type can be determined as follows. When finding $\eta_{\mathrm{opt}}(\mathbf{p}_j)$ by the maximizing task (6.64) and closed-form formula (6.69), the solution $\eta_{\mathrm{opt}}(\mathbf{p}_j)$ is not unique, if the dimension r of span$(\mathbf{S}_{p_j})$ is greater than one. That dimension is equal to the dimension of the eigenvalue space of the matrix $\mathbf{K}$ in equation (6.72) corresponding to the largest eigenvalue d, and so there are r equal largest eigenvalues. Therefore, if two or three of the largest eigenvalues are equal, then the dipole at $\mathbf{p}_j$ is rotating. In practice, two (or three) non-rotating dipoles may appear as a single rotating one if they are sufficiently close to each other. Usually, on the basis of the measured data, there is no way to distinguish between these two cases.

6.11 Iterative Beamformers

In iterative beamforming, sources are found one after another, and in each iteration step, one either projects out [128] or null-constrains [208] the signals of the previously found dipoles from the measurement data and, using the truncated data, searches for the remaining dipoles.

Iterative beamforming has two main benefits. After each iteration step, the remaining dipoles show better than in the earlier steps. In particular, this is true for strongly correlated, or even synchronous, source dipoles. The second great benefit is that the iterative technique replaces the search for several local maxima of the localizer with a much easier search for the global maximum in each iteration step. In particular, this simplifies a beamformer search for sources in 3-D volume, and also facilitates designing an automated beamformer.

The two iterative beamformers we present are the RAP beamformer [128] and the iterative *multiple constrained minimum variance* (MCMV) beamformer [208].

6.12 Iterative RAP Beamformer with Fixed-Oriented Dipoles

The idea of the *recursively applied and projected* (RAP) *beamformer* has been taken from the RAP-MUSIC algorithm [210] and adapted to beamformers.

We start with the case when the dipoles in the ROI are *fixed-oriented,* so that the localizer $\tau(\mathbf{p})$ has the form

$$\tau(\mathbf{p}) = \frac{\mathbf{l}(\mathbf{p})^{\mathrm{T}}\mathbf{F}\,\mathbf{l}(\mathbf{p})}{\mathbf{l}(\mathbf{p})^{\mathrm{T}}\mathbf{G}\,\mathbf{l}(\mathbf{p})}, \tag{6.78}$$

where $\mathbf{F}$ and $\mathbf{G}$, for different localizer types, are as in equations (6.55) to (6.57). Let the data matrix $\mathbf{Y}$ be due to the source dipoles $(\mathbf{p}_1, \eta_1), \ldots, (\mathbf{p}_n, \eta_n)$ and the time-course matrix $\mathbf{S}$, with time courses $\mathbf{S}_j = \mathbf{S}(j, :)$, $j = 1, \ldots, n$, and to additive noise $\boldsymbol{\varepsilon}$, that is,

$$\mathbf{Y} = \mathbf{A}\mathbf{S} + \boldsymbol{\varepsilon} \tag{6.79}$$

with the $m \times n$ mixing matrix $\mathbf{A} = [\mathbf{l}(\mathbf{p}_1), \ldots, \mathbf{l}(\mathbf{p}_n)]$.

The iteration is started by finding the first location $\widehat{\mathbf{p}}_1$, say $\widehat{\mathbf{p}}_1 \simeq \mathbf{p}_1$, as the global maximum point of the localizer $\tau(\mathbf{p})$.

After having found $\widehat{\mathbf{p}}_1, \ldots, \widehat{\mathbf{p}}_{k-1}$, in the kth iteration step, we find $\widehat{\mathbf{p}}_k$ as follows. We may assume that $\widehat{\mathbf{p}}_j \simeq \mathbf{p}_j$ for $j = 1, \ldots, k-1$. Compute the topographies $\mathbf{l}(\widehat{\mathbf{p}}_j)$, $j = 1, \ldots, k-1$, and denote $\mathbf{B}_k = [\mathbf{l}(\widehat{\mathbf{p}}_1), \ldots, \mathbf{l}(\widehat{\mathbf{p}}_{k-1})]$. Next, we form the orthogonal projection

$$\mathbf{Q}_k = \mathbf{I} - \mathbf{B}_k\mathbf{B}_k^{\dagger}, \tag{6.80}$$

which projects $\mathbb{R}^m$ onto the orthospace of span$(\mathbf{B}_k)$. Thus $\mathbf{Q}_k\mathbf{l}(\widehat{\mathbf{p}}_j) = 0 \simeq \mathbf{Q}_k\mathbf{l}(\mathbf{p}_j)$ for $j = 1, \ldots, k-1$. Then, we apply $\mathbf{Q}_k$ to the data matrix $\mathbf{Y}$, and it (approximately) projects topographies $\mathbf{l}(\mathbf{p}_j)$, $j = 1, \ldots, k-1$, out of $\mathbf{Y}$. The resulting transformed, or updated, data matrix $\mathbf{Q}_k\mathbf{Y}$ is

$$\begin{aligned} \mathbf{Q}_k\mathbf{Y} &= \mathbf{Q}_k\mathbf{A}\mathbf{S} + \mathbf{Q}_k\boldsymbol{\varepsilon} = \sum_{j=1}^{n} \mathbf{Q}_k\mathbf{l}(\mathbf{p}_j)\mathbf{S}_j + \mathbf{Q}_k\boldsymbol{\varepsilon} \\ &\simeq \sum_{j=k}^{n} \mathbf{Q}_k\mathbf{l}(\mathbf{p}_j)\mathbf{S}_j + \mathbf{Q}_k\boldsymbol{\varepsilon} \end{aligned} \tag{6.81}$$

$$= [\mathbf{Q}_k\mathbf{l}(\mathbf{p}_k), \ldots, \mathbf{Q}_k\mathbf{l}(\mathbf{p}_n)] \begin{bmatrix} \mathbf{S}_k \\ \vdots \\ \mathbf{S}_n \end{bmatrix} + \mathbf{Q}_k\boldsymbol{\varepsilon}. \tag{6.82}$$

So, the transformed data $\mathbf{Q}_k\mathbf{Y}$ are, besides some noise, due to the remaining source dipoles at $\mathbf{p}_k, \ldots, \mathbf{p}_n$ with the transformed topographies $\mathbf{Q}_k\mathbf{l}(\mathbf{p}_k), \ldots, \mathbf{Q}_k\mathbf{l}(\mathbf{p}_n)$ and the original time courses $\mathbf{S}_k, \ldots, \mathbf{S}_n$. The transformed data and noise covariance matrices are

$$\mathbf{C}_k = \mathbf{Q}_k\mathbf{Y}(\mathbf{Q}_k\mathbf{Y})^{\mathrm{T}} = \mathbf{Q}_k\mathbf{Y}\mathbf{Y}^{\mathrm{T}}\mathbf{Q}_k{}^{\mathrm{T}} = \mathbf{Q}_k\mathbf{C}\mathbf{Q}_k, \text{ and} \tag{6.83}$$

$$\mathbf{C}_{\varepsilon,k} = (\mathbf{Q}_k\boldsymbol{\varepsilon})(\mathbf{Q}_k\boldsymbol{\varepsilon})^{\mathrm{T}} = \mathbf{Q}_k\boldsymbol{\varepsilon}\boldsymbol{\varepsilon}^{\mathrm{T}}\mathbf{Q}_k{}^{\mathrm{T}} = \mathbf{Q}_k\mathbf{C}_\varepsilon\mathbf{Q}_k, \tag{6.84}$$

where we have used the fact that $\mathbf{Q}_k$ is symmetric, that is, $\mathbf{Q}_k{}^{\mathrm{T}} = \mathbf{Q}_k$. Next, we apply the beamformer to the transformed data equation (6.82) with the transformed topographies $\mathbf{Q}_k\mathbf{l}(\mathbf{p}_k), \ldots, \mathbf{Q}_k\mathbf{l}(\mathbf{p}_n)$ and transformed $\mathbf{C}_k$. We get the transformed localizer $\tau_k(\mathbf{p})$ for the kth iteration step, *the RAP localizer for fixed-oriented dipoles,* as

$$\tau_k(\mathbf{p}) = \frac{\mathbf{l}(\mathbf{p})^{\mathrm{T}}\mathbf{F}_k\mathbf{l}(\mathbf{p})}{\mathbf{l}(\mathbf{p})^{\mathrm{T}}\mathbf{G}_k\mathbf{l}(\mathbf{p})}, \tag{6.85}$$

where the transformed $\mathbf{F}_k$ and $\mathbf{G}_k$ for the three beamformer types are

$$\mathbf{F}_k = \mathbf{Q}_k, \quad \mathbf{G}_k = \mathbf{C}_k^{\dagger} \quad \text{for AG,} \tag{6.86}$$

$$\mathbf{F}_k = \left(\mathbf{Q}_k\mathbf{C}_\varepsilon\mathbf{Q}_k\right)^{\dagger}, \quad \mathbf{G}_k = \mathbf{C}_k^{\dagger} \quad \text{for AI,} \tag{6.87}$$

$$\mathbf{F}_k = \mathbf{C}_k^{\dagger}, \quad \mathbf{G}_k = \mathbf{C}_k^{\dagger}\mathbf{C}_{\boldsymbol{\varepsilon}}\mathbf{C}_k^{\dagger} \quad \text{for PZ.} \tag{6.88}$$

Notice that above we have used the the pseudoinverses of $\mathbf{C}_k$ and $\mathbf{Q}_k\mathbf{C}_\varepsilon\mathbf{Q}_k$ because they are not invertible (ranks are less than m). The above equations are easily derived by observing that

$$\mathbf{Q}_k{}^{\mathrm{T}} = \mathbf{Q}_k \text{ and } \mathbf{Q}_k{}^{\mathrm{T}}\mathbf{Q}_k = \mathbf{Q}_k, \tag{6.89}$$

because $\mathbf{Q}_k$ is an orthogonal projection, and

$$\mathbf{Q}_k(\mathbf{Q}_k\mathbf{C}\mathbf{Q}_k)^{\dagger}\mathbf{Q}_k = (\mathbf{Q}_k\mathbf{C}\mathbf{Q}_k)^{\dagger}, \tag{6.90}$$

and similarly for $\mathbf{C}_\varepsilon$, due to the identities

$$\mathbf{P}(\mathbf{P}\mathbf{Z}\mathbf{P})^{\dagger} = (\mathbf{P}\mathbf{Z}\mathbf{P})^{\dagger} = (\mathbf{P}\mathbf{Z}\mathbf{P})^{\dagger}\mathbf{P} \tag{6.91}$$

for any $m \times m$ orthogonal projection $\mathbf{P}$ and any $m \times m$ matrix $\mathbf{Z}$.

The location $\widehat{\mathbf{p}}_k$ is now found as the *global maximum* point of $\tau_k(\mathbf{p})$, and the kth iteration step is completed.

Note that for $\mathbf{p}_j$, $j=k,\ldots,n$, like in equation (6.59),

$$\tau_k(\mathbf{p}_j)=c_k(\mathbf{p}_j)||\widetilde{\mathbf{S}}_j^{\{k\}}||^2 \tag{6.92}$$

where $c_k(\mathbf{p})$ is the transformed weight, and $\widetilde{\mathbf{S}}_j^{\{k\}}=(\mathbf{P}_j^{\{k\}}\mathbf{S}_j{}^{\mathrm{T}})^{\mathrm{T}}$ and $\mathbf{P}_j^{\{k\}}=\mathbf{I}-\mathbf{R}_j^{\{k\}}(\mathbf{R}_j^{\{k\}})^{\dagger}$ is the orthogonal projection onto the orthospace of $\mathrm{span}(\mathbf{R}_j^{\{k\}})$ with

$$\mathbf{R}_j^{\{k\}}=[\mathbf{S}_k{}^{\mathrm{T}},\ldots,\mathbf{S}_{j-1}{}^{\mathrm{T}},\mathbf{S}_{j+1}{}^{\mathrm{T}},\ldots,\mathbf{S}_n{}^{\mathrm{T}}]. \tag{6.93}$$

For fixed j, $k\leq j<n$, the subspace $\mathrm{span}(\mathbf{R}_j^{\{k\}})$ becomes smaller, and so its orthospace gets larger as k increases. Therefore, $||\widetilde{\mathbf{S}}_j^{\{k\}}||$ increases as k increases, as long as $k<j$. Because $c_k(\mathbf{p}_j)$ (as numerical simulations show) decreases much more slowly, the "visibility" of the source dipole at $\mathbf{p}_j$ improves at every iteration step as long as that dipole remains among the non-found source dipoles.

In particular, if some source dipoles are synchronous (or highly correlated), and thus poorly visible in the first iteration steps, they then become visible later. Namely, after all non-synchronous source dipoles have been found and projected out, at least one synchronous source becomes visible, and after it has been projected out, the rest of the synchronous ones become visible because their time courses are (usually) no longer linearly dependent.

The iteration is continued until all source locations are found. It is, however, difficult to know when this happens, because the number n of the source dipoles is unknown. One stopping rule for the iteration is provided by the maximum of $\tau_k(\mathbf{p})$ over all $\mathbf{p}$ in the ROI (or rather over the grid of the ROI). Namely, that maximum usually decreases as the iteration proceeds, and clearly drops after all true source locations have been found, because the next found locations $\mathbf{p}\notin\{\mathbf{p}_1,\ldots,\mathbf{p}_n\}$ and so $\mathbf{S}_{\mathbf{p}}=0$ (see figure 6.6).

One can still take some extra steps to make sure that all true source locations are among the found locations $\{\widehat{\mathbf{p}}_1,\ldots,\widehat{\mathbf{p}}_{k-1}\}$. Note that if there are synchronous source dipoles, $\max_{\mathbf{p}\in\mathrm{ROI}}\tau_k(\mathbf{p})$ decreases until all non-synchronous ones have been found, and then it jumps at the next iteration and after that continues to decrease to the final drop when all synchronous ones have also been found.

After having found all true source locations and possibly some false ones, we can identify the true locations and get estimates for the corresponding time courses as in section 6.8.

6.13 Iterative RAP Beamformer with Freely-Oriented Dipoles

Next we consider the case where the dipoles in the ROI are *freely oriented*, and use the RAP-beamformer technique with *optimal orientations* in the following way.

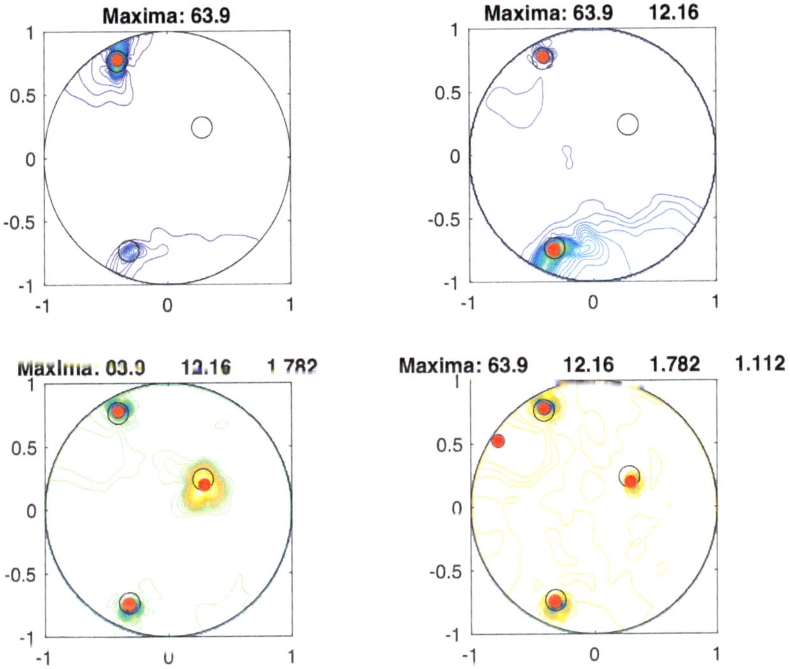

Figure 6.6
RAP beamformer simulation with AI localizer and EEG data in a three-layer spherical model. Three source dipoles lie on the innermost upper hemisphere with random orientations. The localizer contour plots are projected onto a unit disc. The simulated data $\mathbf{Y} = \mathbf{A}\,\mathbf{S} + \boldsymbol{\varepsilon}$ contain colored noise with SNR = 1. Two of the dipoles are synchronous and the third one is correlated with them but linearly independent.

Four iteration steps are taken and results presented in subfigures. The true dipole locations are marked by black circles and the found locations by red balls. The localizer contour plots and the global maxima of the localizer in the first and second steps are shown in the upper subfigures from left to right, and the maxima in the third and fourth steps in the lower subfigures from left to right.

Upper left. The contour plot shows two dipoles, while one of the synchronous dipoles is invisible. The non-synchronous dipole shows clearly as the global maximum, and it is the first found location. *Upper right.* After the first found topography has been out projected, the global maximum drops very low, because the two synchronous dipoles only remain. One shows clearly as the global maximum, and it is the second found location. *Lower left.* After the two found topographies have been out-projected, the last remaining dipole shows as the global maximum and it becomes the third found location. *Lower right.* After the three found topographies have been out-projected, the global maximum drops to its lowest value and the next found location is a false one.

We find the first estimated source location $\widehat{\mathbf{p}}_1$ as the global maximum point of the localizer $\tau_{\text{opt}}(\mathbf{p})$, and the corresponding estimated topography is $\widehat{\mathbf{l}}_1 = \mathbf{L}(\widehat{\mathbf{p}}_1)\eta_{\text{opt}}(\widehat{\mathbf{p}}_1)$; see equations (6.69) and (6.70).

After having found $\widehat{\mathbf{p}}_1, \ldots, \widehat{\mathbf{p}}_{k-1}$ and the estimated topographies $\widehat{\mathbf{l}}_1, \ldots, \widehat{\mathbf{l}}_{k-1}$, we find $\widehat{\mathbf{p}}_k$ and $\widehat{\mathbf{l}}_k$ in the kth iteration step in the following way. As with the fixed-orientation case, with $\mathbf{B}_k = [\widehat{\mathbf{l}}_1, \ldots, \widehat{\mathbf{l}}_{k-1}]$ we form the out-projector $\mathbf{Q}_k = \mathbf{I} - \mathbf{B}_k \mathbf{B}_k^{\dagger}$ and the transformed data matrix $\mathbf{Q}_k \mathbf{Y}$ as in equation (6.82). We form the transformed $\mathbf{F}_k$ and $\mathbf{G}_k$ as in equations (6.86) to (6.88).

Because optimal orientations depend on the data via $\mathbf{F}_k$ and $\mathbf{G}_k$ matrices, we must update them with the transformed $\mathbf{F}_k$ and $\mathbf{G}_k$. We get

$$\eta_k^{\text{opt}}(\mathbf{p}) = \arg\max_{\|\eta\|=1} \frac{\eta^{\mathrm{T}} \mathbf{L}(\mathbf{p})^{\mathrm{T}} \mathbf{F}_k \mathbf{L}(\mathbf{p}) \eta}{\eta^{\mathrm{T}} \mathbf{L}(\mathbf{p})^{\mathrm{T}} \mathbf{G}_k \mathbf{L}(\mathbf{p}) \eta} \tag{6.94}$$

and, as in equation (6.70), the transformed localizer, *the RAP localizer for freely-oriented dipoles*, is

$$\tau_k^{\text{opt}}(\mathbf{p}) = \max_{\|\eta\|=1} \frac{\eta^{\mathrm{T}} \mathbf{L}(\mathbf{p})^{\mathrm{T}} \mathbf{F}_k \mathbf{L}(\mathbf{p}) \eta}{\eta^{\mathrm{T}} \mathbf{L}(\mathbf{p})^{\mathrm{T}} \mathbf{G}_k \mathbf{L}(\mathbf{p}) \eta} = d, \tag{6.95}$$

and

$$\eta_k^{\text{opt}}(\mathbf{p}) = \mathbf{E}_k \mathbf{u} / \|\mathbf{E}_k \mathbf{u}\|, \tag{6.96}$$

where

$$\mathbf{E}_k = \left(\mathbf{L}(\mathbf{p})^{\mathrm{T}} \mathbf{G}_k \mathbf{L}(\mathbf{p})\right)^{-1/2},$$

and d is the largest eigenvalue and $\mathbf{u}$ is the corresponding eigenvector of the matrix $\mathbf{K} = \mathbf{E}_k \mathbf{L}(\mathbf{p})^{\mathrm{T}} \mathbf{F}_k \mathbf{L}(\mathbf{p}) \mathbf{E}_k$. Alternatively, we can use the generalized eigenvalue technique of equations (6.74) and (6.75) with $\mathbf{F}$ and $\mathbf{G}$ replaced by $\mathbf{F}_k$ and $\mathbf{G}_k$.

To complete the kth iteration step, we find $\widehat{\mathbf{p}}_k$ as the global maximum point of $\tau_k^{\text{opt}}(\mathbf{p})$ and set

$$\widehat{\mathbf{l}}_k^{\text{opt}} = \mathbf{L}(\widehat{\mathbf{p}}_k) \eta_k^{\text{opt}}(\widehat{\mathbf{p}}_k), \tag{6.97}$$

where $\eta_k^{\text{opt}}(\widehat{\mathbf{p}}_k)$ is computed by equation (6.96) with $\mathbf{p} = \mathbf{p}_k$ or, alternatively, by the generalized eigenvalue technique.

The stopping rule for the iterations is as in the case of fixed orientations. The true source locations are identified among the found locations and their time courses estimated similarly as earlier.

Finally, we note that the RAP beamformer can also handle a case where some, or all, source dipoles are rotating, that is, with time-depending orientations, only one change to the algorithm is needed: in forming the out-projector $\mathbf{Q}_k$, we must use $\mathbf{B}_k = [\mathbf{L}(\widehat{\mathbf{p}}_1), \ldots, \mathbf{L}(\widehat{\mathbf{p}}_{k-1})]$ instead of $[\widehat{\mathbf{l}}_1, \ldots, \widehat{\mathbf{l}}_{k-1}]$, because rotating dipoles have no time-independent topographies. After having found dipole location $\widehat{\mathbf{p}}_k$, the kind of dipole at that point can be studied by finding the number of the largest equal eigenvalues in forming the maximum in equation (6.95) with $\mathbf{p} = \widehat{\mathbf{p}}_k$, as explained in section 6.10.

6.14 Out-Projecting and Null-Constraining

In the RAP beamformer, the topographies of dipole sources are removed by out-projecting them from the measurement data. A conventional way to carry out a similar operation is the *null-constraining* used with the multiple (linearly) constrained minimum variance (MCMV or LCMC) beamformers; see, for example, Sekihara and Nagarajan [282] or Moiseev et al. [208]. We show here that the two operations, although they look quite different, actually lead to the same weight vectors and output powers.

Let $\mathbf{Y} = \mathbf{A}\,\mathbf{S} + \boldsymbol{\varepsilon}$ be a noisy data matrix, where $\mathbf{A} = [\mathbf{l}_1, \ldots, \mathbf{l}_n]$ with topographies $\mathbf{l}_1, \ldots, \mathbf{l}_n$ of source dipoles at locations $\mathbf{p}_1, \ldots, \mathbf{p}_n$. We first out-project topographies $\mathbf{l}_1, \ldots, \mathbf{l}_{k-1}$, $2 \leq k \leq n$. Let $\mathbf{Q} = \mathbf{I} - \mathbf{B}\,\mathbf{B}^\dagger$ be the out-projector with $\mathbf{B} = [\mathbf{l}_1, \ldots, \mathbf{l}_{k-1}]$. We apply a scalar beamformer to the transformed data equation $\mathbf{Q}\mathbf{Y} = \mathbf{Q}\mathbf{A}\,\mathbf{S} + \mathbf{Q}\,\boldsymbol{\varepsilon}$. Let $\mathbf{p}$ be any location in the ROI and $\mathbf{l}_p = \mathbf{l}(\mathbf{p})$ its topography if dipoles are fixed-oriented, or $\mathbf{l}_p = \mathbf{L}(\mathbf{p})\,\eta_{\text{opt}}(\mathbf{p})$ if they are freely oriented. We form the weight vector $\mathbf{w}_p$ for the transformed data equation with transformed topography $\mathbf{Q}\,\mathbf{l}_p$ and the transformed data covariance matrix $\mathbf{C}_Q = \mathbf{Q}\,\mathbf{C}\,\mathbf{Q}$, and by equation (6.40) we get

$$\mathbf{w}_p = \underset{(\mathbf{Q}\,\mathbf{l}_p)^{\mathrm{T}}\mathbf{w}=1}{\arg\min}\ \mathbf{w}^{\mathrm{T}}\mathbf{C}_Q\mathbf{w} = \frac{\mathbf{C}_Q^\dagger\mathbf{Q}\,\mathbf{l}_p}{(\mathbf{Q}\,\mathbf{l}_p)^{\mathrm{T}}\mathbf{C}_Q^\dagger\mathbf{Q}\,\mathbf{l}_p} = \frac{\mathbf{C}_Q^\dagger\mathbf{l}_p}{\mathbf{l}_p{}^{\mathrm{T}}\mathbf{C}_Q^\dagger\mathbf{l}_p}, \tag{6.98}$$

where we have used the identities (6.91), and by equation (6.41), the corresponding output power is

$$\mu(\mathbf{p}) = \mathbf{w}_p{}^{\mathrm{T}}\mathbf{C}_Q\mathbf{w}_p = \frac{1}{(\mathbf{Q}\,\mathbf{l}_p)^{\mathrm{T}}\mathbf{C}_Q^\dagger\mathbf{Q}\,\mathbf{l}_p} = \frac{1}{\mathbf{l}_p{}^{\mathrm{T}}\mathbf{C}_Q^\dagger\mathbf{l}_p}, \tag{6.99}$$

where we have also used the identities (6.91).

Next, we consider an MCMV beamformer, and form the weight vector $\mathbf{v}_p$ for the data equation $\mathbf{Y} = \mathbf{A}\,\mathbf{S} + \boldsymbol{\varepsilon}$ with topographies $\mathbf{l}_1, \ldots, \mathbf{l}_{k-1}$, $2 \leq k \leq n$ being null-constrained from the data. To that end, let $\mathbf{p}$ be any location in the ROI. In the null-constraining, the weight vector $\mathbf{v}_p$ is found via minimizing $\mathbf{w}^{\mathrm{T}}\mathbf{C}\,\mathbf{w}$ with constraints, but now in addition to the constraint $\mathbf{l}_p{}^{\mathrm{T}}\mathbf{w} = 1$, one sets extra constraints $\mathbf{l}_j{}^{\mathrm{T}}\mathbf{w} = 0$ for $j = 1, \ldots, k-1$, and define $\mathbf{v}_p$ by

$$\mathbf{v}_p = \underset{\substack{\mathbf{l}_j^T\mathbf{w}=0,\, j=1,\ldots,k-1 \\ \mathbf{l}_p^T\mathbf{w}=1}}{\arg\min} \mathbf{w}^T\mathbf{C}\,\mathbf{w}. \tag{6.100}$$

The closed form for $\mathbf{v}_p$ is found as follows. The above constraints can be written in a concise form as $\mathbf{H}^T\mathbf{w} = \mathbf{e}_k = [0, \ldots, 0, 1]^T \in \mathbb{R}^k$ with

$$\mathbf{H} = [\mathbf{l}_1, \ldots, \mathbf{l}_{k-1}, \mathbf{l}_p]. \tag{6.101}$$

By making the substitution $\mathbf{z} = \mathbf{C}^{1/2}\mathbf{w}$ in $\mathbf{w}^T\mathbf{C}\,\mathbf{w} = ||\mathbf{C}^{1/2}\mathbf{w}||^2$ or, reversely, $\mathbf{w} = \mathbf{C}^{-1/2}\mathbf{z}$, we get

$$\mathbf{v}_p = \underset{\mathbf{H}^T\mathbf{C}^{-1/2}\mathbf{z}=\mathbf{e}_k}{\arg\min} ||\mathbf{z}||^2. \tag{6.102}$$

Here, the minimizing $\widehat{\mathbf{z}}$ is the minimum-norm solution of the underdetermined equation $\mathbf{F}\,\mathbf{z} = \mathbf{e}_k$, with $\mathbf{F} = \mathbf{H}^T\mathbf{C}^{-1/2}$, and so $\widehat{\mathbf{z}} = \mathbf{F}^T(\mathbf{F}\,\mathbf{F}^T)^{-1}\mathbf{e}_k$ and

$$\mathbf{v}_p = \mathbf{C}^{-1/2}\widehat{\mathbf{z}} = \mathbf{C}^{-1}\mathbf{H}(\mathbf{H}^T\mathbf{C}^{-1}\mathbf{H})^{-1}\mathbf{e}_k, \tag{6.103}$$

and the output power is

$$\xi_k(\mathbf{p}) = \mathbf{v}_p{}^T\mathbf{C}\,\mathbf{v}_p = \mathbf{e}_k{}^T(\mathbf{H}^T\mathbf{C}^{-1}\mathbf{H})^{-1}\mathbf{e}_k. \tag{6.104}$$

To show that the weight vectors $\mathbf{w}_p$ and $\mathbf{v}_p$ are equal and also that the output powers coincide, we observe that for any $\mathbf{v} \in \mathbb{R}^m$, the constraints $\mathbf{l}_j{}^T\mathbf{v} = 0$ for $j = 1, \ldots, k-1$ are equivalent to the equation $\mathbf{v} \in \operatorname{span}(\mathbf{B})^\perp$ with $\mathbf{B} = [\mathbf{l}_1, \ldots, \mathbf{l}_{k-1}]$, which further is equivalent to the equation $\mathbf{v} = \mathbf{Q}\,\mathbf{w}$, $\mathbf{w} \in \mathbb{R}^m$, where $\mathbf{Q} = \mathbf{I} - \mathbf{B}\,\mathbf{B}^\dagger$. It follows that

$$\begin{aligned} \xi_k(\mathbf{p}) = \mathbf{v}_p{}^T\mathbf{C}\,\mathbf{v}_p &= \min_{\substack{\mathbf{l}_j{}^T\mathbf{v}=0,\, j=1,\ldots,k-1 \\ \mathbf{l}_p{}^T\mathbf{v}=1}} \mathbf{v}^T\mathbf{C}\,\mathbf{v} = \min_{\mathbf{l}_p{}^T\mathbf{Q}\mathbf{w}=1} (\mathbf{Q}\mathbf{w})^T\mathbf{C}\,\mathbf{Q}\mathbf{w} \\ &= \min_{(\mathbf{Q}\,\mathbf{l}_p)^T\mathbf{w}=1} \mathbf{w}^T(\mathbf{Q}\mathbf{C}\mathbf{Q})\mathbf{w} = \mathbf{w}_p{}^T\mathbf{C}_Q\mathbf{w}_p = \mu(\mathbf{p}), \end{aligned} \tag{6.105}$$

and so

$$\mathbf{e}_k^T\left(\mathbf{H}^T\mathbf{C}^{-1}\mathbf{H}\right)^{-1}\mathbf{e}_k = \xi_k(\mathbf{p}) = \mu(\mathbf{p}) = \frac{1}{\mathbf{l}_p^T\mathbf{C}_Q^\dagger\mathbf{l}_p}, \tag{6.106}$$

that is, the output powers coincide. The above reasoning also shows that the first minimum in equation (6.105) is reached by $\mathbf{v}_p = \mathbf{Q}\,\mathbf{w}_p$. By equation (6.98) and identities (6.91), we get

$$\mathbf{w}_p = \frac{1}{\mathbf{l}_p{}^T\mathbf{C}_Q^\dagger\mathbf{l}_p}\,\mathbf{C}_Q^\dagger\mathbf{l}_p = \frac{1}{\mathbf{l}_p{}^T\mathbf{C}_Q^\dagger\mathbf{l}_p}\,\mathbf{Q}\,\mathbf{C}_Q^\dagger\mathbf{l}_p. \tag{6.107}$$

Therefore, $\mathbf{w}_p \in \text{span}(\mathbf{Q})$, which implies that $\mathbf{Q}\,\mathbf{w}_p = \mathbf{w}_p$, and so

$$\mathbf{v}_p = \mathbf{Q}\,\mathbf{w}_p = \mathbf{w}_p. \tag{6.108}$$

The claims have been proved.

We can extend equations (6.99) and (6.106) to the cases where $\mathbf{C}$ is replaced by $\mathbf{I}$ or $\mathbf{C}_\varepsilon = \boldsymbol{\varepsilon}\,\boldsymbol{\varepsilon}^{\mathrm{T}}$ by using (the end of) Appendix 6.21, the above reasoning, and the identity $\mathbf{Q}^\dagger = \mathbf{Q}$ (valid for all orthogonal projections), and we get

$$\mathbf{e}_k{}^{\mathrm{T}}(\mathbf{H}^{\mathrm{T}}\mathbf{H})^{-1}\mathbf{e}_k = \frac{1}{\mathbf{l}(\mathbf{p})^{\mathrm{T}}\mathbf{Q}\,\mathbf{l}(\mathbf{p})}, \text{ and} \tag{6.109}$$

$$\mathbf{e}_k{}^{\mathrm{T}}(\mathbf{H}^{\mathrm{T}}\mathbf{C}_\varepsilon^{-1}\mathbf{H})^{-1}\mathbf{e}_k = \frac{1}{\mathbf{l}(\mathbf{p})^{\mathrm{T}}(\mathbf{Q}\mathbf{C}_\varepsilon\mathbf{Q})^\dagger\mathbf{l}(\mathbf{p})}. \tag{6.110}$$

With these equations, we can also write for fixed-oriented dipoles the iterative localizers (6.85) in MCMV form; for instance, for the iterative AI localizer, the RAP and MCMV forms are

$$\tau_k(\mathbf{p}) = \frac{\mathbf{l}(\mathbf{p})^{\mathrm{T}}(\mathbf{Q}_k\mathbf{C}_\varepsilon\mathbf{Q}_k)^\dagger\mathbf{l}(\mathbf{p})}{\mathbf{l}(\mathbf{p})^{\mathrm{T}}(\mathbf{Q}_k\mathbf{C}\mathbf{Q}_k)^\dagger\mathbf{l}(\mathbf{p})} \tag{6.111}$$

$$= \frac{\mathbf{e}_k{}^{\mathrm{T}}(\mathbf{H}^{\mathrm{T}}\mathbf{C}^{-1}\mathbf{H})^{-1}\mathbf{e}_k}{\mathbf{e}_k{}^{\mathrm{T}}(\mathbf{H}^{\mathrm{T}}\mathbf{C}_\varepsilon^{-1}\mathbf{H})^{-1}\mathbf{e}_k}, \tag{6.112}$$

where $\mathbf{Q}_k = \mathbf{I} - \mathbf{B}_k\mathbf{B}_k^\dagger$, $\mathbf{B}_k = [\mathbf{l}(\widehat{\mathbf{p}}_1), \ldots, \mathbf{l}(\widehat{\mathbf{p}}_{k-1})]$ and $\mathbf{H} = [\mathbf{B}_k, \mathbf{l}(\mathbf{p})]$.

Note that the above MCMV form of $\tau_k(\mathbf{p})$ for fixed-oriented dipoles is simpler and faster to compute than the RAP form, because no $\mathbf{Q}_k$ is needed and both $\mathbf{H}^{\mathrm{T}}\mathbf{C}^{-1}\mathbf{H}$ and $\mathbf{H}^{\mathrm{T}}\mathbf{C}_\varepsilon^{-1}\mathbf{H}$ are only $k \times k$ matrices while $\mathbf{Q}_k\mathbf{C}\mathbf{Q}_k$ and $\mathbf{Q}_k\mathbf{C}_\varepsilon\mathbf{Q}_k$ are $m \times m$ matrices.

However, with freely-oriented dipoles, the RAP-type iterative localizers (6.95) are more practical than the corresponding MCMV forms because they can be directly formed with $\mathbf{F}_k$ and $\mathbf{G}_k$, given by equations (6.86) to (6.88).

The extension of the iterative MCMV localizers from fixed oriented to freely-oriented dipoles is more involved. The extension can be formally done easily by replacing $\mathbf{l}(\mathbf{p})$ by $\mathbf{L}(\mathbf{p})\,\eta$ in the fixed-orientation localizers and by taking the maximum over all η with $||\eta|| = 1$. However, the difficulty is in getting that maximum in a closed form. One way to do that is to write the expression to be maximized in terms of some $\mathbf{F}_k$ and $\mathbf{G}_k$ matrices as in equation (6.95). But finding appropriate $\mathbf{F}_k$ and $\mathbf{G}_k$, starting from expressions like equation (6.112), is not easy. However, that can be done with some involved matrix algebra, for instance, with the technique of Appendix B in Moiseev et al. [200], and get the wanted (lengthy) expression for $\mathbf{F}_k$ and $\mathbf{G}_k$.

In Moiseev et al. [208], the iterative MCMV localizers are introduced differently from the way suggested above. The resulting localizers are called iterative *multi-source localizers*, and they lead to somewhat different beamformers from the RAP ones. We discuss this alternative approach in the next two sections.

6.15 Iterative Multi-Source AI and PZ Beamformers with Fixed-Oriented Dipoles

Next we describe the iterative *multi-source AI*, briefly MAI, and *multi-source PZ*, briefly MPZ, *beamformers* with fixed-oriented dipoles. Let again the $m \times n$ data matrix

$$\mathbf{Y} = \mathbf{A}\mathbf{S} + \boldsymbol{\varepsilon} \tag{6.113}$$

be due to source dipoles at locations $\mathbf{p}_1, \ldots, \mathbf{p}_n$ with topographies $\mathbf{l}(\mathbf{p}_1), \ldots, \mathbf{l}(\mathbf{p}_n)$, mixing matrix $\mathbf{A} = [\mathbf{l}(\mathbf{p}_1), \ldots, \mathbf{l}(\mathbf{p}_n)]$, time-course matrix $\mathbf{S}$, and (white or colored) noise $\boldsymbol{\varepsilon}$.

The first iteration step is taken with the usual AI or PZ localizer $\tau(\mathbf{p})$ in equations (6.49) or (6.50) depending on whether the type of beamformer is MAI or MPZ. The first found location $\widehat{\mathbf{p}}_1$ is the global maximum point of $\tau(\mathbf{p})$ over ROI, and the corresponding topography is $\widehat{\mathbf{l}}_1 = \mathbf{l}(\widehat{\mathbf{p}}_1)$.

Next assume that source dipole locations $\widehat{\mathbf{p}}_1, \ldots, \widehat{\mathbf{p}}_{k-1}$ and corresponding topographies $\widehat{\mathbf{l}}_1, \ldots, \widehat{\mathbf{l}}_{k-1}$ have been found. Set

$$\mathbf{B}_k = [\widehat{\mathbf{l}}_1, \ldots, \widehat{\mathbf{l}}_{k-1}] \text{ and } \mathbf{H} = [\mathbf{B}_k, \mathbf{l}(\mathbf{p})], \tag{6.114}$$

where $\mathbf{p}$ is any point in the ROI. The iterative MAI and MPZ localizers in Moiseev et al. [208] are given by

$$\tau_k^{\mathrm{MAI}}(\mathbf{p}) = \mathrm{trace}\Big(\mathbf{H}^{\mathrm{T}}\mathbf{C}_{\varepsilon}^{-1}\mathbf{H}\,(\mathbf{H}^{\mathrm{T}}\mathbf{C}^{-1}\mathbf{H})^{-1}\Big), \tag{6.115}$$

$$\tau_k^{\mathrm{MPZ}}(\mathbf{p}) = \mathrm{trace}\Big(\mathbf{H}^{\mathrm{T}}\mathbf{C}^{-1}\mathbf{H}\,(\mathbf{H}^{\mathrm{T}}\mathbf{C}^{-1}\mathbf{C}_{\varepsilon}\mathbf{C}^{-1}\mathbf{H})^{-1}\Big), \tag{6.116}$$

where $\mathbf{C} = \mathbf{Y}\mathbf{Y}^{\mathrm{T}}$ and $\mathbf{C}_{\varepsilon} = \boldsymbol{\varepsilon}\,\boldsymbol{\varepsilon}^{\mathrm{T}}$ are the data and noise covariance matrices. The kth found location $\widehat{\mathbf{p}}_k$ is the global maximum point of $\tau_k^{\mathrm{MAI}}(\mathbf{p})$, or $\tau_k^{\mathrm{MPZ}}(\mathbf{p})$, over the ROI, and the corresponding topography is $\widehat{\mathbf{l}}_k = \mathbf{l}(\widehat{\mathbf{p}}_k)$. The kth iteration step is completed. The iteration is continued until all source locations are found.

In Moiseev et al. [208], it is shown that $\tau_{\mathrm{AI}}(\mathbf{p})$ and $\tau_{\mathrm{PZ}}(\mathbf{p})$ have no location bias, that is, $\widehat{\mathbf{p}}_k = \mathbf{p}_k$, if $k = n$ and $\widehat{\mathbf{l}}_i$ are exactly equal to $\mathbf{l}_i$ for all $i = 1, \ldots, k-1$. Although this result does not imply that $\widehat{\mathbf{p}}_k = \mathbf{p}_k$ also for $k < n$, the iterative MAI and MPZ localizers can be expected to perform well in scanning for source locations if noise is not large. In fact, numerical simulations suggest that both iterative beambormers, RAP and MCMV ones, perform about equally well.

Furthermore, numerical simulations show that if the localizer values are normalized to vary between 0 and 1, the RAP AI and MAI localizers are close to each other, and so are the RAP PZ and MPZ ones. Therefore, in iteration, as with the RAP localizers, and also with MAI and MPZ ones, time-correlated sources become more visible as the iteration proceeds. Also, the same stopping rule for iteration can be used with MAI and PMZ beamformers as with the RAP ones.

6.16 Iterative MAI and MPZ Beamformers with Freely-Oriented Dipoles

The iterative MAI and MPZ beamformers of Moiseev et al. [208] for freely-oriented dipoles work as follows. Let again $\mathbf{Y}=\mathbf{A}\mathbf{S}+\boldsymbol{\varepsilon}$ be the data matrix with $\mathbf{A}=[\mathbf{l}_1,\ldots,\mathbf{l}_n]$, where $\mathbf{l}_1,\ldots,\mathbf{l}_n$ are the topographies of the source dipoles at $\mathbf{p}_1,\ldots,\mathbf{p}_n$.

The iteration is started by finding the first estimated dipole location $\widehat{\mathbf{p}}_1$ as the global maximum point of the optimal orientation AI or PZ localizer $\tau_{\text{opt}}(\mathbf{p})$ of equation (6.67), computed by equation (6.70) or with the generalized eigenvalue technique; see (6.74) and the text after that. Let $\widehat{\mathbf{l}}_1=\mathbf{L}(\widehat{\mathbf{p}}_1)\eta_{\text{opt}}(\widehat{\mathbf{p}}_1)$ be the corresponding estimated topography with $\eta_{\text{opt}}(\widehat{\mathbf{p}}_1)$; see equation (6.70) and the text after (6.74).

After having found locations $\widehat{\mathbf{p}}_1,\ldots,\widehat{\mathbf{p}}_{k-1}$ and topographies $\widehat{\mathbf{l}}_1,\ldots,\widehat{\mathbf{l}}_{k-1}$, we find the $\widehat{\mathbf{p}}_k$ and $\widehat{\mathbf{l}}_k$ in the kth iteration step as follows. Let $\mathbf{B}_k=[\widehat{\mathbf{l}}_1,\ldots,\widehat{\mathbf{l}}_{k-1}]$. For any $\mathbf{p}$ in the ROI, we define the iterative MAI and PMZ localizers in the usual way by replacing $\mathbf{l}(\mathbf{p})$ by $\mathbf{L}(\mathbf{p})\eta$ in the corresponding localizers for fixed-oriented dipoles (6.115) and (6.116), and maximizing with respect to η, that is,

$$\tau_k^{\text{opt}}(\mathbf{p})=\max_{||\eta||=1}\tau_k(\mathbf{p},\eta), \tag{6.117}$$

where $\tau_k(\mathbf{p},\eta)$ is either $\tau_k^{\text{AI}}(\mathbf{p})$ or $\tau_k^{\text{PZ}}(\mathbf{p})$ with $\mathbf{H}=[\mathbf{B}_k,\mathbf{L}(\mathbf{p})\eta]$.

Although $\tau_k^{\text{AI}}(\mathbf{p})$ and $\tau_k^{\text{PZ}}(\mathbf{p})$ are given in (6.115) and (6.116) in concise forms, it is quite demanding to find closed forms for $\tau_k^{\text{opt}}(\mathbf{p})$ in equation (6.117). It is, however, done in Moiseev et al. [208], and the resulting localizers (up to additive constants independent of $\mathbf{p}$) are

$$\tau_k^{\text{opt}}(\mathbf{p})=\max_{||\eta||=1}\frac{\eta^{\text{T}}\mathbf{L}(\mathbf{p})^{\text{T}}\mathbf{F}\,\mathbf{L}(\mathbf{p})\,\eta}{\eta^{\text{T}}\mathbf{L}(\mathbf{p})^{\text{T}}\mathbf{G}\,\mathbf{L}(\mathbf{p})\,\eta}, \tag{6.118}$$

where $\mathbf{F}$ and $\mathbf{G}$ for MAI and MPZ types are

$$\mathbf{F}=\mathbf{Z}_1\mathbf{W}\,\mathbf{Z}_2\mathbf{W}\,\mathbf{Z}_1-\mathbf{Z}_1\mathbf{W}\,\mathbf{Z}_2-\mathbf{Z}_2\mathbf{W}\,\mathbf{Z}_1+\mathbf{Z}_2, \tag{6.119}$$

$$\mathbf{G}=\mathbf{Z}_1-\mathbf{Z}_1\mathbf{W}\,\mathbf{Z}_1, \tag{6.120}$$

where

$$\mathbf{W} = \mathbf{B}_k(\mathbf{B}_k{}^{\mathrm{T}}\mathbf{Z}_1\mathbf{B}_k)^{-1}\mathbf{B}_k{}^{\mathrm{T}} \text{ with } \mathbf{B}_k = [\widehat{\mathbf{l}}_1, \ldots, \widehat{\mathbf{l}}_{k-1}], \tag{6.121}$$

and

$$\mathbf{Z}_1 = \mathbf{C}^{-1}, \qquad \mathbf{Z}_2 = \mathbf{C}_\varepsilon^{-1} \qquad \text{for MAI, and} \tag{6.122}$$

$$\mathbf{Z}_1 = \mathbf{C}^{-1}\mathbf{C}_\varepsilon\mathbf{C}^{-1}, \qquad \mathbf{Z}_2 = \mathbf{C}^{-1} \qquad \text{for MPZ.} \tag{6.123}$$

The optimal orientation $\eta_k^{\mathrm{opt}}(\mathbf{p})$ is the η that yields the maximum in equation (6.118). Both $\tau_k^{\mathrm{opt}}(\mathbf{p})$ and $\eta_k^{\mathrm{opt}}(\mathbf{p})$ are obtained in closed forms by equations (6.69) and (6.70) or, alternatively, by the generalized eigenvalue technique.

Now, $\widehat{\mathbf{p}}_k$ is found as the global maximum point of $\tau_k^{\mathrm{opt}}(\mathbf{p})$, and $\widehat{\mathbf{l}}_k = \mathbf{L}(\widehat{\mathbf{p}}_k)\eta_k^{\mathrm{opt}}(\widehat{\mathbf{p}}_k)$. This completes the kth iteration step.

The stopping rule for the iterations and the identification of the true source locations among the found ones can be the same as with the RAP beamformer, or the stopping rule in Moiseev et al. [208] can be used.

6.17 Iterative MAI and PMZ Beamformers and Time-Dependent Orientations

Like iterative RAP beamformers, iterative MAI and PMZ beamformers can also handle time-dependent orientations (i.e., rotating dipoles) with the following change in the algorithm: in equation (6.121), set $\mathbf{B}_k = [\mathbf{L}(\widehat{\mathbf{p}}_1), \ldots, \mathbf{L}(\widehat{\mathbf{p}}_{k-1})]$.

After having found dipole location $\widehat{\mathbf{p}}_k$, as with the RAP beamformer, the kind of the dipole at that point can be studied by observing the number of the largest equal eigenvalues in finding the maximum in (6.118) with $\mathbf{p} = \widehat{\mathbf{p}}_k$, as explained in section 6.10.

6.18 Vector Beamformers

Vector beamformers are designed for freely-oriented dipoles. First, we present the traditional (non-iterative) *vector beamformer* [315]. Its localizer has a compact, easily computed form, but it is location-biased as numerical simulations show (see section 6.7 for location bias). Then, we present an improved *unbiased vector beamformer* and its *iterative* version; both of these beamformers are presented for the first time in this book.

We start with the traditional vector beamformer and consider the $m \times N$ data matrix

$$\mathbf{Y} = \mathbf{A}\,\mathbf{S} + \boldsymbol{\varepsilon} \tag{6.124}$$

due to n source dipoles at locations $\mathbf{p}_1, \ldots, \mathbf{p}_n$ with orientations $\eta_1, \ldots, \eta_n$, time-course matrix $\mathbf{S}$, and noise $\boldsymbol{\varepsilon}$. We briefly write $\mathbf{L_p} = \mathbf{L}(\mathbf{p})$ for the $m \times 3$ lead-field matrices at locations $\mathbf{p} \in$ ROI.

We can write $\mathbf{Y}$ in the form

$$\mathbf{Y} = \sum_{j=1}^{n} \mathbf{L}_{\mathbf{p}_j} \eta_j \mathbf{S}(j,:) + \boldsymbol{\varepsilon} \tag{6.125}$$

$$= \sum_{j=1}^{n} \sum_{k=1}^{3} \mathbf{L}_{\mathbf{p}_j} \mathbf{e}_k \eta_j(k)\, \mathbf{S}(j,:) + \boldsymbol{\varepsilon}, \tag{6.126}$$

where $\mathbf{e}_1$, $\mathbf{e}_2$, and $\mathbf{e}_3$ are the coordinate unit vectors in the 3-space. This presentation shows that data $\mathbf{Y}$ can also be considered due to $3n$ source dipoles such that at every source location $\mathbf{p}_j$, there are three dipoles with fixed orientations $\mathbf{e}_1$, $\mathbf{e}_2$, and $\mathbf{e}_3$; topographies $\mathbf{L}_{\mathbf{p}_j}\mathbf{e}_k$; and linearly dependent time courses $\eta_j(k)\, \mathbf{S}(j,:)$, $k = 1, 2, 3$. With this presentation, we cannot, however, apply the fixed-orientation beamformer localizers for searching the sources, because the linear dependence of time courses spoils the scanning.

In the traditional vector beamformer, this problem is overcome by out-projecting two of the dipoles assigned to location $\mathbf{p}_j$, or rather their topographies $\mathbf{L}_{\mathbf{p}_j}\mathbf{e}_k = \mathbf{L}_{\mathbf{p}_j}(:,k)$, $k = 1, 2, 3$, and so removing the linear dependence. Let the needed out-projectors be the orthogonal matrices $\mathbf{Q}_{\mathbf{p},i}$, $i = 1, 2, 3$ defined by

$$\mathbf{Q}_{\mathbf{p},1} = \mathbf{I} - [\mathbf{L_p}(:,2), \mathbf{L_p}(:,3)][\mathbf{L_p}(:,2), \mathbf{L_p}(:,3)]^{\dagger}, \tag{6.127}$$

$$\mathbf{Q}_{\mathbf{p},2} = \mathbf{I} - [\mathbf{L_p}(:,3), \mathbf{L_p}(:,1)][\mathbf{L_p}(:,3), \mathbf{L_p}(:,1)]^{\dagger}, \tag{6.128}$$

$$\mathbf{Q}_{\mathbf{p},3} = \mathbf{I} - [\mathbf{L_p}(:,1), \mathbf{L_p}(:,2)][\mathbf{L_p}(:,1), \mathbf{L_p}(:,2)]^{\dagger}, \tag{6.129}$$

so that $\mathbf{Q}_{\mathbf{p},1}$ projects out $\mathbf{L_p}(:,2)$ and $\mathbf{L_p}(:,3)$, and so on, at every $\mathbf{p} \in$ ROI. Next we apply the out-projector $\mathbf{Q}_{\mathbf{p},i}$ to both sides of the data equation (6.126) and get the transformed data equation

$$\mathbf{Q}_{\mathbf{p},i}\mathbf{Y} = \sum_{j=1}^{n} \sum_{k=1}^{3} \mathbf{Q}_{\mathbf{p},i}\mathbf{L}_{\mathbf{p}_j}(:,k)\, \eta_j(k)\, \mathbf{S}(j,:) + \mathbf{Q}_{\mathbf{p},i}\, \boldsymbol{\varepsilon}. \tag{6.130}$$

We see that the transformed data covariance matrix is $\mathbf{Q}_{\mathbf{p},i}\mathbf{C}\,\mathbf{Q}_{\mathbf{p},i}$, with $\mathbf{C} = \mathbf{Y}\mathbf{Y}^{\mathrm{T}}$, and the transformed topographies are $\mathbf{Q}_{\mathbf{p},i}\mathbf{L_p}(:,k)$, $\mathbf{p} \in$ROI and $k = 1, 2, 3$. Note that $\mathbf{Q}_{\mathbf{p},i}\mathbf{L_p}(:,k) = 0$ for $k \neq i$. Using the non-zero topographies $\mathbf{Q}_{\mathbf{p},i}\mathbf{L_p}(:,i)$, we form an AG-type localizer $\mu_i^{\mathrm{AG}}(\mathbf{p})$ for the transformed data equation by

$$\mu_i^{\mathrm{AG}}(\mathbf{p}) = \frac{1}{(\mathbf{Q}_{\mathbf{p},i}\mathbf{L_p}(:,i))^{\mathrm{T}} (\mathbf{Q}_{\mathbf{p},i}\mathbf{C}\,\mathbf{Q}_{\mathbf{p},i})^{\dagger}\, \mathbf{Q}_{\mathbf{p},i}\mathbf{L_p}(:,i)}, \tag{6.131}$$

for $i=1,2,3$. If in the proof of equation (6.27) in Appendix 6.20, we replace $\mathbf{C}$ and $\mathbf{l}(\mathbf{p})$ by $\mathbf{Q}_{\mathbf{p},i}\mathbf{C}\,\mathbf{Q}_{\mathbf{p},i}$ and $\mathbf{Q}_{\mathbf{p},i}\mathbf{L}_{\mathbf{p}}(:,i)$, and take into account equation

$$\mathbf{Q}_{\mathbf{p}_j,i}\mathbf{L}_{\mathbf{p}_j}\boldsymbol{\eta}_j=\boldsymbol{\eta}_j(i)\,\mathbf{Q}_{\mathbf{p}_j,i}\mathbf{L}_{\mathbf{p}_j}(:,i) \tag{6.132}$$

and equations (6.41), (6.36), and (6.37), we see that

$$\mu_i^{\mathrm{AG}}(\mathbf{p})\simeq\begin{cases}|\boldsymbol{\eta}_j(i)|^2||\widetilde{\mathbf{S}}(j,:)||^2 & \text{if } \mathbf{p}=\mathbf{p}_j, \text{ and}\\ 0 & \text{if } \mathbf{p}\notin\{\mathbf{p}_1,\ldots,\mathbf{p}_n\},\end{cases} \tag{6.133}$$

where $\widetilde{\mathbf{S}}(j,:)=\widetilde{\mathbf{S}}_j$ is as in (6.27). Therefore, $\mu_i^{\mathrm{AG}}(\mathbf{p})$, $i=1,2,3$, works as a localizer, but it is location-biased like the scalar beamformer AG localizer.

On the other hand, as in the proof of equation (6.106), with small modifications, we see that

$$\mu_i^{\mathrm{AG}}(\mathbf{p})=\mathbf{e}_i{}^{\mathrm{T}}(\mathbf{L}_{\mathbf{p}}{}^{\mathrm{T}}\mathbf{C}^{-1}\mathbf{L}_{\mathbf{p}})^{-1}\,\mathbf{e}_i,\quad i=1,2,3. \tag{6.134}$$

These equations, with $\sum_{i=1}^{3}|\boldsymbol{\eta}_j(i)|^2=1$, yield the (vector) $\mu_{\mathrm{AG}}(\mathbf{p})$ localizer

$$\mu_{\mathrm{AG}}(\mathbf{p})=\mathrm{trace}\big((\mathbf{L}_{\mathbf{p}}{}^{\mathrm{T}}\mathbf{C}^{-1}\mathbf{L}_{\mathbf{p}})^{-1}\big)\simeq\begin{cases}||\widetilde{\mathbf{S}}(j,:)||^2 & \text{if } \mathbf{p}=\mathbf{p}_j, \text{ and}\\ 0 & \text{if } \mathbf{p}\notin\{\mathbf{p}_1,\ldots,\mathbf{p}_n\}.\end{cases} \tag{6.135}$$

It follows that the localizer $\mu_{\mathrm{AG}}(\mathbf{p})$ is also location-biased.

In the literature [282, 315], the $\mu_{\mathrm{AG}}(\mathbf{p})$ localizer is often presented as the minimum

$$\mu_{\mathrm{AG}}(\mathbf{p})=\min_{\mathbf{W}^{\mathrm{T}}\mathbf{L}_p=\mathbf{I}}\ \mathrm{trace}(\mathbf{W}^{\mathrm{T}}\mathbf{C}\,\mathbf{W}), \tag{6.136}$$

where $\mathbf{W}$ varies over all $\mathbf{W}\in\mathbb{R}^{m\times 3}$ and the minimum is reached by

$$\mathbf{W}_p=\mathbf{C}^{-1}\mathbf{L}_p\big(\mathbf{L}_p{}^{\mathrm{T}}\mathbf{C}^{-1}\mathbf{L}_p\big)^{-1}, \tag{6.137}$$

which (for noiseless data) yields

$$\mathbf{W}_p{}^{\mathrm{T}}\mathbf{Y}=\eta_p\widetilde{\mathbf{S}}_p, \tag{6.138}$$

where $\mathbf{S}_p=\mathbf{S}(j,:)$ if $\mathbf{p}=\mathbf{p}_j$, and $\mathbf{S}_p=0$ otherwise. For proving equations (6.136) and (6.137), first observe that

$$\mathrm{trace}(\mathbf{W}^{\mathrm{T}}\mathbf{C}\,\mathbf{W})=\sum_{i=1}^{3}\mathbf{W}(:,i)^{\mathrm{T}}\mathbf{C}\mathbf{W}(:,i), \tag{6.139}$$

and so the minimizing in equation (6.136) can be performed separately in each term of the above sum for $\mathbf{w}_i=\mathbf{W}(:,i)$ with constraint $\mathbf{w}_i{}^{\mathrm{T}}\mathbf{L}_p=\mathbf{e}_i{}^{\mathrm{T}}$. Equations (6.136) and (6.137) are now derived by modifying the proof of (6.103), and observing the consequences of the resulting equation. Equation (6.138) is obtained, for instance, by the techniques in this section and section 6.14.

The problem of the location bias of the $\mu_{\mathrm{AG}}(\mathbf{p})$ localizer has been alleviated by weighting it with (an AI-type) weight $1/\mathrm{trace}\big((\mathbf{L_p C}_\varepsilon^{-1}\mathbf{L_p})^{-1}\big)$, where $\mathbf{C}_\varepsilon$ is the noise covariance matrix. The weighting yields the traditional *vector beamformer* localizer [315],

$$\mu_{\mathrm{trad}}(\mathbf{p}) = \frac{\mathrm{trace}\big((\mathbf{L_p}^{\mathrm{T}}\mathbf{C}^{-1}\mathbf{L_p})^{-1}\big)}{\mathrm{trace}\big((\mathbf{L_p}^{\mathrm{T}}\mathbf{C}_\varepsilon^{-1}\mathbf{L_p})^{-1}\big)}\,. \tag{6.140}$$

This localizer, however, is still biased, as numerical simulations show.

A better remedy is found by starting again with $\mu_i^{\mathrm{AG}}(\mathbf{p})$ and weighting it with (an AI-type) weight $1/\big(\mathbf{e}_i^{\mathrm{T}}(\mathbf{L_p}^{\mathrm{T}}\mathbf{C}_\varepsilon^{-1}\mathbf{L_p})^{-1}\mathbf{e}_i\big)$, yielding the localizer

$$\mu_i^{\mathrm{AI}}(\mathbf{p}) = \frac{\mathbf{e}_i^{\mathrm{T}}(\mathbf{L_p C}^{-1}\mathbf{L_p})^{-1}\,\mathbf{e}_i}{\mathbf{e}_i^{\mathrm{T}}(\mathbf{L_p C}_\varepsilon^{-1}\mathbf{L_p})^{-1}\,\mathbf{e}_i}\,, \tag{6.141}$$

which can be shown to be location-unbiased by modifying the proof in Appendix 6.24 for the unbiasedness of the scalar AI beamformer. It follows that the sum $\sum_{i=1}^{3}\mu_i^{\mathrm{AI}}(\mathbf{p})$ is also unbiased, which is just the improved *unbiased vector beamformer* localizer

$$\mu_{\mathrm{vec}}(\mathbf{p}) = \sum_{i=1}^{3}\frac{\mathbf{F}(i,i)}{\mathbf{G}(i,i)}, \quad \text{where} \tag{6.142}$$

$$\mathbf{F} = (\mathbf{L_p}^{\mathrm{T}}\mathbf{C}^{-1}\mathbf{L_p})^{-1} \text{ and } \mathbf{G} = (\mathbf{L_p}^{\mathrm{T}}\mathbf{C}_\varepsilon^{-1}\mathbf{L_p})^{-1}\,.$$

Numerical simulations show that $\mu_{\mathrm{vec}}(\mathbf{p})$, in general, performs better than $\mu_{\mathrm{trad}}(\mathbf{p})$. Furthermore, the performance of $\mu_{\mathrm{vec}}(\mathbf{p})$ is about as good as that of the scalar beamformer (for freely-oriented dipoles) with moderate noise, say SNR $> 1/2$, but it is often worse for SNR $< 1/2$.

The *iterative* version of the unbiased vector beamformer is obtained as a direct generalization of the non-iterative one, as follows. In the first iteration step, one uses $\mu_{\mathrm{vec}}(\mathbf{p})$, and the first source location estimate $\widehat{\mathbf{p}}_1$ is the global maximum point of $\mu_{\mathrm{vec}}(\mathbf{p})$. After location estimates $\widehat{\mathbf{p}}_1,\ldots,\widehat{\mathbf{p}}_{k-1}$ have been found, one forms an *unbiased localizer* for the kth iteration step by

$$\mu_k^{\mathrm{vec}}(\mathbf{p}) = \sum_{i=1}^{3}\frac{\mathbf{F}_k(3(k-1)+i,3(k-1)+i)}{\mathbf{G}_k(3(k-1)+i,3(k-1)+i)} \tag{6.143}$$

where

$$\mathbf{F}_k = (\mathbf{H}^{\mathrm{T}}\mathbf{C}^{-1}\mathbf{H})^{-1} \text{ and } \mathbf{G}_k = (\mathbf{H}^{\mathrm{T}}\mathbf{C}_\varepsilon^{-1}\mathbf{H})^{-1} \text{ and } \mathbf{H} = [\mathbf{L}_{\widehat{\mathbf{p}}_1},\ldots,\mathbf{L}_{\widehat{\mathbf{p}}_{k-1}},\mathbf{L_p}].$$

The kth location estimate $\widehat{\mathbf{p}}_k$ is the global maximum point of $\mu_k^{\mathrm{vec}}(\mathbf{p})$.

Iteration is continued until all source locations have been found; one can use the rule that the iteration is stopped at that k after which a clear drop in $\max_{\mathbf{p}} \mu_k^{\mathrm{vec}}(\mathbf{p})$ happens (if such a drop is observed). The estimates $\widehat{\eta}_j$ for orientations can be computed with the AI RAP, or MAI, localizers, as earlier, and an estimate for the time-course matrix $\mathbf{S}$ is found as with the AI RAP or MAI beamformer.

6.19 Summary on Beamformers

The $m \times N$ measurement data matrix $\mathbf{Y} = \mathbf{A}\,\mathbf{S} + \boldsymbol{\varepsilon}$ has been obtained. We want to estimate the number n of dipolar sources, as well as their locations, orientations, and time courses. That is done in the following steps.

- Time-center the data matrix $\mathbf{Y}$ by replacing each element $\mathbf{Y}(i,k)$ by $\mathbf{Y}(i,k) - N^{-1}\sum_{j=1}^{N}\mathbf{Y}(i,j)$, $i = 1,\ldots,m$.
- If the data covariance matrix $\mathbf{C} = \mathbf{Y}\,\mathbf{Y}^{\mathrm{T}}$ is under-ranked ($\mathrm{rank}(\mathbf{C}) < m$), or ill-posed (e.g., due to filtering in the time domain, or setting the potential zero level in EEG), it must be regularized. This can be done, e.g., by replacing $\mathbf{C}$ by $\mathbf{C} + \gamma\,\mathbf{I}$ with $\gamma = p^2 m^{-1}\mathrm{trace}(\mathbf{C})$. This is equivalent to adding $100p$ percent white noise to the measured data $\mathbf{Y}$, that is, $\left(\mathrm{trace}(\gamma\,\mathbf{I})/\mathrm{trace}(\mathbf{C})\right)^{-1/2} = p$. If $\mathbf{C}_\varepsilon$ (or its good estimate) is known and it will be used in localizers, then $\mathbf{C}_\varepsilon$ must similarly be replaced by $\mathbf{C}_\varepsilon + \gamma\,\mathbf{I}$. The parameter $p > 0$ controls the amount of regularizing. The value $p = 0.05$ can be used, or an appropriate p is found by experimenting.
- Fix the scanning grid in the ROI.
- Choose a non-iterative or iterative beamformer and its type. If a non-iterative beamformer is chosen, run only the first iterative step of the following iterative beamformers.

Iterative beamformer for fixed-oriented dipoles

The first iteration step:

- Choose $\mathbf{F}$ and $\mathbf{G}$ in equations (6.55) to (6.57) for RAP according to the chosen type AG, AI, or PZ; for MAI and MPZ, choose types AI and PZ, respectively.
- For every $\mathbf{p}$ in the scanning grid, compute the localizer

$$\mu_1(\mathbf{p}) = \frac{\mathbf{l}(\mathbf{p})^{\mathrm{T}}\mathbf{F}\,\mathbf{l}(\mathbf{p})}{\mathbf{l}(\mathbf{p})^{\mathrm{T}}\mathbf{G}\,\mathbf{l}(\mathbf{p})}. \tag{6.144}$$

- Let $\widehat{\mathbf{p}}_1 = \arg\max_{\mathbf{p}} \mu_1(\mathbf{p})$, as the global maximum point of $\mu_1(\mathbf{p})$, be the first found source location estimate, and let $M_1 = \max_{\mathbf{p}} \mu_1(\mathbf{p})$ be the global maximum.

- If you are running the non-iterative beamformer, stop the iteration here. If feasible, draw a contour plot of the localizer $\mu_1(\mathbf{p})$ and, e.g., by visual inspection, find estimates $\widehat{\mathbf{p}}_1, \ldots, \widehat{\mathbf{p}}_{\hat{n}}$ for source locations as the highest "hill-tops" of the contour plot; here, $\hat{n}$ is an estimate for the true number n of sources. Find estimates for the source dipole orientations as $\widehat{\eta}_i = \eta(\widehat{\mathbf{p}}_i)$, $i = 1, \ldots, \hat{n}$. Estimation of the time courses is presented below.

The kth iteration step:

- Location estimates $\widehat{\mathbf{p}}_1, \ldots, \widehat{\mathbf{p}}_{k-1}$ have been found. Let $\mathbf{B}_k = [\mathbf{l}(\widehat{\mathbf{p}}_1), \ldots, \mathbf{l}(\widehat{\mathbf{p}}_{k-1})]$.
- For AI, AG, and PZ RAP beamformers, set $\mathbf{Q}_k = \mathbf{I} - \mathbf{B}_k \mathbf{B}_k^{\dagger}$. Form $\mathbf{F}_k$ and $\mathbf{G}_k$ with $\mathbf{C}_k = \mathbf{Q}_k \mathbf{C}\, \mathbf{Q}_k$ by equations (6.86) to (6.88), and for all $\mathbf{p}$ in the scanning grid, compute

$$\mu_k(\mathbf{p}) = \frac{\mathbf{l}(\mathbf{p})^{\mathrm{T}} \mathbf{F}_k \, \mathbf{l}(\mathbf{p})}{\mathbf{l}(\mathbf{p})^{\mathrm{T}} \mathbf{G}_k \, \mathbf{l}(\mathbf{p})}. \tag{6.145}$$

For MAI and PMZ beamformers and for every $\mathbf{p}$ in the scanning grid, let $\mathbf{H} = [\mathbf{B}_k, \mathbf{l}(\mathbf{p})]$ and compute $\mu_k(\mathbf{p})$ as in equations (6.115) and (6.116).
- Let

$$\widehat{\mathbf{p}}_k = \arg\max_{\mathbf{p}} \mu_k(\mathbf{p}),$$

$$M_k = \max_{\mathbf{p}} \mu_k(\mathbf{p}),$$

$$\widehat{\eta}_k = \mathbf{l}(\widehat{\mathbf{p}}_k).$$

- Stop the iteration after K steps. Let the estimate for n be the index $K = \hat{n}$ after which there is a clear drop in the sequence $M_1, \ldots, M_K$. If no such drop is observed, continue the iteration for some extra steps to see the drop (which does not always show, even for $K > n$).
- For estimating the time courses $\mathbf{S}_i$, set $\widehat{\mathbf{A}} = [\mathbf{l}(\widehat{\mathbf{p}}_1), \ldots, \mathbf{l}(\widehat{\mathbf{p}}_{\hat{n}})]$. Solve $\mathbf{Y} \simeq \widehat{\mathbf{A}}\mathbf{S}$ for $\mathbf{S}$ in a regularized way, and get an estimate $\widehat{\mathbf{S}}$ for $\mathbf{S}$,

$$\widehat{\mathbf{S}} = (\widehat{\mathbf{A}}^{\mathrm{T}} \widehat{\mathbf{A}} + \lambda \mathbf{I})^{-1} \widehat{\mathbf{A}}^{\mathrm{T}} \mathbf{Y}, \tag{6.146}$$

where $\lambda > 0$ is a regularization parameter, say, $\lambda = 10^{-4}\, c$ where c is the maximum of the diagonal elements of $\widehat{\mathbf{A}}^{\mathrm{T}}\widehat{\mathbf{A}}$. Row vectors $\widehat{\mathbf{S}}(i, :)$ are the desired estimates of $\mathbf{S}(i, \cdot)$, $i = 1, \ldots, \hat{n}$; see also section 6.8.

Iterative beamformer for freely-oriented dipoles

The first iteration step:

- Choose $\mathbf{F}$ and $\mathbf{G}$ in equations (6.55) to (6.57) for RAP according to the chosen type AG, AI, or PZ; for MAI and MPZ, choose types AI and PZ, respectively.

- For every $\mathbf{p}$ in the scanning grid, compute the localizer

$$\mu_1(\mathbf{p}) = \max_{\|\eta\|=1} \frac{\eta^{\mathrm{T}}\mathbf{L}(\mathbf{p})^{\mathrm{T}}\mathbf{F}\,\mathbf{L}(\mathbf{p})\,\eta}{\eta^{\mathrm{T}}\mathbf{L}(\mathbf{p})^{\mathrm{T}}\mathbf{G}\,\mathbf{L}(\mathbf{p})\,\eta}, \tag{6.147}$$

where the maximum can, e.g., be computed with the generalized eigenvalue technique (see section 6.9), which you can perform with MATLAB by using the following command lines:

$[\mathbf{E},\mathbf{D}] = \mathrm{eig}\,(\boldsymbol{Lp}' * \mathbf{F} * \boldsymbol{Lp},\ \boldsymbol{Lp}' * \mathbf{G} * \boldsymbol{Lp});$

$\mathrm{d} = \max\,(\mathrm{diag}\,(\mathbf{D}));$

where $\boldsymbol{Lp} = \mathbf{L}(\mathbf{p})$. This yields $\mu_1(\mathbf{p}) = \mathrm{d}$.

- Let the first estimated location be

$$\widehat{\mathbf{p}}_1 = \arg\max_{\mathbf{p}} \mu_1(\mathbf{p}). \tag{6.148}$$

Next, e.g., with the generalized eigenvalue technique, compute

$$\widehat{\eta}_1 = \arg\max_{\|\eta\|=1} \frac{\eta^{\mathrm{T}}\mathbf{L}(\widehat{\mathbf{p}}_1)^{\mathrm{T}}\mathbf{F}\,\mathbf{L}(\widehat{\mathbf{p}}_1)\,\eta}{\eta^{\mathrm{T}}\mathbf{L}(\widehat{\mathbf{p}}_1)^{\mathrm{T}}\mathbf{G}\,\mathbf{L}(\widehat{\mathbf{p}}_1)\,\eta}, \tag{6.149}$$

which can be done with MATLAB as follows:

$[\mathbf{E},\mathbf{D}] = \mathrm{eig}\,(\boldsymbol{Lp1}' * \mathbf{F} * \boldsymbol{Lp1},\ \boldsymbol{Lp1}' * \mathbf{G} * \boldsymbol{Lp1});$

$[\mathrm{d},\mathrm{i}] = \max\,(\mathrm{diag}\,(\mathbf{D}));$

$\mathbf{a} = \mathbf{E}(:,\mathrm{i})\,,$

and $\widehat{\eta}_1 = \|\mathbf{a}\|^{-1}\mathbf{a}$, where $\boldsymbol{Lp1} = \mathbf{L}(\widehat{\mathbf{p}}_1)$. Set $\widehat{\mathbf{l}}_1 = \mathbf{L}(\widehat{p}_1)\widehat{\eta}_1$. Finally, set

$$M_1 = \max_{\mathbf{p}} \mu_1(\mathbf{p}). \tag{6.150}$$

The first iteration step has been completed.

- If you are running the non-iterative beamformer, stop the iteration here. If feasible, draw a contour plot of the localizer $\mu(\mathbf{p})$ and, e.g., by visual inspection, find estimates $\widehat{\mathbf{p}}_1,\ldots,\widehat{\mathbf{p}}_{\hat{n}}$ for source locations as the highest "hill-tops" of the contour plot; here, $\hat{n}$ is an estimate for the true number n of sources. Find estimates $\widehat{\eta}_i$, $i=1,\ldots,\hat{n}$, for the source dipole orientations, e.g., with the generalized eigenvalue technique as above for $\widehat{\eta}_1$. Set $\widehat{\mathbf{l}}_i = \mathbf{L}(\widehat{\mathbf{p}}_i)\widehat{\eta}_i$, $i=1,\ldots,\hat{n}$, to be estimates for the topographies of the source dipoles. With $\widehat{\mathbf{A}} = [\widehat{\mathbf{l}}_1,\ldots,\widehat{\mathbf{l}}_{\hat{n}}]$, the time courses are estimated as in the case of the iterative beamformer for fixed-oriented dipoles.

The kth iteration step:

- Location estimates $\widehat{\mathbf{p}}_1,\ldots,\widehat{\mathbf{p}}_{k-1}$ and the corresponding topography estimates $\widehat{\mathbf{l}}_1,\ldots,\widehat{\mathbf{l}}_{k-1}$ have been found. Let $\mathbf{B}_k=[\widehat{\mathbf{l}}_1,\ldots,\widehat{\mathbf{l}}_{k-1}]$.
- For AG, AI, and PZ RAP beamformers, set $\mathbf{Q}_k=\mathbf{I}-\mathbf{B}_k\mathbf{B}_k^{\dagger}$, and with $\mathbf{C}_k=\mathbf{Q}_k\mathbf{C}\,\mathbf{Q}_k$, form $\mathbf{F}_k$ and $\mathbf{G}_k$ by equations (6.86) to (6.88). For MAI and PMZ, form $\mathbf{F}_k$ and $\mathbf{G}_k$ by equations (6.119) to (6.123).
- For every $\mathbf{p}$ in the scanning grid (e.g., with the generalized eigenvalue technique as in the first iteration step), compute

$$\mu_k(\mathbf{p})=\max_{\|\eta\|=1}\frac{\eta^{\mathrm{T}}\mathbf{L}(\mathbf{p})^{\mathrm{T}}\mathbf{F}_k\,\mathbf{L}(\mathbf{p})\,\eta}{\eta^{\mathrm{T}}\mathbf{L}(\mathbf{p})^{\mathrm{T}}\mathbf{G}_k\,\mathbf{L}(\mathbf{p})\,\eta}, \tag{6.151}$$

and set

$$\widehat{\mathbf{p}}_k=\arg\max_{\mathbf{p}}\ \mu_k(\mathbf{p}), \tag{6.152}$$

$$\widehat{\eta}_k=\arg\max_{\|\eta\|=1}\frac{\eta^{\mathrm{T}}\mathbf{L}(\widehat{\mathbf{p}}_k)^{\mathrm{T}}\mathbf{F}_k\,\mathbf{L}(\widehat{\mathbf{p}}_k)\,\eta}{\eta^{\mathrm{T}}\mathbf{L}(\widehat{\mathbf{p}}_k)^{\mathrm{T}}\mathbf{G}_k\,\mathbf{L}(\widehat{\mathbf{p}}_k)\,\eta}, \tag{6.153}$$

$$\widehat{\mathbf{l}}_k=\mathbf{L}(\widehat{\mathbf{p}}_k)\,\widehat{\eta}_k, \tag{6.154}$$

$$M_k=\max_{\mathbf{p}}\ \mu_k(\mathbf{p})\,. \tag{6.155}$$

The kth iteration step has been completed.

- The iteration is stopped as done earlier with iterative beamformers for fixed-oriented dipoles. Let $\hat{n}$ be the obtained estimate for the number n of source dipoles. With $\widehat{\mathbf{A}}=[\widehat{\mathbf{l}}_1,\ldots,\widehat{\mathbf{l}}_{\hat{n}}]$, the time courses are estimated as in the case of the iterative beamformer for fixed-oriented dipoles.

Iterative vector beamformer

The first iteration step:

- For every $\mathbf{p}$ in the scanning grid, compute

$$\mu_1(\mathbf{p})=\sum_{i=1}^{3}\frac{\mathbf{F}(i,i)}{\mathbf{G}(i,i)},\quad\text{where} \tag{6.156}$$

$$\mathbf{F}=(\mathbf{L}_{\mathbf{p}}\mathbf{C}^{-1}\mathbf{L}_{\mathbf{p}})^{-1}\ \text{and}\ \mathbf{G}=(\mathbf{L}_{\mathbf{p}}\mathbf{C}_{\varepsilon}^{-1}\mathbf{L}_{\mathbf{p}})^{-1}\,.$$

- Let the first estimated location be

$$\widehat{\mathbf{p}}_1=\arg\max_{\mathbf{p}}\ \mu_1(\mathbf{p}). \tag{6.157}$$

The first iteration step has been completed.

- If you are running the non-iterative (unbiased) vector beamformer, stop the iteration here. If feasible, draw a contour plot of the localizer $\mu_1(\mathbf{p})$ and, e.g., by visual inspection, find estimates $\widehat{\mathbf{p}}_1, \ldots, \widehat{\mathbf{p}}_{\hat{n}}$ for source locations as the highest "hill-tops" of the contour plot; here, $\hat{n}$ is an estimate for the true number n of sources. With the RAP AI or MAI localizer, find estimates $\widehat{\eta}_i$, $i=1,\ldots,\hat{n}$, for the source dipole orientations (e.g., with the generalized eigenvalue technique, as in the case of iterative beamformer for freely-oriented dipoles). Set $\widehat{\mathbf{l}}_i = \mathbf{L}(\widehat{\mathbf{p}}_i)\widehat{\eta}_i$, $i=1,\ldots,\hat{n}$, to be the estimates for the topographies of the source dipoles. With $\widehat{\mathbf{A}} = [\widehat{\mathbf{l}}_1, \ldots, \widehat{\mathbf{l}}_{\hat{n}}]$, the time courses are estimated as in the case of the iterative beamformer for freely-oriented dipoles.

The kth iteration step:

- Location estimates $\widehat{\mathbf{p}}_1, \ldots, \widehat{\mathbf{p}}_{k-1}$ have been found.
- For every $\mathbf{p}$ in the scanning grid, compute

$$\mu_k(\mathbf{p}) = \sum_{i=1}^{3} \frac{\mathbf{F}_k(3(k-1)+i, 3(k-1)+i)}{\mathbf{G}_k(3(k-1)+i, 3(k-1)+i)}, \quad \text{where} \tag{6.158}$$

$$\mathbf{F}_k = (\mathbf{H}^{\mathrm{T}}\mathbf{C}^{-1}\mathbf{H})^{-1} \text{ and } \mathbf{G}_k = (\mathbf{H}^{\mathrm{T}}\mathbf{C}_{\varepsilon}^{-1}\mathbf{H})^{-1}, \text{ and}$$

$$\mathbf{H} = [\mathbf{L}_{\widehat{\mathbf{p}}_1}, \ldots, \mathbf{L}_{\widehat{\mathbf{p}}_{k-1}}, \mathbf{L}_{\mathbf{p}}],$$

and set

$$\widehat{\mathbf{p}}_k = \arg\max_{\mathbf{p}} \mu_k(\mathbf{p}). \tag{6.159}$$

The kth iteration step has been completed.

- The iteration is stopped as with iterative beamformers for freely-oriented dipoles. Let $\hat{n}$ be the obtained estimate for the number n of source dipoles. The estimates $\widehat{\eta}_j$ for orientations and an estimate for the time-course matrix $\mathbf{S}$ can be found as explained at the end of the first iteration step.

6.20 Appendix: Proof of Equation (6.27)

Let $\mathbf{Y} = \mathbf{AS}$, $\mathbf{A} = [\mathbf{l}(\mathbf{p}_1), \ldots, \mathbf{l}(\mathbf{p}_n)]$, $\mathrm{rank}(\mathbf{A}) = n$, and the source-dipole locations $\mathbf{p}_1, \ldots, \mathbf{p}_n$ be as in section 6.4. We want to show that

$$\mathbf{w}_p{}^{\mathrm{T}}\mathbf{Y} = \widetilde{\mathbf{S}}_p \tag{6.160}$$

where

$$\mathbf{w}_p = \arg\min_{\mathbf{l}(\mathbf{p})^{\mathrm{T}}\mathbf{w}=1} \mathbf{w}^{\mathrm{T}}\mathbf{C}\,\mathbf{w} \tag{6.161}$$

and $\mathbf{C}=\mathbf{Y}\mathbf{Y}^{\mathrm{T}}$. Furthermore, $\widetilde{\mathbf{S}}_p=\mathbf{S}_p=0$ if $\mathbf{p}\notin\{\mathbf{p}_1,\ldots,\mathbf{p}_n\}$, and $\widetilde{\mathbf{S}}_p=\widetilde{\mathbf{S}}_k$ for $\mathbf{p}=\mathbf{p}_k$, $1\le k\le n$, and $\widetilde{\mathbf{S}}_k^{\mathrm{T}}=(\mathbf{I}-\mathbf{G}_k\mathbf{G}_k^{\dagger})\mathbf{S}_k^{\mathrm{T}}$ with $\mathbf{G}_k=[\mathbf{S}_1^{\mathrm{T}},\ldots,\mathbf{S}_{k-1}^{\mathrm{T}},\mathbf{S}_{k+1}^{\mathrm{T}},\ldots,\mathbf{S}_n^{\mathrm{T}}]$. We denote the minimum in equation (6.161) by

$$\mu(\mathbf{p})=\min_{\mathbf{l}(\mathbf{p})^{\mathrm{T}}\mathbf{w}=1}\mathbf{w}^{\mathrm{T}}\mathbf{C}\mathbf{w},\quad\text{where}\tag{6.162}$$

$$\mathbf{w}^{\mathrm{T}}\mathbf{C}\mathbf{w}=\mathbf{w}^{\mathrm{T}}\mathbf{A}\mathbf{S}\mathbf{S}^{\mathrm{T}}\mathbf{A}^{\mathrm{T}}\mathbf{w}=||\mathbf{S}^{\mathrm{T}}\mathbf{A}^{\mathrm{T}}\mathbf{w}||^2.\tag{6.163}$$

We first consider the case that $\mathbf{p}\notin\{\mathbf{p}_1,\ldots,\mathbf{p}_n\}$ with $\mathbf{S}_p=\widetilde{\mathbf{S}}_p=0$. We may assume that $\mathbf{l}(\mathbf{p})$ is linearly independent of $\mathbf{l}(\mathbf{p}_1),\ldots,\mathbf{l}(\mathbf{p}_n)$, that is, $\mathbf{l}(\mathbf{p})\notin\mathrm{span}(\mathbf{l}(\mathbf{p}_1),\ldots,\mathbf{l}(\mathbf{p}_n))=\mathrm{span}(\mathbf{A})$. We decompose $\mathbf{l}(\mathbf{p})$ as $\mathbf{l}(\mathbf{p})=\mathbf{l}_1+\mathbf{l}_2$ with $\mathbf{l}_1\in\mathrm{span}(\mathbf{A})$ and $\mathbf{l}_2\in\mathrm{span}(\mathbf{A})^{\perp}$. Then, $\mathbf{l}_2\neq 0$. Let $\mathbf{w}=||\mathbf{l}_2||^{-2}\mathbf{l}_2$. Then, $\mathbf{l}(\mathbf{p})^{\mathrm{T}}\mathbf{w}=\mathbf{l}_2^{\mathrm{T}}\mathbf{w}=1$. Because $\mathbf{l}_2\in\mathrm{span}(\mathbf{A})^{\perp}$, then $\mathbf{A}^{\mathrm{T}}\mathbf{w}=0$, which, by equation (6.163), yields $\mathbf{w}^{\mathrm{T}}\mathbf{C}\mathbf{w}=0$. Therefore, $\mu(\mathbf{p})=0$. So, the minimum in equation (6.161) is reached at $\mathbf{w}_p=||\mathbf{l}_2||^{-2}\mathbf{l}_2$, and $\mathbf{w}_p{}^{\mathrm{T}}\mathbf{Y}=\mathbf{w}_p{}^{\mathrm{T}}\mathbf{A}\mathbf{S}=0=\widetilde{\mathbf{S}}_p$. This proves equation (6.160) for any $\mathbf{p}\notin\{\mathbf{p}_1,\ldots,\mathbf{p}_n\}$.

Next, let $\mathbf{p}=\mathbf{p}_k$ for some k, $1\le k\le n$. Then, $\mathbf{l}(\mathbf{p})=\mathbf{l}(\mathbf{p}_k)=\mathbf{A}(:,k)$. Let $\mathbf{w}\in\mathbb{R}^m$ satisfy the constraint $1=\mathbf{l}(\mathbf{p}_k)^{\mathrm{T}}\mathbf{w}=\mathbf{A}(:,k)^{\mathrm{T}}\mathbf{w}$, and let $\mathbf{z}=[\mathbf{z}(1),\ldots,\mathbf{z}(n)]^{\mathrm{T}}=\mathbf{A}^{\mathrm{T}}\mathbf{w}$. Then, $\mathbf{z}(k)=1$, and for $\mathbf{S}_i=\mathbf{S}(i,:)$, $i=1,\ldots,n$, we get by equation (6.163)

$$\mathbf{w}^{\mathrm{T}}\mathbf{C}\mathbf{w}=||\mathbf{S}^{\mathrm{T}}\mathbf{z}||^2=||\sum_{i=1}^{n}\mathbf{z}(i)\mathbf{S}_i^{\mathrm{T}}||^2=||\mathbf{S}_k^{\mathrm{T}}+\sum_{i\neq k}\mathbf{z}(i)\mathbf{S}_i^{\mathrm{T}}||^2.\tag{6.164}$$

We decompose $\mathbf{S}_k^{\mathrm{T}}$ so that $\mathbf{S}_k^{\mathrm{T}}=\mathbf{u}+\mathbf{v}$ with $\mathbf{u}\in\mathrm{span}(\mathbf{G}_k)^{\perp}$ and $\mathbf{v}\in\mathrm{span}(\mathbf{G}_k)$ with $\mathbf{G}_k=[\mathbf{S}_1^{\mathrm{T}},\ldots,\mathbf{S}_{k-1}^{\mathrm{T}},\mathbf{S}_{k+1}^{\mathrm{T}},\ldots,\mathbf{S}_n^{\mathrm{T}}]$. Also because $\mathbf{y}=\sum_{i\neq k}\mathbf{z}(i)\mathbf{S}_i^{\mathrm{T}}\in\mathrm{span}(\mathbf{G}_k)$, $\mathbf{v}+\mathbf{y}\in\mathrm{span}(\mathbf{G}_k)$, and so $\mathbf{u}$ and $\mathbf{v}+\mathbf{y}$ are orthogonal. Denote $\widetilde{\mathbf{S}}_k^{\mathrm{T}}=\mathbf{u}$. Then, by the Pythagoras theorem (4.39),

$$\mathbf{w}^{\mathrm{T}}\mathbf{C}\mathbf{w}=||\widetilde{\mathbf{S}}_k^{\mathrm{T}}||^2+||\mathbf{v}+\mathbf{y}||^2\geqslant||\widetilde{\mathbf{S}}_k^{\mathrm{T}}||^2,\tag{6.165}$$

implying that $\mu(\mathbf{p}_k)\geqslant||\widetilde{\mathbf{S}}_k^{\mathrm{T}}||^2$.

We next show that $\mu(\mathbf{p}_k)\le||\widetilde{\mathbf{S}}_k^{\mathrm{T}}||^2$. Because $\mathbf{v}\in\mathrm{span}(\mathbf{G}_k)$, there are $\alpha_1,\ldots,\alpha_{k-1},\alpha_{k+1},\ldots,\alpha_n\in\mathbb{R}$ so that $\sum_{i\neq k}\alpha_i\mathbf{S}_i^{\mathrm{T}}=-\mathbf{v}$. Let $\widehat{\mathbf{w}}\in\mathbb{R}^m$ be a solution of the equation

$$\mathbf{A}^{\mathrm{T}}\widehat{\mathbf{w}}=[\alpha_1,\ldots,\alpha_{k-1},1,\alpha_{k+1},\ldots,\alpha_n]^{\mathrm{T}},\tag{6.166}$$

for which a solution exists because $\mathrm{rank}(\mathbf{A})=n$. Then, $\mathbf{l}(\mathbf{p}_k)^{\mathrm{T}}\widehat{\mathbf{w}}=\mathbf{A}(:,k)^{\mathrm{T}}\widehat{\mathbf{w}}=1$, and for $\widehat{\mathbf{z}}=\mathbf{A}^{\mathrm{T}}\widehat{\mathbf{w}}$ we obtain

$$\mathbf{S}^{\mathrm{T}}\mathbf{A}^{\mathrm{T}}\widehat{\mathbf{w}}=\mathbf{S}^{\mathrm{T}}\widehat{\mathbf{z}}=\mathbf{S}_k^{\mathrm{T}}+\sum_{i\neq k}\alpha(i)\,\mathbf{S}_i^{\mathrm{T}}=\mathbf{S}_k^{\mathrm{T}}-\mathbf{v}=\mathbf{u}=\widetilde{\mathbf{S}}_k^{\mathrm{T}}.\tag{6.167}$$

By equation (6.163), this implies that $\widehat{\mathbf{w}}^{\mathrm{T}}\mathbf{C}\,\widehat{\mathbf{w}}=||\widetilde{\mathbf{S}}_k^{\mathrm{T}}||^2$, and so $\mu(\mathbf{p})\leqslant||\widetilde{\mathbf{S}}_k^{\mathrm{T}}||^2$. This, with the inequality $\mu(\mathbf{p})\geqslant||\widetilde{\mathbf{S}}_k^{\mathrm{T}}||^2$, implies that $\mu(\mathbf{p})=||\widetilde{\mathbf{S}}_k{}^{\mathrm{T}}||^2$.

Thus, the minimum in equation (6.170) is reached at $\mathbf{w} = \widehat{\mathbf{w}}$, and by equation (6.176),

$$\widehat{\mathbf{w}}^{\mathrm{T}}\mathbf{Y} = \left(\mathbf{S}^{\mathrm{T}}\mathbf{A}^{\mathrm{T}}\widehat{\mathbf{w}}\right)^{\mathrm{T}} = \widetilde{\mathbf{S}}_k. \tag{6.168}$$

The proof of equation (6.160) is now completed.

6.21 Appendix: Proofs of Equations (6.34) and (6.38)–(6.41)

To prove equation (6.34), let $\mathbf{Y} = \mathbf{AS} + \boldsymbol{\varepsilon}$, $\mathbf{C} = \mathbf{Y}\mathbf{Y}^{\mathrm{T}}$ and

$$\mathbf{w}_p = \arg\min_{\mathbf{l}(\mathbf{p})^{\mathrm{T}}\mathbf{w}=1} \mathbf{w}^{\mathrm{T}}\mathbf{C}\mathbf{w} \tag{6.169}$$

for any $\mathbf{p}$ in the ROI. We want to prove that

$$\mathbf{w}_p = \frac{\mathbf{C}^{-1}\mathbf{l}(\mathbf{p})}{\mathbf{l}(\mathbf{p})^{\mathrm{T}}\mathbf{C}^{-1}\mathbf{l}(\mathbf{p})}. \tag{6.170}$$

To that end, we present $\mathbf{C}$ as $\mathbf{C} = \mathbf{C}^{1/2}\mathbf{C}^{1/2}$ (to recall the concept of $\mathbf{C}^{1/2}$, see equation (4.24)). Note that $\mathbf{C}^{1/2}$ is a symmetric matrix. We get

$$\mathbf{w}^{\mathrm{T}}\mathbf{C}\mathbf{w} = \mathbf{w}^{\mathrm{T}}\mathbf{C}^{1/2}\mathbf{C}^{1/2}\mathbf{w} = (\mathbf{C}^{1/2}\mathbf{w})^{\mathrm{T}}(\mathbf{C}^{1/2}\mathbf{w}) = ||\mathbf{C}^{1/2}\mathbf{w}||^2. \tag{6.171}$$

Let

$$\mathbf{z} = \mathbf{C}^{1/2}\mathbf{w} \text{ and, reversely, } \mathbf{w} = \mathbf{C}^{-1/2}\mathbf{z}. \tag{6.172}$$

Then,

$$\min_{\mathbf{l}(\mathbf{p})^{\mathrm{T}}\mathbf{w}=1} \mathbf{w}^{\mathrm{T}}\mathbf{C}\mathbf{w} = \min_{\mathbf{l}(\mathbf{p})^{\mathrm{T}}\mathbf{C}^{-1/2}\mathbf{z}=1} ||\mathbf{z}||^2. \tag{6.173}$$

We see that the latter minimum above is reached by the vector $\widehat{\mathbf{z}}$ that is the minimum-norm solution of the (underdetermined) equation $\mathbf{l}(\mathbf{p})^{\mathrm{T}}\mathbf{C}^{-1/2}\mathbf{z} = 1$; by equation (4.74), the solution, with $\mathbf{B} = \mathbf{l}(\mathbf{p})^{\mathrm{T}}\mathbf{C}^{-1/2}$, is

$$\widehat{\mathbf{z}} = \mathbf{B}^{\mathrm{T}}(\mathbf{B}\mathbf{B}^{\mathrm{T}})^{-1}1 = \frac{\mathbf{C}^{-1/2}\mathbf{l}(\mathbf{p})}{\mathbf{l}(\mathbf{p})^{\mathrm{T}}\mathbf{C}^{-1}\mathbf{l}(\mathbf{p})}, \tag{6.174}$$

where we have used the identity $\mathbf{C}^{-1} = \mathbf{C}^{-1/2}\mathbf{C}^{-1/2}$. By equation (6.172), it follows that

$$\mathbf{w}_p = \mathbf{C}^{-1/2}\widehat{\mathbf{z}} = \frac{\mathbf{C}^{-1}\mathbf{l}(\mathbf{p})}{\mathbf{l}(\mathbf{p})^{\mathrm{T}}\mathbf{C}^{-1}\mathbf{l}(\mathbf{p})} \tag{6.175}$$

as claimed in equation (6.169).

For proving equations (6.38) and (6.39), assume $\mathbf{l}(\mathbf{p}) \notin \text{span}(\mathbf{C})$. Let $\mathbf{P} = \mathbf{C}\mathbf{C}^\dagger$ be an orthogonal projection onto span(**C**) and $\mathbf{Q} = \mathbf{I} - \mathbf{P}$ be the orthogonal projection onto $\text{span}(\mathbf{C})^\perp = \ker(\mathbf{C}^\mathrm{T}) = \ker(\mathbf{C})$. Let $\mathbf{w}_p = \mathbf{Q}\,\mathbf{l}(\mathbf{p})/||\mathbf{Q}\,\mathbf{l}(\mathbf{p})||^2$ as in equation (6.38). Then, $\mathbf{w}_p \in \ker(\mathbf{C})$, and so $\mathbf{w}_p{}^\mathrm{T}\mathbf{C}\,\mathbf{w}_p = 0$. Furthermore, $\mathbf{l}(\mathbf{p}) = \mathbf{P}\,\mathbf{l}(\mathbf{p}) + \mathbf{Q}\,\mathbf{l}(\mathbf{p})$, and so $\mathbf{l}(\mathbf{p})^\mathrm{T}\mathbf{w}_p = (\mathbf{Q}\,\mathbf{l}(\mathbf{p}))^\mathrm{T}\mathbf{w}_p = 1$. Equations (6.38) and (6.39) have been proved.

For proving equations (6.40) and (6.41), let $\mathbf{l}(\mathbf{p}) \in \text{span}(\mathbf{C})$. Because $\text{span}(\mathbf{C}) = \text{span}(\mathbf{C}^{1/2})$, **P** is also an orthogonal projection onto $\text{span}(\mathbf{C}^{1/2})$. Because $\text{span}(\mathbf{C}^{1/2}) = \ker(\mathbf{C}^{1/2})^\perp$, **P** is also an orthogonal projection onto $\ker(\mathbf{C}^{1/2})^\perp$. Then,

$$\mathbf{w}^\mathrm{T}\mathbf{C}\,\mathbf{w} = ||\mathbf{C}^{1/2}\mathbf{w}||^2 = ||\mathbf{C}^{1/2}\mathbf{P}\,\mathbf{w}||^2, \tag{6.176}$$

where we can make the substitution $\mathbf{P}\,\mathbf{w} = (\mathbf{C}^{1/2})^\dagger\mathbf{z}$ for $\mathbf{z} \in \mathbb{R}^m$. Furthermore, $\mathbf{l}(\mathbf{p}) \in \text{span}(\mathbf{C}) = \text{span}(\mathbf{C}^{1/2})$, and so $\mathbf{l}(\mathbf{p}) = \mathbf{P}\,\mathbf{l}(\mathbf{p})$. This implies that $\mathbf{l}(\mathbf{p})^\mathrm{T}\mathbf{w} = (\mathbf{P}\,\mathbf{l}(\mathbf{p}))^\mathrm{T}\mathbf{w} = \mathbf{l}(\mathbf{p})^\mathrm{T}\mathbf{P}\,\mathbf{w}$, and we get

$$\min_{\mathbf{l}(\mathbf{p})^\mathrm{T}\mathbf{w}=1} \mathbf{w}^\mathrm{T}\mathbf{C}\,\mathbf{w} = \min_{\mathbf{l}(\mathbf{p})^\mathrm{T}\mathbf{P}\,\mathbf{w}=1} ||\mathbf{C}^{1/2}\mathbf{P}\,\mathbf{w}||^2 = \min_{\mathbf{l}(\mathbf{p})^\mathrm{T}(\mathbf{C}^\dagger)^{1/2}\mathbf{z}=1} ||\mathbf{z}||^2, \tag{6.177}$$

where relations $\mathbf{C}^{1/2}(\mathbf{C}^{1/2})^\dagger = \mathbf{P}$ and $||\mathbf{P}\,\mathbf{z}|| \le ||\mathbf{z}||$ have been used. We see that the minimizing $\widehat{\mathbf{z}}$ is the minimum-norm solution of the equation $\mathbf{l}(\mathbf{p})^\mathrm{T}(\mathbf{C}^\dagger)^{1/2}\mathbf{z} = 1$, and we get

$$\widehat{\mathbf{z}} = \frac{(\mathbf{C}^\dagger)^{1/2}\mathbf{l}(\mathbf{p})}{\mathbf{l}(\mathbf{p})^\mathrm{T}\mathbf{C}^\dagger\mathbf{l}(\mathbf{p})}, \tag{6.178}$$

which implies that the minimizing **w** in equation (6.177) is

$$\mathbf{w}_p = (\mathbf{C}^\dagger)^{1/2}\widehat{\mathbf{z}} = \frac{\mathbf{C}^\dagger\mathbf{l}(\mathbf{p})}{\mathbf{l}(\mathbf{p})^\mathrm{T}\mathbf{C}^\dagger\mathbf{l}(\mathbf{p})}. \tag{6.179}$$

This yields that $\mathbf{w}_p{}^\mathrm{T}\mathbf{C}\,\mathbf{w}_p = 1/\left(\mathbf{l}(\mathbf{p})^\mathrm{T}\mathbf{C}^\dagger\mathbf{l}(\mathbf{p})\right)$. We are done with the proofs.

In fact, the following more general result holds: if **B** is any positive semi-definite non-zero $m \times m$ matrix and $\mathbf{a} \in \text{span}(\mathbf{B})$ with $\mathbf{a} \neq 0$, then

$$\mathbf{w}_\mathrm{a} = \arg\min_{\mathbf{a}^\mathrm{T}\mathbf{w}=1} \mathbf{w}^\mathrm{T}\mathbf{B}\,\mathbf{w} = \frac{\mathbf{B}^\dagger\mathbf{a}}{\mathbf{a}^\mathrm{T}\mathbf{B}^\dagger\mathbf{a}}, \text{ and} \tag{6.180}$$

$$\mathbf{w}_\mathrm{a}{}^\mathrm{T}\mathbf{B}\,\mathbf{w}_\mathrm{a} = \frac{1}{\mathbf{a}^\mathrm{T}\mathbf{B}^\dagger\mathbf{a}}. \tag{6.181}$$

To verify these equations, we only need to replace $\mathbf{l}(\mathbf{p})$ and **C** by **a** and **B** in the above proof.

6.22 Appendix: Approximations (6.36) and (6.37)

Consider $\mathbf{Y} = \mathbf{AS} + \boldsymbol{\varepsilon}$ with $\mathbf{A} = [\mathbf{l}(\mathbf{p}_1), \ldots, \mathbf{l}(\mathbf{p}_n)]$ and a small amount of noise, where $\mathbf{p}_1, \ldots, \mathbf{p}_n$ are the active source-dipole locations. Let $\mathbf{C}_0 = \mathbf{ASS}^{\mathrm{T}}\mathbf{A}^{\mathrm{T}}$ denote the noiseless data covariance matrix, and

$$\mu_0(\mathbf{p}) = \underset{\mathbf{l}(\mathbf{p})^{\mathrm{T}}\mathbf{w}=1}{\arg\min}\ \mathbf{w}^{\mathrm{T}}\mathbf{C}_0\mathbf{w} \tag{6.182}$$

the corresponding output power. If $\mathbf{p} \in \{\mathbf{p}_1, \ldots, \mathbf{p}_n\}$, then by equations (6.27) and (6.41),

$$\mu_0(\mathbf{p}) = \frac{1}{\mathbf{l}(\mathbf{p})^{\mathrm{T}}\mathbf{C}_0^{\dagger}\,\mathbf{l}(\mathbf{p})} = ||\widetilde{\mathbf{S}}_p||^2. \tag{6.183}$$

Let $\mathbf{C}_0 = \mathbf{U}\,\mathbf{diag}\,(d_1, \ldots, d_n, 0, \ldots, 0)\,\mathbf{U}^{\mathrm{T}}$ be the eigenvalue decomposition EVD of $\mathbf{C}_0$ with eigenvalues $d_1 \geq \cdots \geq d_n > d_{n+1} = \cdots = d_m = 0$.

Assume first that noise $\boldsymbol{\varepsilon}$ is white with $\boldsymbol{\varepsilon}\,\boldsymbol{\varepsilon}^{\mathrm{T}} = \sigma^2\mathbf{I}$ and $\sigma > 0$ is small. Then,

$$\begin{aligned} \mathbf{C} = \mathbf{YY}^{\mathrm{T}} = \mathbf{C}_0 + \sigma^2\mathbf{I} &= \mathbf{U}\left(\mathbf{diag}\,(d_1, \ldots, d_n, 0, \ldots, 0) + \sigma^2\mathbf{I}\right)\mathbf{U}^{\mathrm{T}} \\ &= \mathbf{U}\,\mathbf{diag}\,(d_1 + \sigma^2, \ldots, d_n + \sigma^2, \sigma^2, \ldots, \sigma^2)\mathbf{U}^{\mathrm{T}}, \end{aligned} \tag{6.184}$$

and so

$$\begin{aligned} \mathbf{C}^{-1} &= \mathbf{U}\,\mathbf{diag}\Big((d_1 + \sigma^2)^{-1}, \ldots, (d_n + \sigma^2)^{-1}, \sigma^{-2}, \ldots, \sigma^{-2}\Big)\mathbf{U}^{\mathrm{T}} \\ &= \mathbf{U}_s\,\mathbf{diag}\Big((d_1 + \sigma^2)^{-1}, \ldots, (d_n + \sigma^2)^{-1}\Big)\mathbf{U}_s{}^{\mathrm{T}} + \sigma^{-2}\mathbf{U}_{ns}\mathbf{U}_{ns}{}^{\mathrm{T}}, \end{aligned} \tag{6.185}$$

where $\mathbf{U}_s = \mathbf{U}(:, 1:n)$ and $\mathbf{U}_{ns} = \mathbf{U}(:, n+1:m)$. Notice that $\mathrm{span}(\mathbf{U}_s) = \mathrm{span}(\mathbf{A})$ and $\mathrm{span}(\mathbf{U}_{ns}) = \mathrm{span}(\mathbf{A})^{\perp}$, and $\mathbf{P}_{ns} = \mathbf{U}_{ns}\mathbf{U}_{ns}{}^{\mathrm{T}}$ is the orthogonal projection onto $\mathrm{span}(\mathbf{A})^{\perp}$. Furthermore, $\mathbf{C}_0^{\dagger} = \mathbf{U}_s\,\mathbf{diag}(d_1^{-1}, \ldots, d_n^{-1})\,\mathbf{U}_s{}^{\mathrm{T}}$. Now let $\mathbf{p} \in \{\mathbf{p}_1, \ldots, \mathbf{p}_n\}$. We get the following useful presentation:

$$\begin{aligned} \mu(\mathbf{p}) &= \frac{1}{\mathbf{l}(\mathbf{p})^{\mathrm{T}}\mathbf{C}^{-1}\mathbf{l}(\mathbf{p})} \\ &= \frac{1}{\mathbf{l}(\mathbf{p})^{\mathrm{T}}\mathbf{U}_s \begin{bmatrix} \frac{1}{d_1+\sigma^2} & & \\ & \ddots & \\ & & \frac{1}{d_n+\sigma^2} \end{bmatrix} \mathbf{U}_s^{\mathrm{T}}\mathbf{l}(\mathbf{p}) + \frac{1}{\sigma^2}\,||\mathbf{U}_{ns}\mathbf{U}_{ns}^{\mathrm{T}}||^2}. \end{aligned} \tag{6.186}$$

Assume now that $\mathbf{l}(\mathbf{p}) \in \text{span}(\mathbf{A})$, that is, $\mathbf{p} \in \{\mathbf{p}_1, \ldots, \mathbf{p}_n\}$. Then $\mathbf{P}_{ns}\mathbf{l}(\mathbf{p}) = 0$ and

$$\mu(\mathbf{p}) = \frac{1}{\mathbf{l}(\mathbf{p})^{\mathrm{T}}\mathbf{C}^{-1}\mathbf{l}(\mathbf{p})}$$

$$= \frac{1}{\mathbf{l}(\mathbf{p})^{\mathrm{T}}\mathbf{U}_s \begin{bmatrix} \frac{1}{d_1+\sigma^2} & & \\ & \ddots & \\ & & \frac{1}{d_n+\sigma^2} \end{bmatrix} \mathbf{U}_s^{\mathrm{T}}\mathbf{l}(\mathbf{p})} \simeq \frac{1}{\mathbf{l}(\mathbf{p})^{\mathrm{T}}\mathbf{C}_0^{\dagger}\mathbf{l}(\mathbf{p})}.$$

for a small $\sigma > 0$, and the second part of equation (6.36) holds. Similarly, with equations (6.40) and (6.27), we see that the first part of equation (6.36) holds for $\mathbf{p} \in \{\mathbf{p}_1, \ldots, \mathbf{p}_n\}$.

Next, assume that $\mathbf{p} \notin \{\mathbf{p}_1, \ldots, \mathbf{p}_n\}$. Then, with a small $\sigma > 0$, we get

$$\mu(\mathbf{p}) = \frac{\sigma^2}{||\mathbf{P}_{ns}\mathbf{l}(\mathbf{p})||^2} \simeq 0,$$

and so equation (6.37) holds.

Let next noise $\boldsymbol{\varepsilon}$ be colored with $\mathbf{C}_\varepsilon = \boldsymbol{\varepsilon}\,\boldsymbol{\varepsilon}^{\mathrm{T}}$. We transform the data equation $\mathbf{Y} = \mathbf{AS} + \boldsymbol{\varepsilon}$ by multiplying its both sides by the whitening matrix $\mathbf{B} = \mathbf{C}_\varepsilon^{-1/2}$, and get the transformed data equation $\mathbf{BY} = \mathbf{BAS} + \boldsymbol{\nu}$, where $\boldsymbol{\nu} = \mathbf{B}\boldsymbol{\varepsilon}$ is white noise because $\boldsymbol{\nu}\boldsymbol{\nu}^{\mathrm{T}} = \mathbf{I}$. By the above results, equations (6.36) and (6.37) are valid with $\mathbf{w}_p$ and $\mu(\mathbf{p})$ replaced by the transformed weight vector $\mathbf{w}_{B,p}$ and power function $\mu_B(\mathbf{p})$ given in equations (6.42) and (6.43). But then, by equations (6.44) and (6.43), equations (6.36) and (6.37) are also valid with $\mathbf{w}_p$ and $\mu(\mathbf{p})$, as claimed.

Later, with the iterative RAP beamformer, we need the results of this appendix extended to a transformed data equation $\mathbf{QY} = \mathbf{QAS} + \mathbf{Q}\boldsymbol{\varepsilon}$, where $\mathbf{Q}$ is an orthogonal projection. This extension can be done by first observing that in the SVD $\mathbf{UDU}^{\mathrm{T}} = \mathbf{QASS}^{\mathrm{T}}\mathbf{A}^{\mathrm{T}}\mathbf{Q}$, the orthogonal matrix $\mathbf{U}$ can be chosen so that $\text{span}(\mathbf{U}(:, 1:k_1)) = \text{span}(\mathbf{QA})$ and $\text{span}(\mathbf{U}(:, 1:k_2)) = \text{span}(\mathbf{Q})$ where $k_1 = \text{rank}(\mathbf{QA})$ and $k_2 = \text{rank}(\mathbf{Q})$.

6.23 Appendix: Local Maxima of Localizers $\tau(\mathbf{p})$

We first observe that

$$\tau(\mathbf{p}) = \frac{\mathbf{l}(\mathbf{p})^{\mathrm{T}}\mathbf{C}^{-1}\mathbf{l}(\mathbf{p})}{\mathbf{l}(\mathbf{p})^{\mathrm{T}}\mathbf{C}^{-2}\mathbf{l}(\mathbf{p})} \tag{6.187}$$

is the whitened version of the PZ localizer, and the whitened version of the AI localizer is

$$\tau(\mathbf{p}) = \frac{||\mathbf{l}(\mathbf{p})||^2}{\mathbf{l}(\mathbf{p})^{\mathrm{T}}\mathbf{C}^{-1}\mathbf{l}(\mathbf{p})}, \tag{6.188}$$

which is also the AG localizer. Therefore, to show that with a small amount of noise, source locations are approximately local maximum points of the AI and AZ localizers, we may assume that noise is white and consider localizers (6.187) and (6.188).

We first consider the AI localizer $\tau(\mathbf{p})$ in equation (6.188). As in Appendix 6.22, by equation (6.186), and with small $\sigma > 0$, we see that $\tau(\mathbf{p}) \simeq ||\mathbf{l}(\mathbf{p})||^2 ||\widetilde{\mathbf{S}}_p||^2$ if $\mathbf{p}$ is one of the source locations $\mathbf{p}_1, \ldots, \mathbf{p}_n$, and $\tau(\mathbf{p}) \simeq 0$ otherwise. This proves the claim for AG and AI localizers with white noise.

Next, we consider the (whitened) PZ localizer $\tau(\mathbf{p})$ in equation (6.187). By equation (6.184), we get

$$\mathbf{C}^{-2} = \mathbf{U}_s \mathbf{diag}\left((d_1+\sigma^2)^{-2}, \ldots, (d_n+\sigma^2)^{-2}\right) \mathbf{U}_s^{\mathrm{T}} + \sigma^{-4}\mathbf{P}_{ns}, \tag{6.189}$$

where $\mathbf{P}_{ns} = \mathbf{U}_{ns}\mathbf{U}_{ns}^{\mathrm{T}}$. With this equation and equation (6.184), we get

$$\tau(\mathbf{p}) = \frac{\mathbf{l}(\mathbf{p})^{\mathrm{T}}\mathbf{C}^{-1}\mathbf{l}(\mathbf{p})}{\mathbf{l}(\mathbf{p})^{\mathrm{T}}\mathbf{C}^{-2}\mathbf{l}(\mathbf{p})} \tag{6.190}$$

$$= \frac{\mathbf{l}(\mathbf{p})^{\mathrm{T}}\mathbf{U}_s \begin{bmatrix} (d_1+\sigma^2)^{-1} & & \\ & \ddots & \\ & & (d_n+\sigma^2)^{-1} \end{bmatrix} \mathbf{U}_s^{\mathrm{T}}\mathbf{l}(\mathbf{p}) + \sigma^{-2}||\mathbf{P}_{ns}\mathbf{l}(\mathbf{p})||^2}{\mathbf{l}(\mathbf{p})^{\mathrm{T}}\mathbf{U}_s \begin{bmatrix} (d_1+\sigma^2)^{-2} & & \\ & \ddots & \\ & & (d_n+\sigma^2)^{-2} \end{bmatrix} \mathbf{U}_s^{\mathrm{T}}\mathbf{l}(\mathbf{p}) + \sigma^{-4}||\mathbf{P}_{ns}\mathbf{l}(\mathbf{p})||^2}. \tag{6.191}$$

With this equation and small $\sigma > 0$, we conclude that

$$\tau(\mathbf{p}) = \frac{\left(\mathbf{l}(\mathbf{p})^{\mathrm{T}}\mathbf{C}^{\dagger}\mathbf{l}(\mathbf{p})\right)^2}{\mathbf{l}(\mathbf{p})^{\mathrm{T}}\left(\mathbf{C}^{\dagger}\right)^2\mathbf{l}(\mathbf{p})} ||\widehat{\mathbf{S}}_p||^2 \tag{6.192}$$

if $\mathbf{p} \in \{\mathbf{p}_1, \ldots, \mathbf{p}_n\}$; otherwise,

$$\tau(\mathbf{p}) \simeq \frac{\sigma^{-2}||\mathbf{P}_{ns}\mathbf{l}(\mathbf{p})||^2}{\sigma^{-4}||\mathbf{P}_{ns}\mathbf{l}(\mathbf{p})||^2} = \sigma^2 \simeq 0. \tag{6.193}$$

These results prove the claims for the PZ localizer.

Finally, we note that the claim (6.66) can be proved, and the above results and proofs can be extended to localizers $\tau_{\text{opt}}(\mathbf{p})$ with optimal orientations by observing that $\mathbf{L}(\mathbf{p})\eta$ is in span(A) only if $\mathbf{p} = \mathbf{p}_i$ for some i, $1 \le i \le n$ and $\eta = \pm\eta_i$, where η_i is the orientation of the source dipole at $\mathbf{p}_i$.

6.24 Appendix: Unbiasedness of Localizers $\tau(\mathbf{p})$

As in the previous appendix, it is sufficient to prove the unbiasedness for whitened localizers. Let $\mathbf{Y} = \mathbf{l}_0\,\mathbf{S} + \boldsymbol{\varepsilon}$ be the data matrix due to a single dipole source at $\mathbf{p}_0$ with topography $\mathbf{l}_0 = \mathbf{l}(\mathbf{p}_0) \in \mathbb{R}^m$ and $1 \times N$ time course $\mathbf{S}$. Let $\boldsymbol{\varepsilon}\boldsymbol{\varepsilon}^{\mathrm{T}} = \sigma^2\mathbf{I}$. Then the data covariance matrix is

$$\mathbf{C} = \mathbf{Y}\mathbf{Y}^{\mathrm{T}} = \mathbf{C}_0 + \sigma^2\mathbf{I}, \tag{6.194}$$

where

$$\mathbf{C}_0 = \mathbf{l}_0\mathbf{S}\,\mathbf{S}^{\mathrm{T}}\mathbf{l}_0{}^{\mathrm{T}} = s^2\mathbf{l}_0\mathbf{l}_0{}^{\mathrm{T}} = s^2\,||\mathbf{l}_0||^2\mathbf{u}_0\mathbf{u}_0{}^{\mathrm{T}}, \tag{6.195}$$

where $s^2 = \mathbf{S}\,\mathbf{S}^{\mathrm{T}} > 0$ and $\mathbf{u}_0$ is the unit vector $||\mathbf{l}_0||^{-1}\mathbf{l}_0$. Let $\mathbf{u}_0\mathbf{u}_0{}^{\mathrm{T}} = \mathbf{U}\,\mathbf{diag}\,(1, 0, \ldots, 0)\mathbf{U}^{\mathrm{T}}$ be the EVD of $\mathbf{u}_0\mathbf{u}_0{}^{\mathrm{T}}$ with $\mathbf{U}(., 1) = \mathbf{u}_0$. Then,

$$\mathbf{C} = d\,\mathbf{U}\,\mathbf{diag}\,(1, 0, \ldots, 0)\,\mathbf{U}^{\mathrm{T}} + \sigma^2\mathbf{I} = \mathbf{U}\begin{bmatrix} d+\sigma^2 & & & \\ & \sigma^2 & & \\ & & \ddots & \\ & & & \sigma^2 \end{bmatrix}\mathbf{U}^{\mathrm{T}}, \tag{6.196}$$

where $d = s^2\,||\mathbf{l}_0||^2$, and so

$$\mathbf{C}^{-1} = \mathbf{U}\begin{bmatrix} (d+\sigma^2)^{-1} & & & \\ & \sigma^{-2} & & \\ & & \ddots & \\ & & & \sigma^{-2} \end{bmatrix}\mathbf{U}^{\mathrm{T}}$$

$$= \frac{1}{d+\sigma^2}\,\mathbf{u}_0\mathbf{u}_0{}^{\mathrm{T}} + \frac{1}{\sigma^2}\,(\mathbf{I} - \mathbf{u}_0\mathbf{u}_0{}^{\mathrm{T}}). \tag{6.197}$$

Let now $\mathbf{p}$ be any location in the ROI and $\mathbf{l}_p = \mathbf{l}(\mathbf{p})$ its topography and $\mathbf{v} = ||\mathbf{l}_0||^{-1}\mathbf{l}_0$. We consider first the (whitened) AI localizer $\tau(\mathbf{p})$ in equation (6.188). We get

$$\tau(\mathbf{p}) = \frac{||\mathbf{l}_p||^2}{\mathbf{l}_p{}^{\mathrm{T}}\mathbf{C}^{-1}\mathbf{l}_p} = \frac{1}{\mathbf{v}^{\mathrm{T}}\mathbf{C}^{-1}\mathbf{v}} = \frac{1}{\frac{1}{d+\sigma^2}\,||\mathbf{u}_0\mathbf{u}_0{}^{\mathrm{T}}\mathbf{v}||^2 + \frac{1}{\sigma^2}\,||(\mathbf{I} - \mathbf{u}_0\mathbf{u}_0{}^{\mathrm{T}})\,\mathbf{v}||^2}. \tag{6.198}$$

Because $||\mathbf{u}_0\mathbf{u}_0{}^{\mathrm{T}}\mathbf{v}||^2 + ||(\mathbf{I} - \mathbf{u}_0\mathbf{u}_0{}^{\mathrm{T}})\,\mathbf{v}||^2 = ||\mathbf{v}||^2 = 1$, with $\gamma = (d+\sigma^2)\sigma^{-2} > 1$, we get

$$(d+\sigma^2)^{-1}\tau(\mathbf{p}) = \frac{1}{||\mathbf{u}_0\mathbf{u}_0{}^{\mathrm{T}}\mathbf{v}||^2 + \gamma\,||(\mathbf{I} - \mathbf{u}_0\mathbf{u}_0{}^{\mathrm{T}})\,\mathbf{v}||^2}$$
$$= \frac{1}{1 + (\gamma - 1)\,||(\mathbf{I} - \mathbf{u}_0\mathbf{u}_0{}^{\mathrm{T}})\,\mathbf{v}||^2} \le 1, \tag{6.199}$$

where the equality is valid if $(\mathbf{I} - \mathbf{u}_0\mathbf{u}_0{}^{\mathrm{T}})\,\mathbf{v} = 0$, that is, $\mathbf{v} = \pm\mathbf{u}_0$ or, equivalently, $\mathbf{p} = \mathbf{p}_0$. Thus, $\tau(\mathbf{p})$ reaches its maximum at $\mathbf{p}_0$, and so it has no location bias.

Next we consider the PZ localizer $\tau(\mathbf{p})$ in equation (6.187). By equation (6.197), we get

$$\mathbf{C}^{-2} = \mathbf{U}\begin{bmatrix} (d+\sigma^2)^{-2} & & & \\ & \sigma^{-4} & & \\ & & \ddots & \\ & & & \sigma^{-4} \end{bmatrix}\mathbf{U}^{\mathrm{T}}$$
$$= \frac{1}{(d+\sigma^2)^2}\,\mathbf{u}_0\mathbf{u}_0{}^{\mathrm{T}} + \frac{1}{\sigma^4}\,(\mathbf{I} - \mathbf{u}_0\mathbf{u}_0{}^{\mathrm{T}})\,. \tag{6.200}$$

With this equation and equation (6.197), as in equation (6.199), we obtain

$$\tau(\mathbf{p}) = \frac{\mathbf{l}_p{}^{\mathrm{T}}\mathbf{C}^{-1}\mathbf{l}_p}{\mathbf{l}_p{}^{\mathrm{T}}\mathbf{C}^{-2}\mathbf{l}_p} = (d+\sigma^2)\,\frac{1 + (\gamma - 1)\,||(\mathbf{I} - \mathbf{u}_0\mathbf{u}_0{}^{\mathrm{T}})\,\mathbf{v}||^2}{1 + (\gamma - 1)(1+\gamma)\,||(\mathbf{I} - \mathbf{u}_0\mathbf{u}_0{}^{\mathrm{T}})\,\mathbf{v}||^2} \le 1, \tag{6.201}$$

where the equality is valid if $(\mathbf{I} - \mathbf{u}_0\mathbf{u}_0{}^{\mathrm{T}})\,\mathbf{v} = 0$, that is, $\mathbf{v} = \pm\mathbf{u}_0$ or, equivalently, $\mathbf{p} = \mathbf{p}_0$. So, $\tau(\mathbf{p})$ has no location bias.

Also, as in the previous appendix, we note that the above results and proofs can be extended to localizers $\tau_{\mathrm{opt}}(\mathbf{p})$ by observing that $\mathbf{L}(\mathbf{p})\boldsymbol{\eta}$ is in span($\mathbf{A}$) only if $\mathbf{p} = \mathbf{p}_i$ for some i, $1 \le i \le n$ and $\boldsymbol{\eta} = \pm\boldsymbol{\eta}_i$, where $\boldsymbol{\eta}_i$ is the orientation of the source dipole at $\mathbf{p}_i$.

7 MUSIC Algorithm for EEG and MEG

The MUSIC (Multiple Signal Classification) algorithm is a classical data-analysis method from the 1980s [276]. It became widely known in the MEG/EEG community in the late 1990s (see, e.g., Mosher and Leahy [210]).

The MUSIC algorithm is a *spatiotemporal* data-analysis method, which can be applied to a *time series* of MEG/EEG measurement data arising from a *limited number* of sources that can be approximated by current dipoles, here briefly referred to as *electric dipolar sources*, or *source dipoles*. The method also requires a *forward field solver* with a given *head model*.

MUSIC is closely related to beamformers, but it has a simpler structure and is more robust with respect to unknown colored noise. Here we first introduce the traditional MUSIC algorithm and its iterative RAP (recursively applied and projected) version, for both fixed- and freely-oriented source dipoles.

RAP-MUSIC has a hidden structural weakness, called the *RAP dilemma*, which makes it fail with noiseless data or data with nearly white noise [197]. Although the consequences of the RAP dilemma are difficult to observe in practice, because noise always is somewhat colored, it weakens the performance and, in particular, prevents one with RAP-MUSIC to accurately estimate the true number of the active source dipoles that generate the measured data. The RAP dilemma can be removed by a modification of RAP-MUSIC, called TRAP-MUSIC (truncated RAP-MUSIC) [197], which we also present here.

7.1 Measurement Data for MUSIC

Assumptions regarding the measurement data for MUSIC are similar to those for beamformers, and here we briefly recall them. The data for MUSIC must be due to finitely many source dipoles, say, at locations $\mathbf{p}_1, \ldots, \mathbf{p}_n$ in the region of interest (ROI) with orientations $\eta_1, \dots \eta_n$ and topographies

$$\mathbf{l}_j = \mathbf{l}(\mathbf{p}_j, \eta_j) = \mathbf{L}(\mathbf{p}_j)\eta_j, \; j = 1, \dots, n.$$

The $m \times N$ measurement data matrix $\mathbf{Y}$, with m sensors and N time points, is

$$\mathbf{Y} = \mathbf{AS} + \boldsymbol{\varepsilon}, \tag{7.1}$$

where $\mathbf{A} = [\mathbf{l}_1, \dots, \mathbf{l}_n]$ is the $m \times n$ mixing matrix, $\mathbf{S}$ is the source $n \times N$ time course matrix and $\boldsymbol{\varepsilon}$ is the $m \times N$ noise matrix. We assume that $\mathbf{l}_1, \dots, \mathbf{l}_n$ are linearly independent as well as the rows of $\mathbf{S}$, that is, $\text{rank}(\mathbf{A}) = \text{rank}(\mathbf{S}) = n \leq m$. We also assume that $\mathbf{Y}$ is time-centered, that is, $\sum_{j=1}^{N} \mathbf{Y}(i, j) = 0$ for $i = 1, \dots, m$. This also makes $\mathbf{S}$ time-centered, because we may assume that $\boldsymbol{\varepsilon}$ is time-centered as noise.

The n-dimensional subspace $\text{span}(\mathbf{A}) \subset \mathbb{R}^m$ is called the *signal space* and its orthospace $\text{span}(\mathbf{A})^{\perp}$ the *noise (only) space*.

7.2 MUSIC with Fixed-Oriented Source Dipoles

In this section, we assume that dipoles in the ROI are fixed-oriented so that the topography $\mathbf{l}(\mathbf{p})$ depends only on the location $\mathbf{p}$. The search in MUSIC for source dipoles is based on probing whether $\mathbf{l}(\mathbf{p})$ belongs to the signal space or not, for $\mathbf{p}$ in the ROI. The localizer $\mu(\mathbf{p})$ for the search is constructed as follows.

Consider the measurement data matrix $\mathbf{Y}$. Let us assume that we know the number n of the source dipoles and that the data are noiseless, that is, $\boldsymbol{\varepsilon} = 0$. We form the SVD of the noiseless data covariance matrix $\mathbf{C}_0 = \mathbf{Y}\mathbf{Y}^{\mathrm{T}}$, and get

$$\begin{aligned} \mathbf{C}_0 &= \mathbf{U}\, \mathbf{diag}_{m\times m}(d_1, \dots, d_n, 0, \dots, 0)\, \mathbf{U}^{\mathrm{T}} \\ &= \mathbf{U}(:, 1:n)\, \mathbf{diag}_{n\times n}(d_1, \dots, d_n)\, \mathbf{U}(:, 1:n)^{\mathrm{T}}, \end{aligned} \tag{7.2}$$

where $d_1 \geq \cdots \geq d_n > 0$. Because $\text{span}(\mathbf{C}_0) = \text{span}(\mathbf{A})$, this shows that

$$\text{span}(\mathbf{A}) = \text{span}\big(\mathbf{U}(:, 1:n)\big), \tag{7.3}$$

and so

$$\mathbf{P}_{\text{sg}} = \mathbf{U}(:, 1:n)\mathbf{U}(:, 1:n)^{\mathrm{T}} = \mathbf{A}\mathbf{A}^{\dagger} \tag{7.4}$$

is the orthogonal projection from $\mathbb{R}^m$ onto the signal space $\text{span}(\mathbf{A})$ (see figure 7.1).

For MUSIC, as well as for beamformers, we assume that for any $\mathbf{p} \notin \{\mathbf{p}_1, \dots, \mathbf{p}_n\}$ the topographies $\mathbf{l}(\mathbf{p}), \mathbf{l}(\mathbf{p}_1), \dots, \mathbf{l}(\mathbf{p}_n)$ are linearly independent. Then, it follows that ($\Leftrightarrow$ means "if and only if")

$$\mathbf{p} \in \{\mathbf{p}_1, \dots, \mathbf{p}_n\} \;\Leftrightarrow$$

$$\mathbf{l}(\mathbf{p}) \in \text{span}(\mathbf{l}_1, \dots, \mathbf{l}_n) \;\Leftrightarrow$$

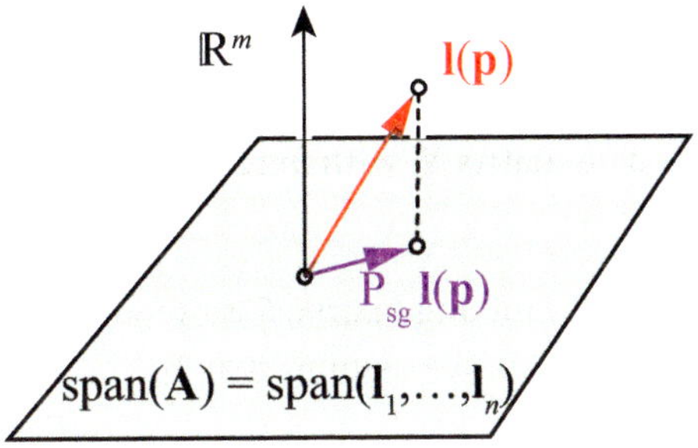

Figure 7.1
Orthogonal projection $\mathbf{P}_{sg}\mathbf{l}(\mathbf{p})$ of topography $\mathbf{l}(\mathbf{p})$ to the signal space $\mathrm{span}(\mathbf{A}) = \mathrm{span}(\mathbf{l}_1, \ldots, \mathbf{l}_n)$.

$$\mathbf{P}_{sg}\,\mathbf{l}(\mathbf{p}) = \mathbf{l}(\mathbf{p}) \;\Leftrightarrow$$
$$||\mathbf{P}_{sg}\,\mathbf{l}(\mathbf{p})|| = ||\mathbf{l}(\mathbf{p})||.$$

This implies that

$$\mathbf{p} \notin \{\mathbf{p}_1, \ldots, \mathbf{p}_n\} \;\Leftrightarrow\; ||\mathbf{P}_{sg}\,\mathbf{l}(\mathbf{p})|| < ||\mathbf{l}(\mathbf{p})||. \tag{7.5}$$

Therefore, we get the *localizer* $\mu(\mathbf{p})$ with noiseless data for MUSIC by setting

$$\mu(\mathbf{p}) = \frac{||\mathbf{P}_{sg}\,\mathbf{l}(\mathbf{p})||^2}{||\mathbf{l}(\mathbf{p})||^2} \begin{cases} = 1, \text{ if } \mathbf{p} \in \{\mathbf{p}_1, \ldots, \mathbf{p}_n\}, \\ < 1, \text{ if } \mathbf{p} \notin \{\mathbf{p}_1, \ldots, \mathbf{p}_n\}. \end{cases} \tag{7.6}$$

Let us next consider the general case where we do not know n, and the data matrix $\mathbf{Y}$ is noisy, that is, $\boldsymbol{\varepsilon} \neq 0$ in equation (7.1). Then, knowing only $\mathbf{Y}$, we must estimate n. To that end, we form the eigenvalue decomposition (EVD) of $\mathbf{C} = \mathbf{Y}\mathbf{Y}^{\mathrm{T}}$,

$$\mathbf{C} = \mathbf{U}\,\mathbf{diag}\,(d_1, \ldots, d_m)\,\mathbf{U}^{\mathrm{T}}. \tag{7.7}$$

We draw the graph of d_j as a function of j. Let $\tilde{n}$ be the integer after which d_j levels approximately to a constant value; see figure 7.2. Use that $\tilde{n} \geq n$ as an estimate for n, and approximate $\mathrm{span}(\mathbf{A})$ by $\mathrm{span}(\mathbf{U}(:, 1:\tilde{n}))$, where $\mathbf{U}$ is as in equation (7.7).

We can now estimate $\mathbf{P}_{sg}$ by

$$\mathbf{P}_s = \mathbf{U}(:, 1:\tilde{n})\mathbf{U}(:, 1:\tilde{n})^{\mathrm{T}},$$

which is the orthogonal projection onto $\mathrm{span}(\mathbf{U}(:, 1:\tilde{n}))$. By replacing $\mathbf{P}_{sg}$ by $\mathbf{P}_s$ in equation (7.6), we get the *localizer* for the MUSIC algorithm, for fixed-oriented dipoles and with unknown n and white or colored noise, as

$$\mu(\mathbf{p}) = \frac{||\mathbf{P}_s\,\mathbf{l}(\mathbf{p})||^2}{||\mathbf{l}(\mathbf{p})||^2}. \tag{7.8}$$

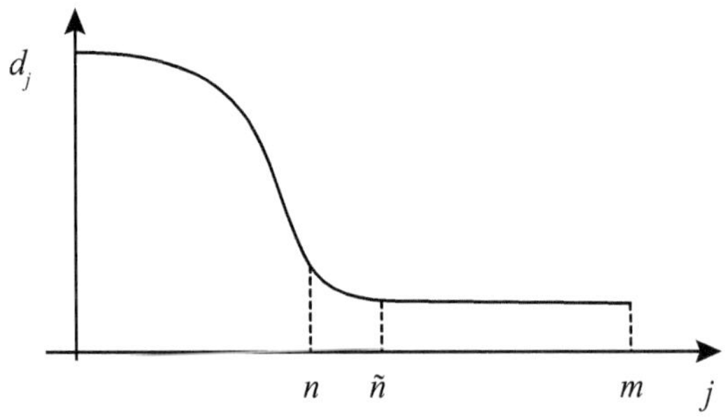

Figure 7.2
Finding an estimate $\tilde{n}$ for the number n of source dipoles by the graph of the singular values d_j of the data matrix $\mathbf{Y}$ drawn as a function of j. The estimate $\tilde{n}$ is found as that index after which the graph settles to a constant level of noise.

Next we study how this localizer works with noisy data. Assume first that $\boldsymbol{\varepsilon} \neq 0$ is *white noise*, that is, $\boldsymbol{\varepsilon}\,\boldsymbol{\varepsilon}^{\mathrm{T}} = \sigma^2\,\mathbf{I}$, $\sigma > 0$. The eigenvalue decomposition (EVD) of $\mathbf{C}$ is then

$$\begin{aligned}\mathbf{C} &= \mathbf{C}_0 + \sigma^2\mathbf{I} = \mathbf{U}\,\mathbf{diag}\,(d_1,\ldots,d_n,0,\ldots,0)\,\mathbf{U}^{\mathrm{T}} + \mathbf{diag}\,(\sigma^2,\ldots,\sigma^2)\\ &= \mathbf{U}\,\mathbf{diag}\,(d_1+\sigma^2,\ldots,d_n+\sigma^2,\sigma^2,\ldots,\sigma^2)\,\mathbf{U}^{\mathrm{T}},\end{aligned} \tag{7.9}$$

where $\mathbf{C}_0 = \mathbf{U}\,\mathbf{diag}\,(d_1,\ldots,d_n)\,\mathbf{U}^{\mathrm{T}}$ is the EVD of the noiseless data covariance matrix $\mathbf{C}_0 = \mathbf{A}^{\mathrm{T}}\mathbf{S}\,\mathbf{S}^{\mathrm{T}}\mathbf{A}$. Furthermore,

$$\mathbf{U}(:,1:\tilde{n}) = [\mathbf{U}(:,1:n),\mathbf{U}(:,n+1:\tilde{n})], \tag{7.10}$$

where $\mathbf{U}(:,j)$, $j = n+1,\ldots,\tilde{n}$, are orthonormal eigenvectors due only to noise and orthogonal to all $\mathbf{U}(:,j)$, $1 \le j \le n$. We get

$$\mathrm{span}(\mathbf{A}) = \mathrm{span}(\mathbf{U}(:,1:n)) \subset \mathrm{span}(\mathbf{U}(:,1:\tilde{n})). \tag{7.11}$$

For $\mathbf{P}_{\mathrm{s}}$ in equation (7.8), this implies two things. First, $||\mathbf{P}_{\mathrm{s}}\mathbf{l}(\mathbf{p})|| = ||\mathbf{l}(\mathbf{p})||$ if $\mathbf{p} \in \{\mathbf{p}_1,\ldots,\mathbf{p}_n\} \subset \mathrm{span}(\mathbf{A})$. Second, if $\mathbf{p} \notin \{\mathbf{p}_1,\ldots,\mathbf{p}_n\}$, then $\mathbf{l}(\mathbf{p}) \notin \mathrm{span}(\mathbf{A})$ and also $\mathbf{l}(\mathbf{p}) \notin \mathrm{span}(\mathbf{U}(:,1:\tilde{n}))$, and so $||\mathbf{P}_{\mathrm{s}}\mathbf{l}(\mathbf{p})|| < ||\mathbf{l}(\mathbf{p})||$. We here ignore the improbable case that $\mathbf{l}(\mathbf{p}) \notin \mathrm{span}(\mathbf{A})$ but $\mathbf{l}(\mathbf{p}) \in \mathrm{span}\big(\mathbf{U}(:,1:\tilde{n})\big)$.

Accordingly,

$$\mu(\mathbf{p}) = \frac{||\mathbf{P}_{\mathrm{s}}\,\mathbf{l}(\mathbf{p})||^2}{||\mathbf{l}(\mathbf{p})||^2} \begin{cases} = 1, \text{ if } \mathbf{p} \in \{\mathbf{p}_1,\ldots,\mathbf{p}_n\},\\ < 1, \text{ otherwise.}\end{cases} \tag{7.12}$$

It follows that the locations of source dipoles are approximately the largest local maximum points of $\mu(\mathbf{p})$, and so the search with this localizer for the source dipoles works as with beamformers.

Let us assume next that the noise is colored. Still $\mathbf{C} = \mathbf{C}_0 + \mathbf{C}_\varepsilon$, with $\mathbf{C}_\varepsilon = \boldsymbol{\varepsilon}\,\boldsymbol{\varepsilon}^{\mathrm{T}}$, but equation (7.11) is only approximately true, because span(**A**) is not excatly equal to span($\mathbf{U}(:, 1:n)$). The localizer $\mu(\mathbf{p})$, however, still approximately satisfies equation (7.12).

MUSIC performs better with white noise than with colored noise, although with low noise, the performances are almost equally good.

7.2.1 MUSIC with Freely-Oriented Source Dipoles

In this subsection, we assume that the dipoles in the ROI are *freely oriented*. As for beamformers with free orientations, we also here assign to every location an orientation determined by the data but in a different way. This procedure allows us to handle MUSIC as with fixed orientations.

We again start by considering the data

$$\mathbf{Y} = \mathbf{A}\mathbf{S} + \boldsymbol{\varepsilon}, \tag{7.13}$$

due to n source dipoles at locations $\mathbf{p}_1, \ldots, \mathbf{p}_n$ with orientations $\eta_1, \ldots, \eta_n$, topographies $\mathbf{l}_j = \mathbf{L}(\mathbf{p}_j)\eta_j$, $j = 1, \ldots, n$, and with the mixing matrix $\mathbf{A} = [\mathbf{l}_1, \ldots, \mathbf{l}_n]$.

Assume first that n is known and $\boldsymbol{\varepsilon} = 0$. Let $\mathbf{C}_0 = \mathbf{U}\mathbf{D}\mathbf{U}^{\mathrm{T}}$ be the EVD of $\mathbf{C}_0 = \mathbf{Y}\,\mathbf{Y}^{\mathrm{T}}$. Then, $\mathrm{span}(\mathbf{A}) = \mathrm{span}(\mathbf{U}(:, 1:n))$ and

$$\mathbf{P}_{sg} = \mathbf{A}\,\mathbf{A}^{\dagger} = \mathbf{U}(:, 1:n)\mathbf{U}(:, 1:n)^{\mathrm{T}} \tag{7.14}$$

is the orthogonal projection onto the signal space span(**A**).

If $\mathbf{p} \notin \{\mathbf{p}_1, \ldots, \mathbf{p}_n\}$, then for every η, $||\eta|| = 1$, $\mathbf{L}(\mathbf{p})\eta \notin \mathrm{span}(\mathbf{A})$, or equivalently, $||\mathbf{P}_{sg}\mathbf{L}(\mathbf{p})\eta|| < ||\mathbf{L}(\mathbf{p})\eta||$, implying that

$$\max_{||\eta||=1} \frac{||\mathbf{P}_{sg}\mathbf{L}(\mathbf{p})\eta||^2}{||\mathbf{L}(\mathbf{p})\eta||^2} < 1. \tag{7.15}$$

On the other hand, if $\mathbf{p} = \mathbf{p}_j$ for some j, $1 \le j \le n$, with the orientation η_j, then $\mathbf{L}(\mathbf{p}_j)\eta_j = \mathbf{l}_j \in \mathrm{span}(\mathbf{A})$ or, equivalently,

$$||\mathbf{P}_{sg}\mathbf{L}(\mathbf{p}_j)\eta_j|| = ||\mathbf{L}(\mathbf{p}_j)\eta_j||,$$

which implies that

$$\max_{||\eta||=1} \frac{||\mathbf{P}_{sg}\mathbf{L}(\mathbf{p}_j)\eta||^2}{||\mathbf{L}(\mathbf{p}_j)\eta||^2} = \frac{||\mathbf{P}_{sg}\mathbf{L}(\mathbf{p}_j)\eta_j||^2}{||\mathbf{L}(\mathbf{p}_j)\eta_j||^2} = 1. \tag{7.16}$$

Due to equations (7.15) and (7.16), we can now form the MUSIC localizer, with known n and $\boldsymbol{\varepsilon} = 0$, as

$$\mu(\mathbf{p}) = \max_{||\eta||=1} \frac{||\mathbf{P}_{sg}\mathbf{L}(\mathbf{p})\eta||^2}{||\mathbf{L}(\mathbf{p})\eta||^2}, \tag{7.17}$$

with $\mu(\mathbf{p}) = 1$, if $\mathbf{p} \in \{\mathbf{p}_1, \ldots, \mathbf{p}_n\}$, and $\mu(\mathbf{p}) < 1$, otherwise.

Next we consider the data equation (7.13) with unknown n and white or colored noise $\boldsymbol{\varepsilon} \neq 0$. We form an estimate $\tilde{n} \geq n$ for n, for example, in the following way. We form the EVD of $\mathbf{C} = \mathbf{Y}\mathbf{Y}^{\mathrm{T}}$,

$$\mathbf{C} = \mathbf{U}\,\mathbf{diag}\,(d_1, \ldots, d_m)\,\mathbf{U}^{\mathrm{T}}, \tag{7.18}$$

draw the graph of d_j as a function of j, and find $\tilde{n}$ as the integer after which d_j is much smaller than d_1 (with colored noise, numbers d_j often keep decreasing for all j and do not settle to a near-constant level). Using $\mathbf{U}$ in equation (7.18), we approximate $\mathbf{P}_{sg}$ by

$$\mathbf{P}_s = \mathbf{U}(:, 1:\tilde{n})\mathbf{U}(:, 1:\tilde{n})^{\mathrm{T}}, \tag{7.19}$$

which is an orthogonal projection onto $\mathrm{span}\big(\mathbf{U}(:, 1:\tilde{n})\big)$.

We now extend the localizer (7.17), with known n and zero noise, by replacing $\mathbf{P}_{sg}$ by $\mathbf{P}_s$ in equation (7.17) and obtain the MUSIC localizer, with freely-oriented dipoles, unknown n, and (white or) colored noise, as

$$\mu(\mathbf{p}) = \max_{||\eta||=1} \frac{||\mathbf{P}_s\mathbf{L}(\mathbf{p})\eta||^2}{||\mathbf{L}(\mathbf{p})\eta||^2}. \tag{7.20}$$

Now, let $\eta_{\mathrm{vec}}(\mathbf{p})$ be the unit vector η that yields the maximum in equation (7.20), that is,

$$\eta_{\mathrm{vec}}(\mathbf{p}) = \arg\max_{||\eta||=1} \frac{||\mathbf{P}_s\mathbf{L}(\mathbf{p})\eta||^2}{||\mathbf{L}(\mathbf{p})\eta||^2}. \tag{7.21}$$

We get the closed forms for $\mu(\mathbf{p})$ and $\eta_{\mathrm{vec}}(\mathbf{p})$, as with the beamformer in equations (6.69) and (6.70), in the following way. Notice first that $||\mathbf{P}_s\mathbf{L}(\mathbf{p})\eta||^2 = \eta^{\mathrm{T}}\mathbf{L}(\mathbf{p})^{\mathrm{T}}\mathbf{P}_s\mathbf{L}(\mathbf{p})\eta$ and $||\mathbf{L}(\mathbf{p})\eta||^2 = \eta^{\mathrm{T}}\mathbf{L}(\mathbf{p})^{\mathrm{T}}\mathbf{L}(\mathbf{p})\eta$. Let d be the largest eigenvalue and $\mathbf{u}$ the corresponding eigenvector of the matrix

$$\mathbf{F}\,\mathbf{L}(\mathbf{p})^{\mathrm{T}}\mathbf{P}_s\mathbf{L}(\mathbf{p})\,\mathbf{F},$$

where

$$\mathbf{F} = \left(\mathbf{L}(\mathbf{p})^{\mathrm{T}}\mathbf{L}(\mathbf{p})\right)^{-1/2}.$$

Then,

$$\mu(\mathbf{p}) = d \text{ and} \tag{7.22}$$

$$\eta_{\mathrm{vec}}(\mathbf{p}) = \frac{1}{||\mathbf{F}\,\mathbf{u}||}\,\mathbf{F}\,\mathbf{u}. \tag{7.23}$$

Alternatively, $\mu(\mathbf{p})$ and $\eta_{\text{vec}}(\mathbf{p})$ are obtained as the largest generalized eigenvalue and the corresponding unit eigenvector of the generalized eigenvalue equation (6.74) with $\mathbf{A} = \mathbf{L}(\mathbf{p})^{\mathrm{T}}\mathbf{P}_{\mathrm{s}}\mathbf{L}(\mathbf{p})$ and $\mathbf{B} = \mathbf{L}(\mathbf{p})^{\mathrm{T}}\mathbf{L}(\mathbf{p})$.

As with beamformers, $\eta_{\text{vec}}(\mathbf{p})$ also recovers the orientations of the source dipoles (up to the sign). Namely, let $\mathbf{p}_j$ be a source-dipole location, with orientation η_j. Because $\mathbf{P}_{\mathrm{s}}$ approximates $\mathbf{P}_{\mathrm{sg}}$, and $||\mathbf{P}_{\mathrm{sg}}\mathbf{L}(\mathbf{p}_j)\eta|| = ||\mathbf{L}(\mathbf{p}_j)\eta||$ only if $\eta = \pm\eta_j$, we can reason that $\eta_{\text{vec}}(\mathbf{p}_j) \simeq \pm\eta_j$.

We notice that the localizer (7.20) actually transforms MUSIC with freely-oriented dipoles back to the fixed-orientations case. Namely, if we assign the orientation $\eta_{\text{vec}}(\mathbf{p})$ and the topography $\mathbf{l}_{\text{vec}}(\mathbf{p}) = \mathbf{L}(\mathbf{p})\,\eta_{\text{vec}}(\mathbf{p})$ to every location $\mathbf{p}$ in the ROI, the localizer (7.20) can be written as

$$\mu(\mathbf{p}) = \frac{||\mathbf{P}_{\mathrm{s}}\mathbf{l}_{\text{vec}}(\mathbf{p})||^2}{||\mathbf{l}_{\text{vec}}(\mathbf{p})||^2}, \tag{7.24}$$

which is just the basic form of MUSIC localizer (7.8) with fixed orientations with $\mathbf{l}(\mathbf{p})$ replaced by $\mathbf{l}_{\text{vec}}(\mathbf{p})$.

7.3 Whitening the Data Equation

As the reasoning in the previous subsection suggests, MUSIC performs better with white than colored noise. However, if the noise is colored, as it usually is, and we have a reasonably good estimate for the noise covariance matrix $\mathbf{C}_{\boldsymbol{\varepsilon}} = \boldsymbol{\varepsilon}\boldsymbol{\varepsilon}^{\mathrm{T}}$, we can transform the situation to that of the white noise by *whitening* the data equation as follows.

We form the *whitening matrix* $\mathbf{B} = \mathbf{C}_{\boldsymbol{\varepsilon}}^{-1/2}$, multiply the data equation $\mathbf{Y} = \mathbf{A}\mathbf{S} + \boldsymbol{\varepsilon}$ by $\mathbf{B}$, and get the whitened equation

$$\mathbf{BY} = \mathbf{BAS} + \boldsymbol{\nu},$$

where $\boldsymbol{\nu} = \mathbf{B}\boldsymbol{\varepsilon}$ is white noise, because

$$\mathrm{Cov}(\boldsymbol{\nu}) = \boldsymbol{\nu}\boldsymbol{\nu}^{\mathrm{T}} = \mathbf{B}\boldsymbol{\varepsilon}\boldsymbol{\varepsilon}^{\mathrm{T}}\mathbf{B} = \mathbf{C}_{\boldsymbol{\varepsilon}}^{-1/2}\mathbf{C}_{\boldsymbol{\varepsilon}}\mathbf{C}_{\boldsymbol{\varepsilon}}^{-1/2} = \mathbf{I}.$$

Because the transformed mixing matrix is

$$\mathbf{BA} = \mathbf{B}\,[\mathbf{l}_1, \ldots, \mathbf{l}_n] = [\mathbf{B}\,\mathbf{l}_1, \ldots, \mathbf{B}\,\mathbf{l}_n],$$

we see that the transformed data

$$\mathbf{BY} = [\mathbf{B}\,\mathbf{l}_1, \ldots, \mathbf{B}\,\mathbf{l}_n]\,\mathbf{S} + \boldsymbol{\nu}$$

are due to dipoles at the original locations $\mathbf{p}_1, \ldots, \mathbf{p}_n$ with the transformed topographies $\mathbf{B}\mathbf{l}_1, \ldots, \mathbf{B}\mathbf{l}_n$ and with the original time courses $\mathbf{S}$.

So, we can apply the MUSIC algorithm to the transformed data with the transformed data covariance matrix $(\mathbf{BY})(\mathbf{BY})^{\mathrm{T}} = \mathbf{BCB}$ and the transformed topographies $\mathbf{B\,l}(\mathbf{p})$, yielding the *transformed localizer*

$$\mu_Z(\mathbf{p}) = \frac{||\mathbf{P}_s^Z\, \mathbf{Bl}(\mathbf{p})||^2}{||\mathbf{Bl}(\mathbf{p})||^2}, \tag{7.25}$$

where $\mathbf{P}_s^Z = \mathbf{U}(:, 1:\tilde{n})\, \mathbf{U}(:, 1:\tilde{n})^{\mathrm{T}}$, and $\mathbf{UDU}^{\mathrm{T}} = \mathbf{BCB}$ is the EVD of $\mathbf{BCB}$. This whitened localizer has an improved performance in the search for the source dipoles.

If we only know an approximation $\mathbf{F}$ of $\mathbf{C}_\varepsilon$, and $\mathbf{F}$ is a symmetric, positive definite matrix, we can still whiten the data with $\mathbf{B} = \mathbf{F}^{-1/2}$. The resulting noise $\boldsymbol{\nu} = \mathbf{F}^{-1/2}\boldsymbol{\varepsilon}$ is not completely white, but MUSIC is still expected to perform better with the transformed data equation than with the original one with (more strongly) colored noise.

If $\mathbf{C}_\varepsilon$ is under-ranked (i.e., $\mathrm{rank}(\mathbf{C}_\varepsilon) < m$) or ill-posed, before whitening, it should be regularized, for instance, by replacing $\mathbf{C}_\varepsilon$ by $\mathbf{C}_\varepsilon + \gamma\,\mathbf{I}$ with $\gamma = 10^{-3}c$, where c is the maximum of the diagonal elements of $\boldsymbol{C}_\varepsilon$.

Finally, we note that, as simulations show, MUSIC is rather robust with respect to colored noise even if $\mathbf{Y}$ is not whitened.

7.4 RAP-MUSIC for Fixed-Oriented Dipoles

The idea of the iterative RAP-MUSIC algorithm is similar to that of the RAP beamformer; one finds the source locations iteratively at each step, projecting out from the data equation topographies of the previously found dipoles before forming the localizer for the current step. By this procedure, one can replace the finding of several local maxima of the noniterative MUSIC localizer by an easier task of finding at each iteration step the global maximum of the localizer of that step.

Let $\mathbf{C} = \mathbf{UDU}^{\mathrm{T}}$ be the SVD of the data covariance matrix $\mathbf{C} = \mathbf{Y}\,\mathbf{Y}^{\mathrm{T}}$ and

$$\mathbf{U}_s = \mathbf{U}(:, 1:\tilde{n}), \tag{7.26}$$

where $\tilde{n} \geq n$ is the estimate for n, obtained as explained in section 7.2. Let $\mathbf{p}_1, \ldots, \mathbf{p}_n$ be the locations of the source dipoles.

The iteration is started by finding the first source-location estimate $\widehat{\mathbf{p}}_1$ as the global maximum point of the (non-iterative) MUSIC localizer $\mu(\mathbf{p})$ in equation (7.8).

After finding locations $\widehat{\mathbf{p}}_1, \ldots, \widehat{\mathbf{p}}_{k-1}$, one finds the location $\widehat{\mathbf{p}}_k$ in the kth iterative step as follows. We may assume that $\widehat{\mathbf{p}}_j \simeq \mathbf{p}_j$, and so $\mathbf{l}(\widehat{\mathbf{p}}_j) \simeq \mathbf{l}(\mathbf{p}_j)$ for $j = 1, \ldots, k-1$.

We compute the topographies $\mathbf{l}(\widehat{\mathbf{p}}_1), \ldots, \mathbf{l}(\widehat{\mathbf{p}}_{k-1})$ and form the out-projector

$$\mathbf{Q}_k = \mathbf{I} - \mathbf{B}_k\mathbf{B}_k^{\dagger}$$

with $\mathbf{B}_k = [\mathbf{l}(\widehat{\mathbf{p}}_1), \ldots, \mathbf{l}(\widehat{\mathbf{p}}_{k-1})]$. Then,

$$0 = \mathbf{Q}_k \mathbf{l}(\widehat{\mathbf{p}}_j) \simeq \mathbf{Q}_k \mathbf{l}(\mathbf{p}_j),\ j = 1, \ldots, k-1.$$

We apply $\mathbf{Q}_k$ to the data equation $\mathbf{Y} = \mathbf{A}\mathbf{S} + \boldsymbol{\varepsilon}$. We denote the rows of $\mathbf{S}$ as $\mathbf{S}_j = \mathbf{S}(j, :)$, $j = 1, \ldots, n$, and get the transformed data equation

$$\begin{aligned} \mathbf{Q}_k \mathbf{Y} &= \mathbf{Q}_k \mathbf{A}\,\mathbf{S} + \mathbf{Q}_k \boldsymbol{\varepsilon} \\ &= [\mathbf{Q}_k \mathbf{l}(\mathbf{p}_1), \ldots, \mathbf{Q}_k \mathbf{l}(\mathbf{p}_n)]\mathbf{S} + \mathbf{Q}_k \boldsymbol{\varepsilon} \\ &= \sum_{j=1}^{n} \mathbf{Q}_k \mathbf{l}(\mathbf{p}_j)\mathbf{S}_j + \mathbf{Q}_k \boldsymbol{\varepsilon} \simeq \sum_{j=k}^{n} \mathbf{Q}_k \mathbf{l}(\mathbf{p}_j)\mathbf{S}_j + \mathbf{Q}_k \boldsymbol{\varepsilon} \\ &= [\mathbf{Q}_k \mathbf{l}(\mathbf{p}_k), \ldots, \mathbf{Q}_k \mathbf{l}(\mathbf{p}_n)] \begin{bmatrix} \mathbf{S}_k \\ \vdots \\ \mathbf{S}_n \end{bmatrix} + \mathbf{Q}_k \boldsymbol{\varepsilon}. \end{aligned} \tag{7.27}$$

We see that the transformed data $\mathbf{Q}_k \mathbf{Y}$ are due to the dipoles at locations $\mathbf{p}_k, \ldots, \mathbf{p}_n$ with transformed topographies $\mathbf{Q}_k \mathbf{l}(\mathbf{p}_k), \ldots, \mathbf{Q}_k \mathbf{l}(\mathbf{p}_n)$ and with the original time courses $\mathbf{S}_k, \ldots, \mathbf{S}_n$. Accordingly,

$$\mathbf{p} \in \{\mathbf{p}_k, \ldots, \mathbf{p}_n\}, \quad \text{if and only if}$$

$$\mathbf{Q}_k \mathbf{l}(\mathbf{p}) \in \operatorname{span}\big(\mathbf{Q}_k \mathbf{l}(\mathbf{p}_k), \ldots, \mathbf{Q}_k \mathbf{l}(\mathbf{p}_n)\big) \simeq \operatorname{span}\big(\mathbf{Q}_k \mathbf{U}_s\big),$$

where $\mathbf{U}_s$ is as in equation (7.26). Let

$$\mathbf{P}_k = \mathbf{Q}_k \mathbf{U}_s (\mathbf{Q}_k \mathbf{U}_s)^{\dagger} \tag{7.28}$$

be the orthogonal projection onto $\operatorname{span}(\mathbf{Q}_k \mathbf{U}_s)$. Then,

$$||\mathbf{P}_k \mathbf{Q}_k \mathbf{l}(\mathbf{p})|| = ||\mathbf{Q}_k \mathbf{l}(\mathbf{p})|| \text{ if } \mathbf{p} \in \{\mathbf{p}_k, \ldots, \mathbf{p}_n\}$$

and $||\mathbf{P}_k \mathbf{Q}_k \mathbf{l}(\mathbf{p})|| < ||\mathbf{Q}_k \mathbf{l}(\mathbf{p})||$ otherwise.

Thus, the localizer for the kth iteration step is

$$\mu_k(\mathbf{p}) = \frac{||\mathbf{P}_k \mathbf{Q}_k \mathbf{l}(\mathbf{p})||^2}{||\mathbf{Q}_k \mathbf{l}(\mathbf{p})||^2}, \tag{7.29}$$

with $\mu_k(\mathbf{p}) = 1$ if $\mathbf{p} \in \{\mathbf{p}_k, \ldots, \mathbf{p}_n\}$, and $\mu_k(\mathbf{p}) < 1$ otherwise.

We find $\widehat{\mathbf{p}}_k$ as the global maximum point of $\mu_k(\mathbf{p})$ over all $\mathbf{p} \neq \widehat{\mathbf{p}}_1, \ldots, \widehat{\mathbf{p}}_{k-1}$; the exclusion of $\mathbf{p} \neq \widehat{\mathbf{p}}_1, \ldots, \widehat{\mathbf{p}}_{k-1}$ is necessary because $\mathbf{Q}_k \mathbf{l}(\widehat{\mathbf{p}}_j) = 0$ for $j = 1, \ldots, k-1$, which would, without the exclusion, lead to division by zero in equation (7.29).

7.5 RAP-MUSIC for Freely-Oriented Dipoles

The iteration is started by finding the first estimate $\widehat{\mathbf{p}}_1$ for the source locations as the global maximum point of the MUSIC localizer $\mu(\mathbf{p})$ in equation (7.20) for freely-oriented dipoles. The orientation $\widehat{\eta}(\widehat{\mathbf{p}}_1)$ assigned to $\widehat{\mathbf{p}}_1$ is computed by equation (7.23) or by the generalized eigenvalue approach, and we get the corresponding topography $\widehat{\mathbf{l}}_1 = \mathbf{L}(\widehat{\mathbf{p}}_1)\widehat{\eta}(\widehat{\mathbf{p}}_1)$.

After having found locations $\widehat{\mathbf{p}}_1, \ldots, \widehat{\mathbf{p}}_{k-1}$ and the corresponding topographies $\widehat{\mathbf{l}}_1, \ldots, \widehat{\mathbf{l}}_{k-1}$, we find $\widehat{\mathbf{p}}_k$ and $\widehat{\mathbf{l}}_k$ as follows. By reordering if necessary, we assume that $\widehat{\mathbf{p}}_j \simeq \mathbf{p}_j$ and so $\widehat{\mathbf{l}}_j \simeq \mathbf{l}_j$ for $j = 1, \ldots, k-1$.

We form the matrix $\mathbf{B}_k = [\widehat{\mathbf{l}}_1, \ldots, \widehat{\mathbf{l}}_{k-1}]$ and the out-projector

$$\mathbf{Q}_k = \mathbf{I} - \mathbf{B}_k \mathbf{B}_k^{\dagger}.$$

We let

$$\mathbf{P}_k = \mathbf{Q}_k \mathbf{U}_s (\mathbf{Q}_k \mathbf{U}_s)^{\dagger}$$

be the orthogonal projection onto $\mathrm{span}(\mathbf{Q}_k \mathbf{U}_s)$, and we form the localizer for the kth iterative step by

$$\mu_k(\mathbf{p}) = \max_{||\eta||=1} \frac{||\mathbf{P}_k \mathbf{Q}_k \mathbf{L}(\mathbf{p})\eta||^2}{||\mathbf{Q}_k \mathbf{L}(\mathbf{p})\eta||^2}. \tag{7.30}$$

Let $\widehat{\eta}(\mathbf{p})$ be that η that yields the maximum in equation (7.30). We get the closed-form representations for $\mu_k(\mathbf{p})$ and $\widehat{\eta}(\mathbf{p})$ as

$$\mu_k(\mathbf{p}) = d_k, \tag{7.31}$$

$$\widehat{\eta}(\mathbf{p}) = \frac{1}{||\mathbf{F}\,\mathbf{u}_k||} \mathbf{F}\,\mathbf{u}_k, \tag{7.32}$$

where d_k is the largest eigenvalue and $\mathbf{u}_k$ is the corresponding eigenvector of the matrix

$$\mathbf{F}\mathbf{L}(\mathbf{p})^{\mathrm{T}} \mathbf{Q}_k \mathbf{P}_k \mathbf{Q}_k \mathbf{L}(\mathbf{p}) \mathbf{F} \text{ with } \mathbf{F} = \left(\mathbf{L}(\mathbf{p})^{\mathrm{T}} \mathbf{Q}_k \mathbf{L}(\mathbf{p})\right)^{-1/2},$$

or we get the closed forms for $\mu_k(\mathbf{p})$ and $\widehat{\eta}(\mathbf{p})$ with the alternative generalized eigenvalue approach.

We find $\widehat{\mathbf{p}}_k$ as the global maximum point of $\mu_k(\mathbf{p})$ over all $\mathbf{p}$ in the ROI not equal to $\widehat{\mathbf{p}}_1, \ldots, \widehat{\mathbf{p}}_{k-1}$.

7.5.1 The RAP Dilemma

RAP-MUSIC has an undesired property: in the iteration, it tends to leave large localizer values, instead of the expected small ones, in the vicinity of the already found source locations. In particular, with noiseless data or data with white, or almost white, noise, this phenomenon is so strong that it prevents RAP-MUSIC from working properly. We call this phenomenon the *RAP dilemma*. It concerns MUSIC with both fixed- and freely-oriented dipoles. Its origin is explained as follows.

We first assume that the noise $\boldsymbol{\varepsilon}$ is white (or even vanishes). Let $\mathbf{U}_s$ be as in equation (7.26), and let $\mathbf{p}_1, \ldots, \mathbf{p}_n$ be the source locations and $\mathbf{l}_1, \ldots, \mathbf{l}_n$ the corresponding topographies.

Consider the kth iteration step. Let $\widehat{\mathbf{p}}_1, \ldots, \widehat{\mathbf{p}}_{k-1}$ and $\widehat{\mathbf{l}}_1, \ldots, \widehat{\mathbf{l}}_{k-1}$ be the already found estimates for the source locations and topographies. We may assume that $\widehat{\mathbf{p}}_j \simeq \mathbf{p}_j$, but $\widehat{\mathbf{p}}_j \neq \mathbf{p}_j$ due to locating errors; recall that $\widehat{\mathbf{p}}_j$ lies in the scanning grid, while $\mathbf{p}_j$, with high probability, does not lie there, because the scanning grid is independent of the true source locations. This implies that $\widehat{\mathbf{l}}_j \neq \mathbf{l}_j$, for $j = 1, \ldots, k-1$.

Let $\mathbf{Q}_k = \mathbf{B}_k \mathbf{B}_k^\dagger$ be the out-projector with $\mathbf{B}_k = [\widehat{\mathbf{l}}_1, \ldots, \widehat{\mathbf{l}}_{k-1}]$, and let $\mathbf{P}_k = \mathbf{Q}_k \mathbf{U}_s (\mathbf{Q}_k \mathbf{U}_s)^\dagger$ be the orthogonal projection onto $\text{span}(\mathbf{Q}_k \mathbf{U}_s)$. As in equation (7.11), $\text{span}(\mathbf{A}) \subset \text{span}(\mathbf{U}_s)$. Because $\widehat{\mathbf{p}}_j \neq \mathbf{p}_j$, we have

$$0 = \mathbf{Q}_k \widehat{\mathbf{l}}_j \neq \mathbf{Q}_k \mathbf{l}_j, \quad \text{for all } j < k.$$

On the other hand, $\mathbf{l}_j \in \text{span}(\mathbf{A}) \subset \text{span}(\mathbf{U}_s)$. Then, $\mathbf{Q}_k \mathbf{l}_j \in \text{span}(\mathbf{Q}_k \mathbf{U}_s)$, which implies that $\mathbf{P}_k \mathbf{Q}_k \mathbf{l}_j = \mathbf{Q}_k \mathbf{l}_j$, and so

$$\mu_k(\mathbf{p}_j) = \frac{||\mathbf{P}_k \mathbf{Q}_k \mathbf{l}_j||^2}{||\mathbf{Q}_k \mathbf{l}_j||^2} = 1. \tag{7.33}$$

Because $\mu_k(\mathbf{p})$ is a continuous function of $\mathbf{p}$, $\mu_k(\mathbf{p}) \simeq 1$ also in a close vicinity of $\mathbf{p}_j$, for all $j < k$. It follows that the maximum of $\mu_k(\mathbf{p})$, over all $\mathbf{p}$ in the ROI, is equal to 1, and that maximum is also reached at locations $\mathbf{p}_j$, $j < k$. This easily leads to a wrong choice of the global maximum point $\hat{\mathbf{p}}_k$ of $\mu_k(\mathbf{p})$, and the iteration step fails.

With colored noise the RAP dilemma effect is smaller, because the relation $\text{span}(\mathbf{A}) \subset \text{span}(\mathbf{U}_s)$ is only approximately valid; in practice, the effect is usually not seen with a realistic amount of colored noise.

But with colored noise, the RAP dilemma still causes RAP-MUSIC to leave large localizer values around the already found source locations [197]. This is harmful if the drop of the global maximum of the localizer in iterations is used as a stopping rule. Namely, with large localizer values around found source locations, no clear drop of the global maximum of the localizer is seen after all true sources have been found [197]. That may lead to picking up extra false sources and, in particular, prevents estimation of the number n of the true sources by observing the expected drop.

In the next subsection, we introduce a modification of RAP-MUSIC, called TRAP-MUSIC, which does not suffer from the RAP dilemma.

7.5.2 Truncated RAP-MUSIC (TRAP-MUSIC) for Fixed- and Freely-Oriented Dipoles

We use here the same notation as in section 7.4. In RAP-MUSIC at every iteration step one source topography is projected out of the measurement data, leaving $n-k+1$ sources after $k-1$ iteration steps. If there were no estimation error in $\widehat{\mathbf{l}}_j - \mathbf{l}_j$ for $j<k$, the out-projector $\mathbf{Q}_k$ would be exact and topographies $\mathbf{l}_1, \ldots, \mathbf{l}_{k-1}$ would be completely removed and the dimension of the remaining signal space, $\mathrm{span}(\mathbf{Q}_k\mathbf{l}_1, \ldots, \mathbf{Q}_k\mathbf{l}_{k-1})$, would be $n-k+1$. This suggests that the dimension of the approximate remaining subspace $\mathrm{span}(\mathbf{Q}_k\mathbf{U}_s)$ should similarly be lowered by $k-1$. That is just the idea in TRAP-MUSIC, where the $k-1$ principal directions, corresponding to the $k-1$ smallest eigenvalues of $\mathbf{Q}_k\mathbf{U}_s(\mathbf{Q}_k\mathbf{U}_s)^{\mathrm{T}}$, are truncated from $\mathrm{span}(\mathbf{Q}_k\mathbf{U}_s)$. So, TRAP-MUSIC differs from RAP-MUSIC only in the way the localizer is formed in the kth iteration step, which is done as follows.

Let

$$\mathbf{U}_k\mathbf{D}_k\mathbf{U}_k{}^{\mathrm{T}} = \mathbf{Q}_k\mathbf{U}_s(\mathbf{Q}_k\mathbf{U}_s)^{\mathrm{T}} \tag{7.34}$$

be the EVD of $\mathbf{Q}_k\mathbf{U}_s(\mathbf{Q}_k\mathbf{U}_s)^{\mathrm{T}}$, where $\mathbf{U}_k$ is an orthogonal matrix and the eigenvalues $d_j = \mathbf{D}(j,j)$ are given in a descending order $d_1 \geq \cdots \geq d_m$. In TRAP-MUSIC, we let

$$\mathrm{span}(\mathbf{U}_k(:, 1:\tilde{n}-k+1))$$

approximate the remaining signal space, and we set

$$\mathbf{P}_k^{\mathrm{TRAP}} = \mathbf{U}_k(:, 1:\tilde{n}-k+1)\mathbf{U}_k(:, 1:\tilde{n}-k+1)^{\mathrm{T}} \tag{7.35}$$

to be the orthogonal projection onto $\mathrm{span}(\mathbf{U}_k(:, 1:\tilde{n}-k+1))$. With $\mathbf{P}_k^{\mathrm{TRAP}}$, we form the TRAP-MUSIC *localizer* for the kth iteration step as follows. If the dipoles in the ROI are fixed-oriented, then

$$\mu_k^{\mathrm{TRAP}}(\mathbf{p}) = \frac{||\mathbf{P}_k^{\mathrm{TRAP}}\mathbf{Q}_k\mathbf{l}(\mathbf{p})||^2}{||\mathbf{Q}_k\mathbf{l}(\mathbf{p})||^2}. \tag{7.36}$$

If dipoles are freely oriented, then

$$\mu_k^{\mathrm{TRAP}}(\mathbf{p}) = \max_{||\eta||=1} \frac{||\mathbf{P}_k^{\mathrm{TRAP}}\mathbf{Q}_k\mathbf{L}(\mathbf{p})\eta||^2}{||\mathbf{Q}_k\mathbf{L}(\mathbf{p})\eta||^2} \tag{7.37}$$

as in equation (7.30), and $\mu_k^{\mathrm{TRAP}}(\mathbf{p})$ can be given in a closed form as explained after equation (7.30).

The iteration is stopped after all locations $\mathbf{p}_1, \ldots, \mathbf{p}_n$ have been found. Because with TRAP-MUSIC the global maximum of $\mu_k(\mathbf{p})$ drops after all true locations have been found, that drop can be used as the stopping rule (see figure 7.3).

After estimates for all true source locations have been found, and possibly some extra false ones, we can find estimates for the time courses similarly as described for beamformers in section 6.8.

Why TRAP-MUSIC removes the RAP dilemma needs to be explained. Intuitively, it can be explained as follows.

Let us consider TRAP-MUSIC with fixed-oriented dipoles; freely-oriented ones can be treated similarly. Let noise $\boldsymbol{\varepsilon}$ be white or vanishing. Equation (7.9) shows that the columns of $\mathbf{U}_s$ form an orthonormal basis of the subspace $\mathrm{span}(\mathbf{l}_1, \ldots, \mathbf{l}_n, \mathbf{u}_{n+1}, \ldots, \mathbf{u}_{\tilde{n}})$, where columns $\mathbf{u}_j = \mathbf{U}_s(:, j)$, $j = n+1, \ldots, \tilde{n}$, are solely due to noise. Furthermore,

$$\mathrm{span}(\mathbf{Q}_k\mathbf{U}_s) = \mathrm{span}(\mathbf{Q}_k\mathbf{l}_1, \ldots, \mathbf{Q}_k\mathbf{l}_n, \mathbf{Q}_k\mathbf{u}_{n+1}, \ldots, \mathbf{Q}_k\mathbf{u}_{\tilde{n}}). \tag{7.38}$$

The topographies $\mathbf{l}_k, \ldots, \mathbf{l}_n$ and the noise directions $\mathbf{u}_{n+1}, \ldots, \mathbf{u}_{\tilde{n}}$ are not very much affected by the out-projector $\mathbf{Q}_k$, while it almost cancels the topographies $\mathbf{l}_1, \ldots, \mathbf{l}_{k-1}$, that is, $\mathbf{Q}_k\mathbf{l}_j \simeq 0$ for $j < k$. This suggests that $\mathbf{Q}_k\mathbf{l}_k, \ldots, \mathbf{Q}_k\mathbf{l}_n, \mathbf{Q}_k\mathbf{u}_{n+1}, \ldots, \mathbf{Q}_k\mathbf{u}_{\tilde{n}}$ almost entirely span the subspace $\mathrm{span}(\mathbf{U}_k(:, 1:\tilde{n}-k+1))$ corresponding to the $\tilde{n}-k+1$ largest eigenvalues of $\mathbf{Q}_k\mathbf{U}_s(\mathbf{Q}_k\mathbf{U}_s)^{\mathrm{T}}$. Because for $j < k$, the projection $\mathbf{Q}_k\mathbf{l}_j$ very likely spreads into all dimensions of $\mathrm{span}(\mathbf{Q}_k\mathbf{U}_s)$, $\mathbf{Q}_k\mathbf{l}_k$ does not lie entirely in $\mathrm{span}(\mathbf{U}_k(:, 1:\tilde{n}-k+1))$, implying that

$$\mu_k^{\mathrm{TRAP}}(\mathbf{p}_j) = \frac{||\mathbf{P}_k^{\mathrm{TRAP}}\mathbf{Q}_k\mathbf{l}_j||^2}{||\mathbf{Q}_k\mathbf{l}_j||^2} < 1,$$

and so the RAP dilemma does not appear.

In fact, $\mu_k^{\mathrm{TRAP}}(\mathbf{p}_j)$, for $j < k$, is usually much smaller than the above explanation suggests. Namely, a statistical analysis of the structure of the removal of the RAP dilemma by localizer $\mu_k^{\mathrm{TRAP}}(\mathbf{p})$ shows that, on average, $\mu_k^{\mathrm{TRAP}}(\mathbf{p}_j)$, for $j < k$, with small estimation error $\mathbf{l}_j - \widehat{\mathbf{l}}_j$, is about $(\tilde{n}-k+1)/(m-k+1)$, which the numerical simulations also confirm.

In practice, TRAP-MUSIC performs equally well or somewhat better than RAP-MUSIC. However, in the presence of synchronous or highly correlated source dipoles, both algorithms, as well as all previously presented MUSIC algorithms, do not perform well and have difficulties in finding those source dipoles. By synchronous source dipoles, we mean dipoles whose time courses are linearly dependent.

In the next subsection, we present a method, called *double-scanning* MUSIC [196], that can deal much better with synchronous and highly correlated sources.

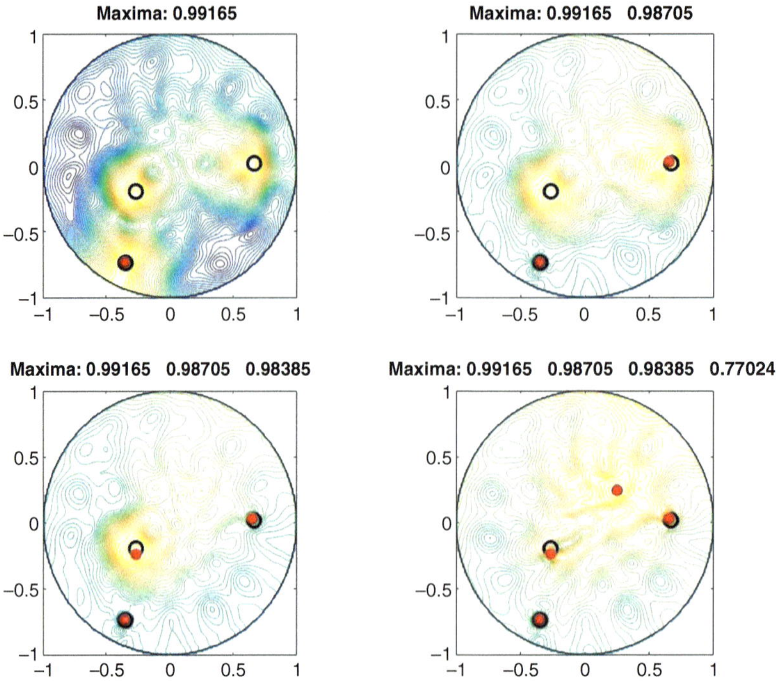

Figure 7.3

TRAP-MUSIC simulation for EEG data in a three-layer spherical model with three correlated source dipoles on the innermost upper hemisphere with z-directed orientations. The localizer contour plots are projected onto a unit disc. The simulated data matrix $\mathbf{Y} = \mathbf{A}\,\mathbf{S} + \boldsymbol{\varepsilon}$ contains colored noise with SNR = 1.

Four iteration steps are taken and results are presented in subfigures. The true dipole locations are marked by black circles and the found locations by red balls. The localizer contour plots and the global maximum points of the localizer in 1 to 4 iteration steps are shown in the subfigures.

Top left: In the initial step, all three dipole locations show and the global maximum point is the first found dipole. *Top right*: After the first found topography has been out-projected, only two remaining dipoles show in the contour plot. *Bottom left*: After three steps, all three locations are found with the global maximum staying close to one. *Bottom right*: After the three found topographies have been out-projected, the global maximum drops clearly lower than the previous ones, and the next found location becomes a false one.

7.6 Double-Scanning (DS-) MUSIC

Synchronous dipolar sources, that is, sources with linearly dependent time courses, or sources with highly correlated time courses are usually difficult to find with beamformers and MUSIC algorithms. Besides iterative beamformers, see chapter 6, several special algorithms have been suggested to deal with such sources. Some successful ones have been based on the null-constraining and prior knowledge of the approximate locations of sources and their synchronous or highly correlated partners [132]. The dual-core technique for beamformers [42, 83, 84] scans through all pairs of locations in the scanning grid, with higher computational cost than scanning only once through the grid, but it does not require any prior knowledge of the source locations. The dual-core algorithms in Brookes et al. [42] and Diwakar et al. [83] work for pairs of synchronous or highly correlated sources, but they often fail with any other mixture of sources. The enhanced dual-core algorithm in Diwakar et al. [84] is designed to improve the dual-core beamformer in Diwakar et al. [83].

Here, we present the *double-scanning* MUSIC algorithm [196] (DS-MUSIC). It is not based on the dual-core technique, and it is designed to work with any mixture of correlated sources, possibly including synchronous or highly correlated ones. It also scans through pairs of locations in a scanning grid and has the same order of computational cost as the dual-core beamformers. Extensive simulations [196] suggest that DS-MUSIC performs better than the enhanced dual-core beamformer.

We also present a *recursive (iterative) version* RDS-MUSIC [196] of DS-MUSIC, which overcomes the problem of identifying the largest local maxima in non-recursive methods.

Let again the data matrix $\mathbf{Y}=\mathbf{AS}+\boldsymbol{\varepsilon}$ be due to source dipoles at locations $\mathbf{p}_1,\ldots,\mathbf{p}_n$ with topographies $\mathbf{l}_1,\ldots,\mathbf{l}_n$, mixing matrix $\mathbf{A}=[\mathbf{l}_1,\ldots,\mathbf{l}_n]$ and time courses $\mathbf{S}_j=\mathbf{S}(j,:)$, $j=1,\ldots,n$.

Dipoles at locations $\mathbf{p}_1,\ldots,\mathbf{p}_k$, $2\leq k\leq n$, are called *synchronous* if their time courses $\mathbf{S}_1,\ldots,\mathbf{S}_k$ are linearly dependent so that there are $\alpha_j\neq 0$, $j=1,\ldots,k$ with $\alpha_1\mathbf{S}_1+\cdots+\alpha_k\mathbf{S}_k=0$. Synchronous dipoles show poorly, or not at all, in MUSIC scanning, as we see in the following.

First consider a synchronous pair of dipoles, that is, $k=2$. Then, $\mathbf{S}_1=\beta\mathbf{S}_2$ with $\beta=-\alpha_2/\alpha_1\neq 0$, and

$$\mathbf{Y}=\mathbf{l}_1\mathbf{S}_1+\cdots+\mathbf{l}_n\mathbf{S}_n+\boldsymbol{\varepsilon}=(\beta\mathbf{l}_1+\mathbf{l}_2)\mathbf{S}_2+\mathbf{l}_3\mathbf{S}_3+\cdots+\mathbf{l}_n\mathbf{S}_n+\boldsymbol{\varepsilon}.$$

So, $\mathbf{l}_1$ and $\mathbf{l}_2$ are not in $\mathrm{span}(\mathbf{Y}_0)=\mathrm{span}(\beta\mathbf{l}_1+\mathbf{l}_2,\mathbf{l}_3,\ldots,\mathbf{l}_n)$, implying that $\mu(\mathbf{p}_j)<1$, $j=1,2$, that is, $\mathbf{p}_1$ and $\mathbf{p}_2$ show poorly, or not at all, in scanning with $\mu(\mathbf{p})$. This reasoning also applies to more than two synchronous dipoles, as can be easily seen. Furthermore, if two time courses are highly correlated, we can reason with a similar analysis that they are easily lost in noise and do not show in the scanning.

Next, we describe the DS-MUSIC algorithm for freely-oriented dipoles; the case for fixed-oriented dipoles is similar, and at the end of this subsection, we briefly explain how it is obtained.

For any pair of locations $\mathbf{p}$ and $\mathbf{q}$ in the ROI, we probe if they are synchronous partners by out-projecting subspace $\mathrm{span}(\mathbf{L}(\mathbf{q}))$ from the data equation by a (wide) out-projector $\mathbf{I}-\mathbf{L}(\mathbf{q})\,\mathbf{L}(\mathbf{q})^{\dagger}$ and form the resulting localizer. By scanning with this localizer, a source location $\mathbf{p}$ shows even if $\mathbf{p}$ and $\mathbf{q}$ are synchronous partners or highly correlated. This idea will be realized in a compact way by forming the *DS-MUSIC localizer* $\mu_{\mathrm{DS}}(\mathbf{p})$ for any location $\mathbf{p}$ in the ROI by

$$\mu_{\mathrm{DS}}(\mathbf{p})=\max_{\mathbf{q}\neq\mathbf{p}}\ \nu_{\mathbf{q}}(\mathbf{p})\quad \text{with} \tag{7.39}$$

$$\nu_{\mathbf{q}}(\mathbf{p})=\max_{\|\eta\|=1}\frac{\|\mathbf{P}_{\mathbf{q}}\mathbf{Q}_{\mathbf{q}}\mathbf{L}(\mathbf{p})\eta\|^2}{\|\mathbf{Q}_{\mathbf{q}}\mathbf{L}(\mathbf{p})\eta\|^2}, \tag{7.40}$$

where $\mathbf{Q}_{\mathbf{q}}=\mathbf{I}-\mathbf{L}(\mathbf{q})\mathbf{L}(\mathbf{q})^{\dagger}$, $\mathbf{P}_{\mathbf{q}}=\mathbf{U}(:,1:\tilde{n})\,\mathbf{U}(:,1:\tilde{n})^{\mathrm{T}}$ with $\mathbf{UDV}^{\mathrm{T}}$ being the SVD of $\mathbf{Q}_{\mathbf{q}}\mathbf{Y}$; $\tilde{n}$ is the estimate of n, $\tilde{n}\geq n$; and $\mathbf{Y}$ is the data matrix. It follows that the locations of source dipoles are (approximately) the largest local maximum points of the localizer $\mu_{\mathrm{DS}}(\mathbf{p})$, as we see in the following.

If $\mathbf{p}$ is a nonsynchronous source location, then $\nu_{\mathbf{q}}(\mathbf{p})\simeq 1$ for all $\mathbf{q}\neq\mathbf{p}$, implying that $\mu_{\mathrm{DS}}(\mathbf{p})\simeq 1$. If a source dipole at $\mathbf{p}$ is the synchronous partner of another dipole at $\mathbf{q}$, then $\nu_{\mathbf{q}}(\mathbf{p})\simeq 1$, due to the out-projecting of $\mathrm{span}(\mathbf{L}(\mathbf{q}))$, which contains $\mathbf{l}_{\mathbf{q}}$, and so the source at $\mathbf{p}$ is no longer synchronous in the data equation transformed by out-projector $\mathbf{Q}_{\mathbf{q}}$. It follows that $\mu_{\mathrm{DS}}(\mathbf{p})=\max_{\mathbf{r}\neq\mathbf{p}}\ \nu_{\mathbf{r}}(\mathbf{p})\simeq 1$. If $\mathbf{p}$ is not a source location, then $\nu_{\mathbf{q}}(\mathbf{p})\simeq 0$ for all $\mathbf{q}\neq\mathbf{p}$, and so $\mu_{\mathrm{DS}}(\mathbf{p})\simeq 0$.

In practice, the ROI is discretized by the scanning grid locations, say, $\mathbf{r}_1,\ldots,\mathbf{r}_M$, and $\nu_{\mathbf{q}}(\mathbf{p})$ is approximated by matrix elements

$$\mathbf{V}(i,j)=\nu_{\mathbf{r}_i}(\mathbf{r}_j),\ \text{for}\quad 1\leq i,j\leq M,\ i\neq j, \tag{7.41}$$

and $\mathbf{V}(i,i)=0$ for $i=1,\ldots,M$. Then, $\mu_{\mathrm{DS}}(\mathbf{r}_j)\simeq\max_i \mathbf{V}(i,j)$ for $j=1,\ldots,M$. We see that the computational load of the DS-MUSIC scanning is of order M^2.

The computation of $\mathbf{V}$ is made faster if it is filled row by row, because then the same matrices $\mathbf{Q}_{\mathbf{r}_i}$ and $\mathbf{P}_{\mathbf{r}_i}$ can be used for computing all $\mathbf{V}(i,j)=\nu_{\mathbf{r}_i}(\mathbf{r}_j)$, $j=1,\ldots,M$. For a compact algorithm for computing $\mu_{\mathrm{DS}}(\mathbf{r}_j)$, $j=1,\ldots,M$; see section 7.7.

After the estimates $\widehat{\mathbf{p}}_1,\ldots,\widehat{\mathbf{p}}_n$ of the source dipole locations have been found as the largest local maximum points of $\mu_{\mathrm{DS}}(\mathbf{p})$, we can find the estimates $\widehat{\eta}_j$ of the corresponding orientations, for instance, by using the "self-consistent" method [284] with wide out-projecting and the optimal orientations as follows.

For each j, $1 \le j \le n$, let

$$\mathbf{B}_j = [\mathbf{L}(\widehat{\mathbf{p}}_1), \ldots, \mathbf{L}(\widehat{\mathbf{p}}_{j-1}), \mathbf{L}(\widehat{\mathbf{p}}_{j+1}), \ldots, \mathbf{L}(\widehat{\mathbf{p}}_n)], \tag{7.42}$$

$$\mathbf{Q}_j = \mathbf{I} - \mathbf{B}_j \mathbf{B}_j^{\dagger}, \tag{7.43}$$

$$\mathbf{P}_j = \mathbf{U}(:, 1:n)\mathbf{U}(:, 1:n)^{\mathrm{T}}, \tag{7.44}$$

where $\mathbf{UDV}^{\mathrm{T}}$ is the SVD of $\mathbf{Q}_j\mathbf{Y}$. Then

$$\widehat{\eta}_j = \arg\max_{\|\eta\|=1} \frac{\|\mathbf{P}_j\mathbf{Q}_j\mathbf{L}(\widehat{\mathbf{p}}_j)\eta\|^2}{\|\mathbf{Q}_j\mathbf{L}(\widehat{\mathbf{p}}_j)\eta\|^2}, \tag{7.45}$$

obtained in a closed form as in equation (7.32), or with the generalized eigenvalue technique.

Finally, the estimate for the time-course matrix $\widehat{\mathbf{S}}$ is found by

$$\widehat{\mathbf{S}} = (\widehat{\mathbf{A}}^{\mathrm{T}}\widehat{\mathbf{A}} + \lambda\mathbf{I})^{-1}\widehat{\mathbf{A}}^{\mathrm{T}}\mathbf{Y}, \tag{7.46}$$

where $\widehat{\mathbf{A}} = [\mathbf{L}(\widehat{\mathbf{p}}_1)\widehat{\eta}_1, \ldots, \mathbf{L}(\widehat{\mathbf{p}}_n)\widehat{\eta}_n]$ and $\lambda > 0$ is an appropriate regularization parameter, say $\lambda = 10^{-4}\bar{d}$, where $\bar{d}$ is the mean of the diagonal elements $d_1, \ldots, d_n$ of $\widehat{\mathbf{A}}^{\mathrm{T}}\widehat{\mathbf{A}}$.

We complete this section by noting that DS-MUSIC for *fixed-oriented dipoles* is obtained similarly as DS-MUSIC for freely-oriented ones by replacing equation (7.40) by

$$\mu_{\mathrm{DS}}(\mathbf{p}) = \max_{\mathbf{q}\neq\mathbf{p}} \frac{||\mathbf{P}_{\mathbf{q}}\mathbf{Q}_{\mathbf{q}}\mathbf{l}(\mathbf{p})||^2}{||\mathbf{Q}_{\mathbf{q}}\mathbf{l}(\mathbf{p})||^2}, \tag{7.47}$$

where $\mathbf{Q}_{\mathbf{q}} = \mathbf{I} - ||\mathbf{l}(\mathbf{q})||^{-2}\mathbf{l}(\mathbf{q})\,\mathbf{l}(\mathbf{q})^{\mathrm{T}}$ and $\mathbf{P}_{\mathbf{q}}$ is as in equation (7.40). Furthermore, equation (7.45) is not needed for $\widehat{\mathbf{A}} = [\mathbf{l}(\widehat{\mathbf{p}}_1), \ldots, \mathbf{l}(\widehat{\mathbf{p}}_n)]$ in equation (7.46).

7.6.1 Recursive Double-Scanning MUSIC (RDS-MUSIC)

We describe the recursive (iterative) DS-MUSIC (RDS-MUSIC) only for freely-oriented dipoles; the case with fixed-oriented dipoles is treated in a similar manner. The computational load of the RDS-MUSIC scanning is of order $n \times M^2$.

As in the previous section, we let the data matrix $\mathbf{Y} = \mathbf{AS} + \boldsymbol{\varepsilon}$ be due to source dipoles at locations $\mathbf{p}_1, \ldots, \mathbf{p}_n$ with topographies $\mathbf{l}_1, \ldots, \mathbf{l}_n$, mixing matrix $\mathbf{A} = [\mathbf{l}_1, \ldots, \mathbf{l}_n]$ and time courses $\mathbf{S}_j = \mathbf{S}(j, :)$, $j = 1, \ldots, n$.

As the first step for RDS-MUSIC, the non-recursive DS-MUSIC is run and the global maximum point $\widehat{\mathbf{p}}_1$ of the DS-MUSIC localizer $\mu_{\mathrm{DS}}(\mathbf{p})$ is the first estimated source location.

After locations $\widehat{\mathbf{p}}_1,\ldots,\widehat{\mathbf{p}}_{k-1}$ have been found, one forms the matrix $\mathbf{B}_k = [\mathbf{L}(\widehat{\mathbf{p}}_1),\ldots,\mathbf{L}(\widehat{\mathbf{p}}_{k-1})]$ and with that one gets the localizer $\mu_k^{\mathrm{RDS}}(\mathbf{p})$ for the kth iteration step as

$$\mu_k^{\mathrm{RDS}}(\mathbf{p}) = \max_{\mathbf{q}\notin\{\widehat{\mathbf{p}}_1,\ldots,\widehat{\mathbf{p}}_{k-1},\mathbf{p}\}} \nu_{\mathbf{q}}(\mathbf{p}), \tag{7.48}$$

where

$$\nu_{\mathbf{q}}(\mathbf{p}) = \max_{\|\boldsymbol{\eta}\|=1} \frac{\|\mathbf{P}_{\mathbf{q}}\mathbf{Q}_{\mathbf{q}}\mathbf{L}(\mathbf{p})\boldsymbol{\eta}\|^2}{\|\mathbf{Q}_{\mathbf{q}}\mathbf{L}(\mathbf{p})\boldsymbol{\eta}\|^2} \tag{7.49}$$

and $\mathbf{Q}_{\mathbf{q}} = \mathbf{I} - \mathbf{B}_{\mathbf{q}}\mathbf{B}_{\mathbf{q}}^{\dagger}$, $\mathbf{B}_{\mathbf{q}} = [\mathbf{B}_k, \mathbf{L}(\mathbf{q})]$, $\mathbf{P}_{\mathbf{q}} = \mathbf{U}(:,1:\tilde{n})\,\mathbf{U}(:,1:\tilde{n})^{\mathrm{T}}$, where $\mathbf{UDV}^{\mathrm{T}}$ is the SVD of $\mathbf{Q}_{\mathbf{q}}\mathbf{Y}$, $\tilde{n}\geq n$ is the estimate of n, and $\mathbf{Y}$ is the data matrix.

In practice, as with the non-recursive RDS-MUSIC, we obtain $\mu_k^{\mathrm{RDS}}(\mathbf{p})$, $k\geq 2$, for scanning-grid locations $\mathbf{r}_1,\ldots,\mathbf{r}_M$ by computing a matrix $\mathbf{V}\in\mathbb{R}^{M\times M}$ as follows. Let the found locations $\widehat{\mathbf{p}}_1,\ldots,\widehat{\mathbf{p}}_{k-1}$ be the grid locations $\mathbf{r}_{j_1},\ldots,\mathbf{r}_{j_{k-1}}$. Set $\mathbf{V}(i,j)=0$ for $i\in\{j_1,\ldots,j_{k-1}\}$ and $j\in\{j_1,\ldots,j_{k-1},i\}$, and set

$$\mathbf{V}(i,j) = \nu_{\mathbf{r}_i}(\mathbf{r}_j) = \max_{\|\boldsymbol{\eta}\|=1} \frac{\|\mathbf{P}_{\mathbf{r}_i}\mathbf{Q}_{\mathbf{r}_i}\mathbf{L}(\mathbf{r}_j)\boldsymbol{\eta}\|^2}{\|\mathbf{Q}_{\mathbf{r}_i}\mathbf{L}(\mathbf{r}_j)\boldsymbol{\eta}\|^2} \tag{7.50}$$

for $i\notin\{j_1,\ldots,j_{k-1}\}$ and $j\notin\{j_1,\ldots,j_{k-1},i\}$, where $\mathbf{Q}_{\mathbf{r}_i}$ and $\mathbf{P}_{\mathbf{r}_i}$ are as in equation (7.49) with $\mathbf{q}$ and $\mathbf{p}$ replaced by $\mathbf{r}_i$ and $\mathbf{r}_j$, respectively. Then $\mu_k^{\mathrm{RDS}}(\mathbf{r}_j) = \max_i \mathbf{V}(i,j)$, $j=1,\ldots,M$.

Again, the computation of $\mathbf{V}$ is made faster if it is filled row by row, so that the same matrices $\mathbf{Q}_{\mathbf{r}_i}$ and $\mathbf{P}_{\mathbf{r}_i}$ can be used for computing all $\mathbf{V}(i,j)=\nu_{\mathbf{r}_i}(\mathbf{r}_j)$, $j=1,\ldots,M$. For a compact algorithm for computing $\mu_k^{\mathrm{RDS}}(\mathbf{r}_j)$, $j=1,\ldots,M$; see section 7.7.

The iteration is continued until all source dipoles have been found, which is often indicated by a significant drop of the global maximum of $\mu_k^{\mathrm{RDS}}(\mathbf{p})$. After (the estimates of) the source dipole locations have been found, the corresponding orientations and time courses can be found as explained in the previous section.

7.7 Summary on MUSIC Algorithm

Let an $m\times N$ measurement data matrix $\mathbf{Y}=\mathbf{A}\,\mathbf{S}+\boldsymbol{\varepsilon}$ be given. We want to estimate the number n of dipolar sources, as well as their locations, orientations, and time courses. This is done in the following steps.

- If noise is colored and the noise covariance matrix $\mathbf{C}_\varepsilon = \boldsymbol{\varepsilon}\,\boldsymbol{\varepsilon}^{\mathrm{T}}$ (or its good estimate) is available, the data matrix $\mathbf{Y}$ can be whitened to improve the performance of MUSIC. Whitening is done by replacing $\mathbf{Y}$ by $\mathbf{C}_\varepsilon^{-1/2}\mathbf{Y}$. If $\mathbf{C}_\varepsilon$ is under-ranked (i.e., $\mathrm{rank}(\mathbf{C}_\varepsilon) < m$) or ill-posed, it must be, before being used in whitening, regularized, for example, by replacing $\mathbf{C}_\varepsilon$ by $\mathbf{C}_\varepsilon + \gamma\,\mathbf{I}$ with $\gamma = 10^{-3}c$, where c is the maximum of the diagonal elements of C_ε. Note that MUSIC is rather robust with respect to colored noise even if $\mathbf{Y}$ is not whitened.

- Fix the scanning grid in the ROI.
- Choose a non-recursive or recursive (iterative) MUSIC. For the recursive MUSIC, choose either the RAP or TRAP type. If a non-recursive MUSIC is chosen, run only the first iterative step in the recursive MUSIC algorithms described below.
- Let $\mathbf{C} = \mathbf{Y}\,\mathbf{Y}^\mathrm{T}$ be the data covariance matrix. Let $\mathbf{U}\,\mathbf{D}\,\mathbf{V}^\mathrm{T} = \mathbf{C}$ be the SVD of $\mathbf{C}$ with singular values $d_1 \geq \cdots \geq d_m$. Form an upper-estimate $\tilde{n} \geq n$, for the (unknown) number n of the dipolar sources, as the index $\tilde{n}$ after which the singular values d_j level to an approximately constant value. Set $\mathbf{U}_s = U(:, 1:\tilde{n})$, and $\mathbf{P}_1 = \mathbf{U}_s\mathbf{U}_s{}^\mathrm{T}$.

Recursive MUSIC for fixed-oriented dipoles

First iteration step:

- For every $\mathbf{p}$ in scanning grid, compute the localizer

$$\mu_1(\mathbf{p}) = \frac{||\mathbf{P}_1\mathbf{l}(\mathbf{p})||^2}{||\mathbf{l}(\mathbf{p})||^2}. \tag{7.51}$$

- Let $\widehat{\mathbf{p}}_1 = \arg\max_\mathbf{p} \mu_1(\mathbf{p})$, as the global maximum point of $\mu_1(\mathbf{p})$, be the first found source location estimate and $\widehat{\mathbf{l}}_1 = \mathbf{l}(\widehat{\mathbf{p}}_1)$. Set $M_1 = \max_\mathbf{p} \mu_1(\mathbf{p})$ be the global maximum.
- If you are running the nonrecursive MUSIC, stop the iteration here. If feasible, draw a contour plot of the localizer $\mu_1(\mathbf{p})$ and, for example, by visual inspection, find estimates $\widehat{\mathbf{p}}_1, \ldots, \widehat{\mathbf{p}}_{\hat{n}}$ for source locations as the highest "hill-tops" of the contour plot; here, $\hat{n}$ is an estimate for the true number n of sources. Find estimates for orientations and topographies of the source dipoles as $\widehat{\eta}_i = \eta(\widehat{\mathbf{p}}_i)$ and $\widehat{\mathbf{l}}_i = \mathbf{l}(\widehat{\mathbf{p}}_i)$, $i = 1, \ldots, \hat{n}$, respectively. Estimation of the time courses is presented below.

The k*th iteration step*:

- Location estimates $\widehat{\mathbf{p}}_1, \ldots, \widehat{\mathbf{p}}_{k-1}$ have been found. Let $\mathbf{B}_k = [\mathbf{l}(\widehat{\mathbf{p}}_1), \ldots, \mathbf{l}(\widehat{\mathbf{p}}_{k-1})]$ and $\mathbf{Q}_k = \mathbf{I} - \mathbf{B}_k\mathbf{B}_k^\dagger$. Let $\mathbf{U}_k\mathbf{D}_k\mathbf{V}_k{}^\mathrm{T} = \mathbf{Q}_k\mathbf{U}_s$ be the SVD of $\mathbf{Q}_k\mathbf{U}_s$.
- Set

$$\mathbf{P}_k = \mathbf{U}_k(:, 1:\tilde{n})\,\mathbf{U}(:, 1:\tilde{n})^\mathrm{T} \quad \text{for RAP-MUSIC, and} \tag{7.52}$$

$$\mathbf{P}_k = \mathbf{U}_k(:, 1:\tilde{n} - k + 1)\,\mathbf{U}(:, 1:\tilde{n} - k + 1)^\mathrm{T} \quad \text{for TRAP-MUSIC.} \tag{7.53}$$

- For all $\mathbf{p}$ in the scanning grid, compute

$$\mu_k(\mathbf{p}) = \frac{||\mathbf{P}_k\mathbf{Q}_k\mathbf{l}(\mathbf{p})||^2}{||\mathbf{Q}_k\mathbf{l}(\mathbf{p})||^2}. \tag{7.54}$$

- Set

$$\widehat{\mathbf{p}}_k = \arg\max_{\mathbf{p}} \mu_k(\mathbf{p}),$$

$$\widehat{\mathbf{l}}_k = \mathbf{l}(\widehat{\mathbf{p}}_k),$$

$$M_k = \max_{\mathbf{p}} \mu_k(\mathbf{p}).$$

- Stop the iteration after $\tilde{n}$ steps. If you are running TRAP-MUSIC, let the estimate for n be the index $\hat{n}$ after which there is a clear drop in the sequence $M_1, \ldots, M_{\tilde{n}}$.
- For estimating the time courses $\mathbf{S}_i$, set $\widehat{\mathbf{A}} = [\widehat{\mathbf{l}}_1, \ldots, \widehat{\mathbf{l}}_{\hat{n}}]$. Solve equation $\mathbf{Y} \simeq \widehat{\mathbf{A}}\,\mathbf{S}$ for $\mathbf{S}$ in a regularized way, and get an estimate $\widehat{\mathbf{S}}$ for $\mathbf{S}$,

$$\widehat{\mathbf{S}} = (\widehat{\mathbf{A}}^{\mathrm{T}}\widehat{\mathbf{A}} + \lambda\mathbf{I})^{-1}\widehat{\mathbf{A}}^{\mathrm{T}}\mathbf{Y}, \tag{7.55}$$

where $\lambda > 0$ is a regularization parameter, say, $\lambda = 10^{-3}\,c$, and c is the maximum of the diagonal elements of $\widehat{\mathbf{A}}^{\mathrm{T}}\widehat{\mathbf{A}}$. Row vectors $\widehat{\mathbf{S}}(i,:)$ are the desired estimates of $\mathbf{S}(i,:)$, $i = 1, \ldots, \hat{n}$. If you do not get a reliable $\hat{n}$ by observing the drop of values of $M_1, \ldots, M_k$, use $\tilde{n}$ instead of $\hat{n}$ and proceed as in section 6.8.

Recursive MUSIC for freely-oriented dipoles

First iteration step:

- For every $\mathbf{p}$ in the scanning grid, compute localizer

$$\mu_1(\mathbf{p}) = \max_{||\eta||=1} \frac{\eta^{\mathrm{T}}\mathbf{L}(\mathbf{p})^{\mathrm{T}}\mathbf{P}_1\mathbf{L}(\mathbf{p})\,\eta}{\eta^{\mathrm{T}}\mathbf{L}(\mathbf{p})^{\mathrm{T}}\mathbf{L}(\mathbf{p})\,\eta}, \tag{7.56}$$

where the maximum can be computed, e.g., with the generalized eigenvalue technique (see section 6.9), which you can do with MATLAB by the following command lines:

```
[E, D] = eig (Lp' * P1 * Lp, Lp' * Lp);
d = max (diag (D));
```

where $\mathbf{P1} = \mathbf{P}_1$ and $\mathbf{Lp} = \mathbf{L}(\mathbf{p})$. This yields $\mu_1(\mathbf{p}) = \mathrm{d}$.

- Let the first estimated location be

$$\widehat{\mathbf{p}}_1 = \arg\max_{\mathbf{p}} \mu_1(\mathbf{p}). \tag{7.57}$$

Next, for example, with the generalized eigenvalue technique, compute

$$\widehat{\eta}_1 = \arg\max_{\|\eta\|=1} \frac{\eta^{\mathrm{T}}\mathbf{L}(\widehat{\mathbf{p}}_1)^{\mathrm{T}}\mathbf{P}_1\,\mathbf{L}(\widehat{\mathbf{p}}_1)\,\eta}{\eta^{\mathrm{T}}\mathbf{L}(\widehat{\mathbf{p}}_1)^{\mathrm{T}}\mathbf{L}(\widehat{\mathbf{p}}_1)\,\eta}, \tag{7.58}$$

which can be done with MATLAB as follows:

```
[E, D] = eig (Lp1' * P1 * Lp1, Lp1' * Lp1);
[d, i] = max (diag (D));
a = E(:, i);
```

where $\mathbf{Lp1} = \mathbf{L}(\widehat{\mathbf{p}}_1)$ and $\mathbf{P1} = \mathbf{P}_1$. Get $\widehat{\eta}_1 = \|\mathbf{a}\|^{-1}\mathbf{a}$. Finally, set

$$\widehat{\mathbf{l}}_1 = \mathbf{L}(\widehat{\mathbf{p}}_1), \text{ and} \tag{7.59}$$

$$M_1 = \max_{\mathbf{p}} \mu_1(\mathbf{p}). \tag{7.60}$$

The first iteration step has been completed.

- If you are running the non-recursive MUSIC, stop iteration here. If feasible, draw a contour plot of the localizer $\mu_1(\mathbf{p})$ and, for example, by visual inspection, find estimates $\widehat{\mathbf{p}}_1, \ldots, \widehat{\mathbf{p}}_{\hat{n}}$ for source locations as the highest "hill-tops" of the contour plot; here, $\hat{n}$ is an estimate for the true number n of sources. Find estimates $\widehat{\eta}_i$, $i = 1, \ldots, \hat{n}$, for the source dipole orientations, e.g., with the generalized eigenvalue technique as above for $\widehat{\eta}_1$. Set $\widehat{\mathbf{l}}_i = \mathbf{L}(\widehat{\mathbf{p}}_i)\widehat{\eta}_i$, $i = 1, \ldots, \hat{n}$, to be estimates for the topographies of source dipoles. With $\widehat{\mathbf{A}} = [\widehat{\mathbf{l}}_1, \ldots, \widehat{\mathbf{l}}_{\hat{n}}]$, the time courses are estimated as with the recursive MUSIC for fixed-oriented dipoles.

The kth iteration step:

- Location estimates $\widehat{\mathbf{p}}_1, \ldots, \widehat{\mathbf{p}}_{k-1}$ and the corresponding topography estimates $\widehat{\mathbf{l}}_1, \ldots, \widehat{\mathbf{l}}_{k-1}$ have been found. Let $\mathbf{B}_k = [\widehat{\mathbf{l}}_1, \ldots, \widehat{\mathbf{l}}_{k-1}]$ and $\mathbf{Q}_k = \mathbf{I} - \mathbf{B}_k\mathbf{B}_k^{\dagger}$. Let

  $$\mathbf{U}_k\mathbf{D}_k{\mathbf{V}_k}^{\mathrm{T}} = \mathbf{Q}_k\mathbf{U}_s$$

  be the SVD of $\mathbf{Q}_k\mathbf{U}_s$.

- Set

  $$\mathbf{P}_k = \mathbf{U}_k(:, 1:\tilde{n})\,\mathbf{U}_k(:, 1:\tilde{n})^{\mathrm{T}} \text{ for RAP-MUSIC, and} \tag{7.61}$$

  $$\mathbf{P}_k = \mathbf{U}_k(:, 1:\tilde{n}-k+1)\,\mathbf{U}_k(:, 1:\tilde{n}-k+1)^{\mathrm{T}} \text{ for TRAP-MUSIC.} \tag{7.62}$$

- For every $\mathbf{p}$ in the scanning grid, for example, with the generalized eigenvalue technique, compute

  $$\mu_k(\mathbf{p}) = \max_{\|\eta\|=1} \frac{\eta^{\mathrm{T}}\mathbf{L}(\mathbf{p})^{\mathrm{T}}\mathbf{Q}_k\mathbf{P}_k\mathbf{Q}_k\mathbf{L}(\mathbf{p})\,\eta}{\eta^{\mathrm{T}}\mathbf{L}(\mathbf{p})^{\mathrm{T}}\mathbf{Q}_k\,\mathbf{L}(\mathbf{p})\,\eta}, \tag{7.63}$$

which you can do with MATLAB by the following command lines:

$[\mathbf{E}, \mathbf{D}] = \text{eig}\,(\mathbf{Lp}' * \mathbf{Qk} * \mathbf{Pk} * \mathbf{Qk} * \mathbf{Lp},\ \mathbf{Lp}' * \mathbf{Qk} * \mathbf{Lp});$

$\text{d} = \max\,(\text{diag}\,(\mathbf{D}));$

where $\mathbf{Pk} = \mathbf{P}_k$, $\mathbf{Qk} = \mathbf{Q}_k$ and $\mathbf{Lp} = \mathbf{L}(\mathbf{p})$. This yields $\mu_k(\mathbf{p}) = \text{d}$. Set

$$\widehat{\mathbf{p}}_k = \arg\max_{\mathbf{p}} \mu_k(\mathbf{p}).$$

Next, for example, with the generalized eigenvalue technique, compute

$$\widehat{\eta}_k = \arg\max_{||\eta||=1} \frac{\eta^{\mathrm{T}} \mathbf{L}(\widehat{\mathbf{p}}_k)^{\mathrm{T}} \mathbf{Q}_k \mathbf{P}_k \mathbf{Q}_k \mathbf{L}(\widehat{\mathbf{p}}_k)\, \eta}{\eta^{\mathrm{T}} \mathbf{L}(\widehat{\mathbf{p}}_k)^{\mathrm{T}} \mathbf{Q}_k\, \mathbf{L}(\widehat{\mathbf{p}}_k)\, \eta},$$

which can be done as follows with MATLAB:

$[\mathbf{E}, \mathbf{D}] = \text{eig}\,(\mathbf{Lpk}' * \mathbf{Qk} * \mathbf{Pk} * \mathbf{Qk} * \mathbf{Lpk},\ \mathbf{Lpk}' * \mathbf{Qk} * \mathbf{Lpk});$

$[\text{d}, \text{i}] = \max\,(\text{diag}\,(\mathbf{D}));$

$\mathbf{a} = \mathbf{E}(:, \text{i});$

where $\mathbf{Lpk} = \mathbf{L}(\widehat{\mathbf{p}}_k)$, $\mathbf{Pk} = \mathbf{P}_k$, and $\mathbf{Qk} = \mathbf{Q}_k$. Get $\widehat{\eta}_k = ||\mathbf{a}||^{-1}\mathbf{a}$. Finally, set

$$\widehat{\mathbf{l}}_k = \mathbf{L}(\widehat{\mathbf{p}}_k)\,\widehat{\eta}_k,$$

$$M_k = \max_{\mathbf{p}} \mu_k(\mathbf{p})\,.$$

The kth iteration step has been completed.

- The iteration is stopped and time courses estimated as earlier with recursive MUSIC for fixed-oriented dipoles.

Double-scanning MUSIC (DS-MUSIC) for freely-oriented dipoles

- Let $\mathbf{Y} = \mathbf{A}\,\mathbf{S} + \boldsymbol{\varepsilon}$ be the $m \times N$ data matrix, possibly whitened, and let $\tilde{n} \geq n$ be the estimate for the number n of sources as in the beginning of this section.
- Let $\mathbf{r}_1, \ldots, \mathbf{r}_M$ be the scanning grid locations in the ROI.
- Below, we use the notation: if A is a set and $B \subset A$, then $A \setminus B$ is the set of all elements of A that do not belong to B.
- Compute $\mu_{\mathrm{DS}}(\mathbf{r}_j)$, $j = 1, \ldots, M$ iteratively as follows.
- Start by setting $\mathbf{u}(j) = 0$, $j = 1, \ldots, M$.
- For $i = 1, \ldots, M$, set

  $\mathbf{Q}_i = \mathbf{I} - \mathbf{L}(\mathbf{r}_i)\mathbf{L}(\mathbf{r}_i)^{\dagger}$ and

  $\mathbf{P}_i = \mathbf{U}(:, 1:\tilde{n})\,\mathbf{U}(:, 1:\tilde{n})^{\mathrm{T}},$

where $\mathbf{U\,D\,W}$ is the SVD of $\mathbf{P}_i\mathbf{Y}$.
Set $\{h_1, \ldots, h_{M-1}\} = \{1, \ldots, M\} \setminus \{i\}$.
For $j = h_1, \ldots, h_{M-1}$, set
$\mathbf{L}_{ij} = \mathbf{Q}_i\mathbf{L}(\mathbf{r}_j)$ and
$v = \max_{||\eta||=1} \frac{\eta^T \mathbf{L}_{ij}{}^T \mathbf{P}_i \mathbf{L}_{ij}\,\eta}{\eta^T \mathbf{L}_{ij}{}^T \mathbf{L}_{ij}\,\eta}$,
where v can be computed by the generalized eigenvalue technique with the following MATLAB lines:

```
E = eig (Lij' * Pi * Lij, Lij' * Lij);
v = max (E);
```

where $\mathbf{Lij} = \mathbf{L}_{ij}$ and $\mathbf{Pi} = \mathbf{P}_i$,
and then substitute
$\mathbf{u}(j) \longleftarrow \max\big(\mathbf{u}(j), v\big)$.
End j-for-loop.
End i-for-loop.

- $\mu_{\mathrm{DS}}(\mathbf{r}_j) = \mathbf{u}(j)$, $j = 1, \ldots, M$.
- The estimated source locations $\widehat{\mathbf{p}}_1, \ldots, \widehat{\mathbf{p}}_n$ are the largest local maximum points of $\mu_{\mathrm{DS}}(\mathbf{p})$.
- For estimating the true orientations η_j of the source dipoles at locations $\mathbf{p}_1, \ldots, \mathbf{p}_n$, and their time courses, proceed as explained in section 7.6 in equations (7.42) and (7.45).

Recursive double-scanning MUSIC (RDS-MUSIC) for freely-oriented dipoles

- Let again $\mathbf{Y} = \mathbf{A}\,\mathbf{S} + \boldsymbol{\varepsilon}$ be the $m \times N$ data matrix, possibly whitened, and with the estimate $\tilde{n} \geq n$ for the number n of sources as in the beginning of this section.
- Let $\mathbf{r}_1, \ldots, \mathbf{r}_M$ be the scanning grid locations in the ROI.

First iteration (recursion) step:

- Let $\mu_1^{\mathrm{RDS}}(\mathbf{p}) = \mu_{\mathrm{DS}}(\mathbf{p})$ be the non-recursive DS-MUSIC localizer.
- Let the first estimated source location be
$\mathbf{r}_{j_1} = \arg\max_j \mu_1^{\mathrm{RDS}}(\mathbf{r}_j)$.
The first iteration step is completed.

The kth iteration (recursion) step:

- We again use the following notation. If A is a set and $B \subset A$, then $A \setminus B$ is the set of all elements of A that do not belong to B.
- Let $\mathbf{r}_{j_1}, \ldots, \mathbf{r}_{j_{k-1}}$ be the found location estimates.

- Set $\mathbf{B} = [\mathbf{L}(\mathbf{r}_{j_1}), \ldots, \mathbf{L}(\mathbf{r}_{j_{k-1}})]$.
- Set $\{h_1, \ldots, h_{M-k+1}\} = \{1, \ldots, M\} \setminus \{j_1, \ldots, j_{k-1}\}$.
- Next we compute $\mu_k^{\mathrm{RDS}}(\mathbf{r}_j)$, $j = 1, \ldots, M$ by a double for-loop. Set $\mathbf{u}(j) = 0$, $j = 1, \ldots, M$.
- For $i = h_1, \ldots, h_{M-k+1}$, set

  $\mathbf{B}_i = [\mathbf{B}, \mathbf{L}(\mathbf{r}_i)]$,

  $\mathbf{Q}_i = \mathbf{I} - \mathbf{B}_i \mathbf{B}_i^{\dagger}$,

  $\mathbf{P}_i = \mathbf{U}(:, 1:\tilde{n})\, \mathbf{U}(:, 1:\tilde{n})^{\mathrm{T}}$,

  where $\mathbf{U\,D\,W}$ is the SVD of $\mathbf{Q}_i\mathbf{Y}$, and set

  $\{g_1, \ldots, g_{M-k+2}\} = \{1, \ldots, M\} \setminus \{j_1, \ldots, j_{k-1}, i\}$.

  For $j = g_1, \ldots, g_{M-k+2}$, set

  $\mathbf{L}_{ij} = \mathbf{Q}_i \mathbf{L}(\mathbf{r}_j)$ and

  $v = \max_{\|\eta\|=1} \frac{\eta^{\mathrm{T}} \mathbf{L}_{ij}{}^{\mathrm{T}} \mathbf{P}_i \mathbf{L}_{ij}\, \eta}{\eta^{\mathrm{T}} \mathbf{L}_{ij}{}^{\mathrm{T}} \mathbf{L}_{ij}\, \eta}$,

  where v can be computed by the generalized eigenvalue technique with the following MATLAB lines:

  ```
  E = eig (Lij' * Pi * Lij, Lij' * Lij);
  v = max (E);
  ```

  where $\mathbf{Lij} = \mathbf{L}_{ij}$ and $\mathbf{Pi} = \mathbf{P}_i$,

  and then substitute

  $\mathbf{u}(j) \longleftarrow \max\big(\mathbf{u}(j), v\big)$.

  End j-for-loop.

  End i-for-loop.
- $\mu_k^{\mathrm{RDS}}(\mathbf{r}_j) = \mathbf{u}(j)$, $j = 1, \ldots, M$.
- The estimated kth source location is $\mathbf{r}_{j_k}$ with $j_k = \arg\max_j \mu_k^{\mathrm{RDS}}(\mathbf{r}_j)$. The kth iteration step is completed.
- Iteration is continued until all source dipoles have been found. That is usually indicated by a clear drop in the global maximum $max_j\, \mu_k^{\mathrm{RDS}}(\mathbf{r}_j)$ of localizer $\mu_k^{\mathrm{RDS}}(\mathbf{p})$ as a function of the iteration step number k.
- For estimating the true orientations $\boldsymbol{\eta}_j$ of the source dipoles at locations $\mathbf{p}_1, \ldots, \mathbf{p}_n$, and their time courses, proceed as explained in section 7.6 with $\widehat{\mathbf{p}}_i$ replaced with $\mathbf{r}_{j_i}$ in equations (7.42) and (7.45).

8 Independent Component Analysis (ICA)

Independent component analysis (ICA) is a blind-source-separation method, which can be used to analyze time series of MEG and EEG measurements. ICA does not require either a head model or a forward field solver, and so it is a solely data-driven method. Several ICA algorithms are available [64, 137] and in common use, often to separate artifacts from brain signals [52, 202]; in this chapter, we present one of them, FastICA [136], which is popular, robust, and fast.

8.1 Measurement Data and the ICA Assumption

We consider the following measurement data:

$$\mathbf{Y} = \mathbf{A}\mathbf{S} + \boldsymbol{\varepsilon}, \tag{8.1}$$

where $\mathbf{A}$ is the $m \times n$ mixing matrix, $\mathbf{S}$ is the $n \times N$ time-course matrix, and $\boldsymbol{\varepsilon}$ is the $m \times N$ additive noise matrix, statistically independent of $\mathbf{S}$. The data are assumed to be due to n unknown sources, which are spatially and temporally fixed-source *current distributions* $\mathbf{J}_i$, $i = 1, \ldots, n$. If $\mathbf{J}_i$ is clustered in a small volume, say with a diameter smaller than 5 mm, it can be considered (or rather modeled) as a current dipole. Here the sources $\mathbf{J}_i$, however, do not need to be dipoles.

Each $\mathbf{J}_i$ gives rise to a field (magnetic in MEG and electric in EEG), which is measured at m sensors yielding the measurement vector $\mathbf{l}_i$, called the *topography* of the source $\mathbf{J}_i$. The topographies $\mathbf{l}_1, \ldots, \mathbf{l}_n$ are the columns of the mixing matrix so that

$$\mathbf{A} = [\mathbf{l}_1, \ldots, \mathbf{l}_n]. \tag{8.2}$$

Each source $\mathbf{J}_i$ has a *time-dependent amplitude* $s_i(t_k)$ at time instants t_k, $k = 1, \ldots, N$, and the amplitudes form the time-course matrix $\mathbf{S}$ so that $\mathbf{S}(i,k) = s_i(t_k)$, $1 \le i \le n$, $1 \le k \le N$. Because of the linearity of the field with respect to amplitudes and the superposition principle, the measurement vector $\mathbf{y}(t_k) = \mathbf{Y}(:, k)$, due to all simultaneously active n sources, is

$$\mathbf{y}(t_k) = \sum_{i=1}^{n} s_i(t_k)\mathbf{l}_i + \boldsymbol{\varepsilon}(:, k), \tag{8.3}$$

as equation (8.1) states.

We assume that the data matrix $\mathbf{Y}$ in (8.1) is time-centered, that is, all rows of $\mathbf{Y}$ have zero mean. It also follows that the rows of the time-course matrix S are time-centered.

For ICA, we must, in addition, assume that the data $\mathbf{Y}$ satisfy the *ICA assumption*, stated in the following way: the time-course matrix $\mathbf{S}$ arises from a *hidden (latent), non-Gaussian random n-vector* $\mathbf{s} = [s_1, \ldots, s_n]^{\mathrm{T}}$ so that the columns $\mathbf{S}(:, k)$, $k = 1, \ldots, N$, are random samples of $\mathbf{s}$, and furthermore, the components $s_1, \ldots, s_n$ are *statistically independent* random variables with

$$E(s_i) = 0 \text{ and } \mathrm{var}(s_i) = E(s_i^2) = 1, \; i = 1, \ldots, n, \tag{8.4}$$

where the linear operator E stands for the expectation value (mean). The name ICA just refers to these independent components $s_1, \ldots, s_n$. Often in the literature, the rows $\mathbf{S}(i, :)$ are also loosely called independent components.

It follows that the columns of the noiseless data matrix $\mathbf{Y}_0 = \mathbf{AS}$ are samples of the random vector

$$\mathbf{y} = \mathbf{As}. \tag{8.5}$$

This equation is the *statistical model* of the noiseless data.

We note that the equations $E(s_i) = 0$ in equation (8.4) follow from the time-centering of $\mathbf{S}$. The latter equations in equation (8.4) are normalizing conditions for the random variables $s_1, \ldots, s_n$. Namely, if $\mathbf{D} = \mathbf{diag}\,(\sigma_1, \ldots, \sigma_n)$ with $\sigma_i^2 = \mathrm{var}(s_i) > 0$, $i = 1, \ldots, n$, and some $\sigma_i \neq 1$, then we can write equation (8.5) in the form $\mathbf{y} = \mathbf{A}\,\mathbf{D}\,\mathbf{D}^{-1}\mathbf{s}$, and if we replace $\mathbf{A}$ by $\mathbf{A}\,\mathbf{D}$, then $\mathbf{s}$ by $\mathbf{D}^{-1}\mathbf{s}$, the conditions in equation (8.4) become valid.

For given data $\mathbf{Y}$, satisfying the ICA assumption, the goal of ICA is, by a given ICA algorithm, to find $\mathbf{A}$ and $\mathbf{S}$, up to scaling and order of the columns $\mathbf{A}(:, i)$ and rows $\mathbf{S}(i, :)$, so that $\mathbf{Y} = \mathbf{AS}$. The found pair $\mathbf{A}$, $\mathbf{S}$ is called the *separation* of $\mathbf{Y}$.

Note that the uniqueness (up to order and scaling) of the separation requires that the independent components $s_1, \ldots, s_n$ are non-Gaussian. Namely, if they are Gaussian, then by the properties of a normalized Gaussian random vector $\mathbf{s}$, for any orthonormal matrix $\mathbf{V}$, the equation $\mathbf{Y} = (\mathbf{AV}^{\mathrm{T}})(\mathbf{VS})$ holds and the pair $\mathbf{AV}^{\mathrm{T}}$ and $\mathbf{VS}$ forms another separation with the hidden random vector $\mathbf{V}[s_1, \ldots, s_n]^{\mathrm{T}}$, which is also a (normalized Gaussian) random vector with statistically independent components.

The ICA assumption is rather implicit, and in practice, it is difficult to verify whether a given data matrix $\mathbf{Y}$ satisfies it. If, however, the sources $\mathbf{J}_i$ are neural centers acting independently and in a stationary way in time, one may with good reason expect that $\mathbf{Y}$ satisfies the ICA assumption.

Most ICA algorithms, regardless of the validity of the ICA assumption, yield a separation of $\mathbf{Y}=\widehat{\mathbf{A}}\,\widehat{\mathbf{S}}$, a correct or false one, and it remains for the user, often by using some additional information, to find out if $\widehat{\mathbf{A}}$ and $\widehat{\mathbf{S}}$ really approximate the true $\mathbf{A}$ and $\mathbf{S}$ (up to scaling and order). Therefore, the ICA method, with MEG or EEG data, is often of an exploratory nature: if the outcome looks reasonable and interesting, it is a promising starting point for further study of the topic under investigation.

8.2 Preprocessing Data for ICA

Let the data $\mathbf{Y}=\mathbf{AS}+\boldsymbol{\varepsilon}$ be as in equation (8.1). We assume that the noiseless data $\mathbf{Y}_0=\mathbf{AS}$ satisfiy the ICA assumption.

The data $\mathbf{Y}$ must be preprocessed before an ICA algorithm is applied to them. The preprocessing steps are *time-centering, compressing,* and *whitening* (note that here whitening is different from the whitening with beamformers and MUSIC). These steps are required for FastICA, but all of them are not necessary for all ICA algorithms.

Time-centering (i.e., row-centering) is needed, because in ICA, we use covariance matrices. In time-centering, we remove the time average of each row $\mathbf{Y}(j,:)$ from its elements and get the time-centered matrix $\mathbf{Y}_{\mathrm{c}}$ by

$$\mathbf{Y}_{\mathrm{c}}(j,k)=\mathbf{Y}(j,k)-\frac{1}{N}\sum_{i=1}^{N}\mathbf{Y}(j,i), \tag{8.6}$$

$1\leq j\leq m,\ 1\leq k\leq N.$

With ICA, as with beamformers, we denote the *unnormalized covariance matrix* of a matrix $\mathbf{F}$ by $\mathrm{Cov}(\mathbf{F})=\mathbf{F}_{\mathrm{c}}\mathbf{F}_{\mathrm{c}}^{\mathrm{T}}$, where $\mathbf{F}_{\mathrm{c}}$ is the row-centered $\mathbf{F}$. Accordingly, the *normalized covariance matrix* of $\mathbf{F}$ is $N^{-1}\mathrm{Cov}(\mathbf{F})=N^{-1}\mathbf{F}_{\mathrm{c}}\mathbf{F}_{\mathrm{c}}^{\mathrm{T}}$ for any $m\times N$ matrix.

Next we compress $\mathbf{Y}_{\mathrm{c}}$. We form its SVD as

$$\mathbf{Y}_{\mathrm{c}}=\mathbf{U}\,\mathbf{diag}_{m\times N}(d_1,\ldots,d_m)\mathbf{V}^{\mathrm{T}}=\sum_{j=1}^{m}d_j\mathbf{u}_j\mathbf{v}_j^{\mathrm{T}}, \tag{8.7}$$

where $d_1\geq\cdots\geq d_m\geq 0$ (we assume that $m<N$) are the singular values and $\mathbf{u}_j$ and $\mathbf{v}_j$ are the columns of $\mathbf{U}$ and $\mathbf{V}$, respectively. The compressed matrix $\mathbf{Y}_{\mathrm{comp}}$ is given by the

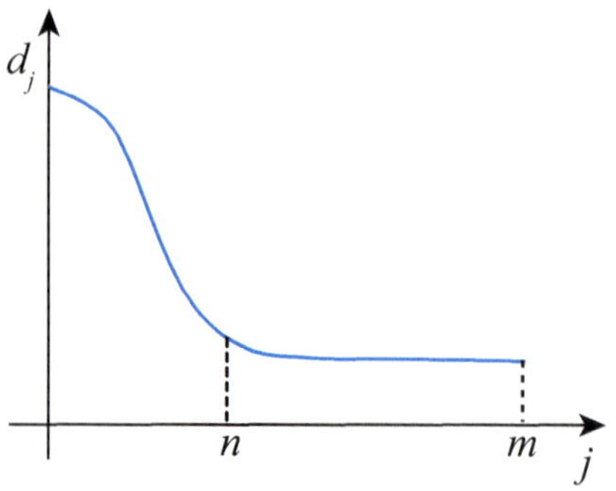

Figure 8.1
Estimation of the number n of hidden components by the graph of the singular values of the time-centered data matrix $\mathbf{Y}_c$ as the function of index j.

truncated singular value decomposition as

$$\begin{aligned}\mathbf{Y}_{\text{comp}} &= \sum_{j=1}^{n} d_j \mathbf{u}_j \mathbf{v}_j^{\mathrm{T}} \\ &= \mathbf{U}(:,1:n)\,\mathbf{diag}(d_1,\ldots,d_n)\,\mathbf{V}(:,1:n)^{\mathrm{T}}. \end{aligned} \tag{8.8}$$

The integer n in equation (8.8) is an estimate of the number of the hidden independent components. This estimation could, for instance, be done by setting n to be the integer after which the sequence $\{d_j\}$ of singular values of $\mathbf{Y}_c$ settles to the constant level of noise so that n estimates the rank of the noiseless data $\mathbf{Y}_0 = \mathbf{AS}$ (see figure 8.1).

The compressing also suppresses noise, because the truncated terms in the SVD of $\mathbf{Y}$ are mostly noise. The compressed data $\mathbf{Y}_{\text{comp}}$ approximate the noiseless data $\mathbf{AS}$, and by ignoring the approximation error, we write

$$\mathbf{Y}_{\text{comp}} = \mathbf{AS} \quad \text{with} \tag{8.9}$$

$$\frac{1}{N}\text{Cov}(\mathbf{S}) = \frac{1}{N}\mathbf{SS}^{\mathrm{T}} = \mathbf{I}. \tag{8.10}$$

Equation (8.10) follows from the fact that $\mathbf{S}(:,j)$, $j = 1,\ldots,N$, are samples of the hidden random vector $\mathbf{s}$, and so $N^{-1}\text{Cov}(\mathbf{S}) \simeq \text{cov}(\mathbf{s}) = \mathbf{I}$ due to equation (8.4) (recall that $\text{cov}(\mathbf{s})$ is the covariance matrix of random vector $\mathbf{s}$).

The last preprocessing step is the whitening of $\mathbf{Y}_{\text{comp}}$, which simplifies the construction of FastICA. The whitening also turns the mixing matrix $\mathbf{A}$ into an orthogonal matrix. Consider the SVD (8.8). We let the $n \times N$ matrix $\mathbf{X}$ denote the whitened $\mathbf{Y}_{\text{comp}}$

matrix and define $\mathbf{X}$ by

$$\mathbf{X} = \mathbf{B}_{\text{wh}} \mathbf{Y}_{\text{comp}} \in \mathbb{R}^{n \times N}, \tag{8.11}$$

where $\mathbf{B}_{\text{wh}}$ is the *whitening matrix*

$$\mathbf{B}_{\text{wh}} = \sqrt{N}\, \mathbf{diag}\, (d_1^{-1}, \ldots, d_n^{-1}) \mathbf{U}(:, 1:n)^{\mathrm{T}}. \tag{8.12}$$

It follows that

$$\mathbf{X} = \sqrt{N}\, \mathbf{V}(:, 1:n)^{\mathrm{T}}, \tag{8.13}$$

because by equation (8.8),

$$\begin{aligned} \mathbf{X} &= \mathbf{B}_{\text{wh}} \mathbf{Y}_{\text{comp}} = \mathbf{B}_{\text{wh}} \mathbf{U}(:, 1:n) \mathbf{diag}\, (d_1, \ldots, d_n) \mathbf{V}(:, 1:n)^{\mathrm{T}} \\ &= \sqrt{N}\, \mathbf{V}(:, 1:n)^{\mathrm{T}}, \end{aligned}$$

since the columns of $\mathbf{U}$ are orthonormal and so $\mathbf{U}(:, 1:n)^{\mathrm{T}} \mathbf{U}(:, 1:n) = \mathbf{I}$. Furthermore, the normalized covariance matrix of $\mathbf{X}$ is

$$\frac{1}{N} \mathrm{Cov}(\mathbf{X}) = \frac{1}{N} \mathbf{X}\mathbf{X}^{\mathrm{T}} = \mathbf{V}(:, 1:n)^{\mathrm{T}} \mathbf{V}(:, 1:n) = \mathbf{I}, \tag{8.14}$$

because the columns of $\mathbf{V}$ are also orthonormal – hence the name "whitened" matrix for $\mathbf{X}$. By equations (8.9) and (8.11), the matrix $\mathbf{X}$ gets the following form:

$$\mathbf{X} = \mathbf{W}\mathbf{S}, \tag{8.15}$$

where

$$\mathbf{W} = \mathbf{B}_{\text{wh}} \mathbf{A} \in \mathbb{R}^{n \times n} \tag{8.16}$$

is called the *weight matrix*. Note that $\mathbf{W}$ is an *orthogonal matrix*, because by equations (8.14), (8.10), and (8.15), we get

$$\mathbf{I} = N^{-1} \mathbf{X}\mathbf{X}^{\mathrm{T}} = N^{-1} \mathbf{W}\mathbf{S}\mathbf{S}^{\mathrm{T}} \mathbf{W}^{\mathrm{T}} = \mathbf{W}\mathbf{I}\mathbf{W}^{\mathrm{T}} = \mathbf{W}\mathbf{W}^{\mathrm{T}},$$

and so $\mathbf{W}^{-1} = \mathbf{W}^{\mathrm{T}}$, showing that $\mathbf{W}$ is orthogonal.

The separation of the data matrix $\mathbf{Y}$ in equation (8.1) with FastICA (or with most other ICA algorithms) proceeds as follows. We time-center, compress, and whiten $\mathbf{Y}$ and get $\mathbf{Y}_{\text{comp}}$ and $\mathbf{X} = \mathbf{W}\mathbf{S}$. We apply FastICA to $\mathbf{X}$ and get an estimate $\widehat{\mathbf{W}}$ for $\mathbf{W}$ up to signs and order of the columns of $\mathbf{W}$. By changing the signs and order of the columns of $\mathbf{A}$ (and of the rows of $\mathbf{S}$), we may assume that $\widehat{\mathbf{W}} \simeq \mathbf{W}$. With $\widehat{\mathbf{W}}$ we get an estimate $\widehat{\mathbf{S}}$ for $\mathbf{S}$ as

$$\widehat{\mathbf{S}} = \widehat{\mathbf{W}}^{\mathrm{T}} \mathbf{X}, \tag{8.17}$$

because by equation (8.15), $\widehat{\mathbf{W}}^{\mathrm{T}}\mathbf{X}=\widehat{\mathbf{W}}^{\mathrm{T}}\mathbf{W}\mathbf{S}\simeq\mathbf{I}\,\mathbf{S}=\mathbf{S}$. And finally, we get an estimate $\widehat{\mathbf{A}}$ for $\mathbf{A}$ as

$$\widehat{\mathbf{A}}=\frac{1}{N}\mathbf{Y}_{\mathrm{comp}}\widehat{\mathbf{S}}^{\mathrm{T}} \tag{8.18}$$

because by equations (8.9), (8.11), and (8.10),

$$N^{-1}\mathbf{Y}_{\mathrm{comp}}\widehat{\mathbf{S}}^{\mathrm{T}}=N^{-1}\mathbf{A}\mathbf{S}\widehat{\mathbf{S}}^{\mathrm{T}}\simeq N^{-1}\mathbf{A}\mathbf{S}\mathbf{S}^{\mathrm{T}}=\mathbf{A}\,\mathbf{I}=\mathbf{A}.$$

We complete the description of the separation of a data matrix $\mathbf{Y}$, which satisfies the ICA assumption, by describing the ICA algorithm FastICA in the next two subsections.

8.3 FastICA for Finding One Weight Vector

Let $\mathbf{X}$ be the compressed and whitened $n \times N$ data matrix with

$$\mathbf{X}=\mathbf{W}\mathbf{S}, \tag{8.19}$$

where $\mathbf{W}=[\mathbf{w}_1,\dots,\mathbf{w}_n]$ is the $n \times n$ orthogonal weight matrix and $\mathbf{S}$ is the $n \times N$ (time-centered) time-course matrix with $N^{-1}\mathbf{S}\mathbf{S}^{\mathrm{T}}=\mathbf{I}$. The columns $\mathbf{w}_j=\mathbf{W}(:,j)\in\mathbb{R}^n$, $j=1,\dots,n$, are called the *weight vectors*.

According to the ICA assumption, the columns $\mathbf{S}(:,k)$, $k=1,\dots,N$, are random samples of a hidden random vector $\mathbf{s}=[s_1,\dots,s_n]^{\mathrm{T}}$, where $s_1,\dots,s_n$ are statistically independent random variables with $E(s_j)=0$ and $\mathrm{var}(s_j)=E(s_j^2)=1$ for all $j=1,\dots,n$. Furthermore, the columns $\mathbf{X}(:,j)$ are random samples of the random variable

$$\mathbf{x}=\mathbf{W}\mathbf{s}. \tag{8.20}$$

This equation implies that

$$\mathbf{x}=\sum_{j=1}^{n} s_j\mathbf{w}_j \quad\text{and}\quad s_j=\mathbf{w}_j{}^{\mathrm{T}}\mathbf{x}, \tag{8.21}$$

because the weight vectors $\mathbf{w}_1,\dots,\mathbf{w}_n$ are orthonormal.

For a given whitened data matrix $\mathbf{X}$, our task is to find $\mathbf{W}$ and $\mathbf{S}$, up to the order and sign of the columns of $\mathbf{W}$ and the rows of $\mathbf{S}$, so that $\mathbf{X}=\mathbf{W}\mathbf{S}$ or, in other words, to separate $\mathbf{X}$ into the product of $\mathbf{W}$ and $\mathbf{S}$.

To that end, we only need to find $\mathbf{w}_1,\dots,\mathbf{w}_n$, up to order and sign, because with them, we get $\mathbf{S}=\mathbf{W}^{\mathrm{T}}\mathbf{X}$ by multiplying equation (8.19) with $\mathbf{W}^{\mathrm{T}}$ and using the identity $\mathbf{W}^{\mathrm{T}}\mathbf{W}=\mathbf{I}$. We seek $\mathbf{w}_1,\dots,\mathbf{w}_n$ with the FastICA algorithm.

The FastICA algorithm is based on *the contrast function method*, which with FastICA works as follows. One first chooses a smooth and even *contrast function* $G(t)\in\mathbb{R}$ for $t\in\mathbb{R}$

("even" means that $G(-t) = G(t)$ for all $t \in \mathbb{R}$). There is a lot of liberty in choosing $G(t)$, but good and popular choices are

$$G(t) = e^{-\frac{1}{2}t^2} \quad \text{and} \quad G(t) = \log\left(\cosh(t)\right). \tag{8.22}$$

After having chosen $G(t)$, we form an *objective function*

$$J(\mathbf{w}) = E\big(G(\mathbf{w}^{\mathrm{T}}\mathbf{x})\big) \simeq \frac{1}{N}\sum_{j=1}^{N} G\big(\mathbf{w}^{\mathrm{T}}\mathbf{X}(:,j)\big) \tag{8.23}$$

for $\mathbf{w} \in \mathbb{R}^n$ with $||\mathbf{w}|| = 1$. Also the form $J(\mathbf{w}) = \left(E\big(G(\mathbf{w}^{\mathrm{T}}\mathbf{x})\big) - c\right)^2$ with $c = E\big(G(u)\big)$, where u is the normalized Gaussian random variable, can be used here, and it leads to the same Lagrangian as below. In this latter form, $J(\mathbf{w})$ can be seen as a "measure of non-Gaussianity." For a motivation for the choice of $J(\mathbf{w})$; see, for instance, Hyvärinen et al. [137].

Next, consider the local extremal points of $J(\mathbf{w})$ (i.e., local maximum, minimum, and saddle points) on the unit sphere $||\mathbf{w}|| = 1$. By Lagrange's method, they are among the roots of Lagrange's equation

$$\nabla J(\mathbf{w}) + \beta\mathbf{w} = 0, \tag{8.24}$$

for $J(\mathbf{w})$ with constraint $||\mathbf{w}||^2 = 1$, where $\nabla J(\mathbf{w})$ is the gradient of function $J(\mathbf{w})$ with respect to $\mathbf{w}$ and β is some constant depending on the root $\mathbf{w}$. In Appendix 8.5, we show that the weight vectors $\mathbf{w}_1, \ldots, \mathbf{w}_n$ and $-\mathbf{w}_1, \ldots, -\mathbf{w}_n$ are among these roots (see figure 8.2).

By differentiating and changing the order of differentiation and the mean operator E, by equation (8.23) we get

$$\nabla J(\mathbf{w}) = E\big(G'(\mathbf{w}^{\mathrm{T}}\mathbf{x})\mathbf{x}\big) \simeq \frac{1}{N}\sum_{j=1}^{N} G'\big(\mathbf{w}^{\mathrm{T}}\mathbf{X}(:,j)\big)\mathbf{X}(:,j). \tag{8.25}$$

Next we proceed as follows. We first derive an iterative FastICA algorithm for finding a single weight vector up to the sign. Thereafter, we derive the proper FastICA algorithm for finding all $\mathbf{w}_1, \ldots, \mathbf{w}_n$ in a joint iterative search.

So, let $\mathbf{w}_k$ be one of the weight vectors $\pm\mathbf{w}_1, \ldots, \pm\mathbf{w}_n$, and we want to find it in an iterative way. The vector $\mathbf{w}_k$ is a root of equation (8.24), and so it is a root of the equation

$$\mathbf{f}(\mathbf{w}) = \nabla J(\mathbf{w}) + \beta_k\mathbf{w} = 0, \tag{8.26}$$

where

$$\beta_k = -\mathbf{w}_k^{\mathrm{T}}\nabla J(\mathbf{w}_k), \tag{8.27}$$

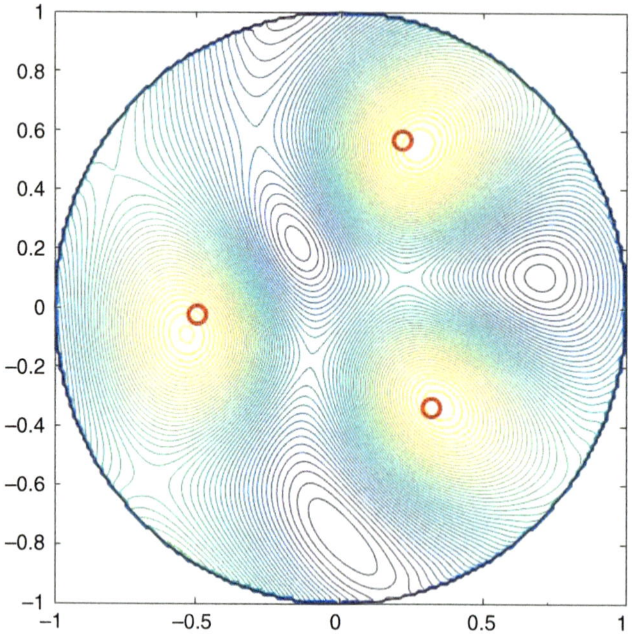

Figure 8.2
Contour plot of $J(\mathbf{w})$ on the upper hemisphere of the unit sphere in $\mathbb{R}^3$ is projected onto the planar unit disc. Here $n=3$, the components s_j of the hidden random vector $\mathbf{s}=[s_1,s_2,s_3]^{\mathrm{T}}$ are independently and uniformly distributed on $-1/2 \le s \le 1/2$, the contrast function is $G(t)=\log(\cosh(t))$, and $N=350$. The weight vectors $\mathbf{w}_1$, $\mathbf{w}_2$, and $\mathbf{w}_3$ are marked on the plot by red circles. Note that the weight vectors are at the local maxima of $J(\mathbf{w})$. The plot also shows local minima and saddle points of $J(\mathbf{w})$.

as we can see by inserting $\mathbf{w}_k$ into equation (8.24) and multiplying the equation by ${\mathbf{w}_k}^{\mathrm{T}}$.

Next, we solve (8.26) for $\mathbf{w}_k$ by the iterative Newton method. Assume that $\mathbf{w}$ is a current approximation of $\mathbf{w}_k$. The case that $\mathbf{w}$ approximates $-\mathbf{w}_k$ is treated likewise. So, $\mathbf{h}=\mathbf{w}_k-\mathbf{w}$ is small and $\mathbf{w}_k=\mathbf{w}+\mathbf{h}$. To form the equation for the iterative step $\mathbf{h}$, we use the differential expansion

$$0=\mathbf{f}(\mathbf{w}+\mathbf{h})=\mathbf{f}(\mathbf{w})+\mathbf{Df}(\mathbf{w})\mathbf{h}+\mathcal{O}\big(||\mathbf{h}||^2\big), \tag{8.28}$$

where $\mathbf{Df}(\mathbf{w})$ is the derivative (Jacobian) matrix of $\mathbf{f}(\mathbf{w})=[f_1(\mathbf{w}),\ldots,f_n(\mathbf{w})]^{\mathrm{T}}$ with $\big(\mathbf{Df}(\mathbf{w})\big)(i,j)=\partial_j f_i(\mathbf{w})$, $1\le i,\, j\le n$, and $\mathcal{O}\big(||\mathbf{h}||^2\big)$ stands for the residual term, which tends to zero at the rate of $||\mathbf{h}||^2$ as $||\mathbf{h}||\to 0$. By differentiating $\mathbf{f}(\mathbf{w})=\nabla J(\mathbf{w})+\beta_k\mathbf{w}$, we get

$$\mathbf{Df}(\mathbf{w}) = \mathbf{H}(\mathbf{w}) + \beta_k \mathbf{I}, \tag{8.29}$$

where

$$\mathbf{H}(\mathbf{w}) = \mathbf{D}(\nabla J)(\mathbf{w}) = E\big(G''(\mathbf{w}^{\mathrm{T}}\mathbf{x})\mathbf{x}\mathbf{x}^{\mathrm{T}}\big) \tag{8.30}$$

is the Hessian matrix of $J(\mathbf{w})$.

The Newton iteration step $\mathbf{h}$ could now be obtained from equation (8.28) by ignoring $\mathcal{O}(||\mathbf{h}||^2)$ and solving the resulting equation for $\mathbf{h}$. The remarkable idea of FastICA is that we do not use that proper Newton step but instead modify equation (8.28) by taking into account that $\mathbf{w}$ is close to $\mathbf{w}_k$ and so $\mathbf{H}(\mathbf{w}) \simeq \mathbf{H}(\mathbf{w}_k)$, and $\mathbf{H}(\mathbf{w}_k)$ has a special structure imposed on it by the fact that $\mathbf{w}_k$ is a weight vector.

The modification has two important benefits. First, it greatly simplifies equation (8.28), and second, it makes $\mathbf{w}_k$ an *attractive* point for the iterative FastICA algorithm, and that applies to all points $\pm\mathbf{w}_1, \ldots, \pm\mathbf{w}_n$, while other roots of Lagrange's equation (8.24) will be *repulsive* for that special iterative step, because for those points, it is a wrong Newton step and makes the algorithm step away from them.

In fact, as we show in Appendix 8.6, it holds that

$$\mathbf{H}(\mathbf{w}_k)\mathbf{h} = \alpha_k \mathbf{h} \ \text{ for any } \mathbf{h} \ \text{ with } \ {\mathbf{w}_k}^{\mathrm{T}}\mathbf{h} = 0, \tag{8.31}$$

where

$$\alpha_k = E\big(G''({\mathbf{w}_k}^{\mathrm{T}}\mathbf{x})\big). \tag{8.32}$$

In equation (8.28), the vector $\mathbf{h} = \mathbf{w}_k - \mathbf{w}$, and so ${\mathbf{w}_k}^{\mathrm{T}}\mathbf{h} \simeq 0$, because $||\mathbf{h}||$ is small and $||\mathbf{w}_k|| = ||\mathbf{w}|| = 1$ (i.e., on the n-dimensional unit sphere, the vector $\mathbf{h} = \mathbf{w}_k - \mathbf{w}$, with small $||\mathbf{h}||$, is almost tangential and so almost perpendicular to the radial vector $\mathbf{w}_k$). Approximating $\mathbf{H}(\mathbf{w}) \simeq \mathbf{H}(\mathbf{w}_k)$, using equations (8.29), (8.31), $\mathbf{f}(\mathbf{w}) = \nabla J(\mathbf{w}) + \beta_k \mathbf{w}$ and ignoring $\mathcal{O}(||\mathbf{h}||^2)$, equation (8.28) gets the form

$$\begin{aligned} 0 &= \nabla J(\mathbf{w}) + \beta_k \mathbf{w} + \alpha_k \mathbf{h} + \beta_k \mathbf{h} \\ &= \nabla J(\mathbf{w}) + \beta_k \mathbf{w} + (\alpha_k + \beta_k)\mathbf{h}. \end{aligned}$$

Solving this equation for $\mathbf{h}$ yields the FastICA iteration step

$$\widehat{\mathbf{h}} = \frac{-1}{(\alpha_k + \beta_k)}\big(\nabla J(\mathbf{w}) + \beta_k \mathbf{w}\big), \tag{8.33}$$

and the updated estimate $\mathbf{w} + \widehat{\mathbf{h}}$ for $\mathbf{w}_k$ gets the form

$$\mathbf{w} + \widehat{\mathbf{h}} = \frac{-1}{(\alpha_k + \beta_k)}\big(\nabla J(\mathbf{w}) - \alpha \mathbf{w}\big), \tag{8.34}$$

where $\alpha \simeq \alpha_k$ with

$$\alpha = E\big(G''(\mathbf{w}^{\mathrm{T}}\mathbf{x})\big). \tag{8.35}$$

Note that in equation (8.34), we have ignored the improbable case that $\alpha_k + \beta_k = 0$.

Assume next that $\alpha_k + \beta_k < 0$, which corresponds to the case that $\mathbf{w}_k$ is a local maximum point of $J(\mathbf{w})$; see Appendix 8.7. If we normalize $\mathbf{w} + \widehat{\mathbf{h}}$ to be a unit vector, we get the FastICA updating $\mathbf{w}_{\text{new}}$ for $\mathbf{w}$ as

$$\mathbf{w}_{\text{new}} = \frac{\nabla J(\mathbf{w}) - \alpha\mathbf{w}}{||\nabla J(\mathbf{w}) - \alpha\mathbf{w}||} \tag{8.36}$$

with

$$\nabla J(\mathbf{w}) \simeq \frac{1}{N}\sum_{j=1}^{N} G'\Big(\mathbf{w}^{\mathrm{T}}\mathbf{X}(:,j)\Big)\mathbf{X}(:,j) \tag{8.37}$$

and

$$\alpha \simeq \frac{1}{N}\sum_{j=1}^{N} G''\Big(\mathbf{w}^{\mathrm{T}}\mathbf{X}(:,j)\Big). \tag{8.38}$$

Because $\widehat{\mathbf{h}}$ is a Newton step for $\mathbf{w}$ close to $\mathbf{w}_k$, $\mathbf{w}_{\text{new}}$ converges to $\mathbf{w}_k$ with quadratic order (which is the convergence order of the Newton iteration) as soon as $\mathbf{w}$ has come sufficiently close to $\mathbf{w}_k$.

Next, we assume that $\alpha_k + \beta_k > 0$, which corresponds to the case that $\mathbf{w}_k$ is a local minimum point of $J(\mathbf{w})$. Then,

$$\frac{\mathbf{w} + \widehat{\mathbf{h}}}{||\mathbf{w} + \widehat{\mathbf{h}}||} = -\frac{\nabla J(\mathbf{w}) - \alpha\mathbf{w}}{||\nabla J(\mathbf{w}) - \alpha\mathbf{w}||} = -\mathbf{w}_{\text{new}}. \tag{8.39}$$

To make the updating the same for both cases $\alpha_k + \beta_k < 0$ and > 0, we ignore the minus sign in equation (8.39), and let the updating for the case $\alpha_k + \beta_k > 0$ be $\mathbf{w}_{\text{new}}$ as in equation (8.36). This means that with $\alpha_k + \beta_k > 0$, the updated $\mathbf{w}_{\text{new}}$ jumps to the vicinity of $-\mathbf{w}_k$ and again back to the vicinity of $\mathbf{w}_k$ at the next iteration step because $\alpha_k + \beta_k$ is the same for $\mathbf{w}_k$ and $-\mathbf{w}_k$, since $G'(t)$ is an odd and $G''(t)$ an even function, making $\alpha_k = -E\big(G''({\mathbf{w}_k}^{\mathrm{T}}\mathbf{x})\big)$ and $\beta_k = {\mathbf{w}_k}^{\mathrm{T}}E\big(G'({\mathbf{w}_k}^{\mathrm{T}}\mathbf{x})\mathbf{x}\big)$ even functions of $\mathbf{w}$.

Although $\mathbf{w}_{\text{new}}$, with $\alpha_k + \beta_k > 0$, jumps back and forth as the iteration proceeds, it converges to $\mathbf{w}_k$ at a quadratic rate up to the sign.

In summary, we conclude that the FastICA algorithm for a single weight vector, when started at a random initial value $\mathbf{w}_0$, with $||\mathbf{w}_0|| = 1$, first moves at random on the sphere $||\mathbf{w}|| = 1$ until it arrives close to one of the vectors $\pm\mathbf{w}_1, \ldots, \pm\mathbf{w}_n$, and then, up to sign, converges to that $\pm\mathbf{w}_k$ at a quadratic rate. This makes the algorithm robust and fast, and if the data are noiseless and the sample errors in equations (8.37) and (8.38) are ignored, we can say that it always converges to one of the vectors $\pm\mathbf{w}_1, \ldots, \pm\mathbf{w}_n$, and not to any other root of Lagrange's equation (8.24), which the proper Newton iteration algorithm would easily do.

8.4 FastICA for Finding All Weight Vectors ("Symmetric Mode")

FastICA can be used in two ways, either by finding one weight vector after another ("deflation mode") or by finding all weight vectors together iteratively ("symmetric mode") [137]. The latter way is more robust and we describe it here.

We need the *symmetric orthogonalization* of matrices, which is given as follows. If $\mathbf{F}$ is a $p \times q$ matrix with $p \geq q$ and $\text{rank}(\mathbf{F}) = q$, then the symmetrically orthogonalized $\mathbf{F}$, denoted by $\mathbf{F}_{\text{sym}}$, is given by

$$\mathbf{F}_{\text{sym}} = \mathbf{F}(\mathbf{F}^{\text{T}}\mathbf{F})^{-\frac{1}{2}} = \mathbf{U}(:, 1:q)\mathbf{V}^{\text{T}}, \tag{8.40}$$

where $\mathbf{F} = \mathbf{UDV}^{\text{T}}$ is the SVD of $\mathbf{F}$. It follows that the columns of $\mathbf{F}_{\text{sym}}$ are orthonormal, as one can check with the equation $\mathbf{F}_{\text{sym}} = \mathbf{U}(:, 1:q)\mathbf{V}^{\text{T}}$. If $p = q$, then $\mathbf{F}_{\text{sym}}$ is an orthogonal matrix. If the columns of $\mathbf{F}$ are orthonormal, $\mathbf{F}_{\text{sym}} = \mathbf{F}$, and if they are nearly orthonormal, then $\mathbf{F}_{\text{sym}} \simeq \mathbf{F}$.

Let $\mathbf{w}_1, \ldots, \mathbf{w}_n \in \mathbb{R}^n$ be the true weight vectors and $\mathbf{W}_{\text{true}} = [\mathbf{w}_1, \ldots, \mathbf{w}_n]$ the corresponding weight matrix. We search for $\mathbf{W}_{\text{true}}$ in an iterative way. We start by choosing an initial orthogonal matrix $\mathbf{W}_0$, for instance, by using symmetric orthogonalization and setting

$$\mathbf{W}_0 = \mathbf{F}(\mathbf{F}^{\text{T}}\mathbf{F})^{-\frac{1}{2}}, \tag{8.41}$$

where $\mathbf{F} = [\mathbf{v}_1, \ldots, \mathbf{v}_n]$ with random vectors $\mathbf{v}_j \in \mathbb{R}^n$. Next, let the orthogonal matrix $\mathbf{W}$ be the current matrix after $k \geq 0$ iteration steps. We update $\mathbf{W}$ by updating each column $\mathbf{u}_j = \mathbf{W}(:, j)$, $j = 1, \ldots, n$, with one FastICA iteration step (8.36) without normalizing it to a unit vector; the symmetric orthogonalization below takes care of that. So,

$$\mathbf{u}_j^{\text{new}} = \nabla J(\mathbf{u}_j) - \alpha_j \mathbf{u}_j, \tag{8.42}$$

where $\nabla J(\mathbf{u}_j)$ and α_j are given by (8.37) and (8.38) with $\mathbf{w} = \mathbf{u}_j$. This yields $\widetilde{\mathbf{W}} = [\mathbf{u}_1^{\text{new}}, \ldots, \mathbf{u}_n^{\text{new}}]$, and by symmetrically orthogonalizing $\widetilde{\mathbf{W}}$, we get the updating

$$\mathbf{W}_{\text{new}} = \widetilde{\mathbf{W}}(\widetilde{\mathbf{W}}^{\text{T}}\widetilde{\mathbf{W}})^{-\frac{1}{2}}. \tag{8.43}$$

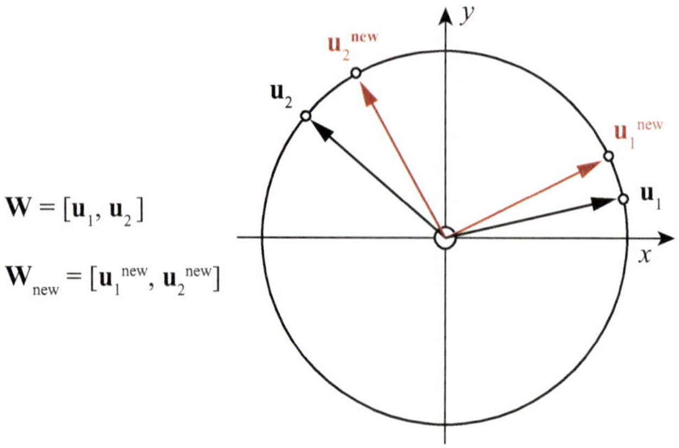

Figure 8.3
The figure illustrates an iteration step in FastICA in $\mathbb{R}^2$. Here, the 2×2 orthogonal matrix $\mathbf{W} = [\mathbf{u}_1, \mathbf{u}_2]$ is the current estimate for the true weight matrix. After upgrading $\mathbf{u}_1$ and $\mathbf{u}_2$ by one FastICA iteration step for a single weight vector, one gets $\widetilde{\mathbf{W}} = [\widetilde{\mathbf{u}}_1, \widetilde{\mathbf{u}}_2]$, and its orthogonal symmetrization yields the upgraded estimate $\mathbf{W}_{\text{new}} = [\mathbf{u}_1^{\text{new}}, \mathbf{u}_2^{\text{new}}]$.

We continue the iteration until $\mathbf{W}_{\text{new}}$, up to the signs of its columns, converges to an orthogonal matrix $\widehat{\mathbf{W}}$, or the number of iterations reaches a preset maximal value. If data $\mathbf{Y}$ are noiseless, $\nabla J(\mathbf{u}_j)$ and α_j contain no sampling error and if FastICA converges, then it converges to $\mathbf{W}_{\text{true}}$ (see figure 8.4).

The failure of convergence is usually due to a too low number N of time points, which causes large sampling errors in $\nabla J(\mathbf{u}_j)$ and α_j.

A practical measure of the convergence is the following distance d_k of two successive estimates $\mathbf{W}_{k-1}$ and $\mathbf{W}_k$ in the iteration steps $k-1$ and k, given by

$$d_k = 1 - \min_i ||\mathbf{C}(i,i)|| \tag{8.44}$$

where $\mathbf{C} = \mathbf{W}_{k-1}{}^{\mathrm{T}}\mathbf{W}_k$. A stopping rule could then be that iterations are continued until $\max(d_{k-1}, d_k) < tol$ or $k > k_{\max}$, where tol is the tolerance, for instance, 10^{-3}, and $k_{\max}$ is the maximal number of iterations.

Note, however, that even a tough stopping rule does not guarantee that the algorithm has converged close to $\mathbf{W}_{\text{true}}$ although it has converged before reaching $k_{\max}$. Therefore, one should always check, possibly by other means, that the resulting separation $\widehat{\mathbf{A}}$ and $\widehat{\mathbf{S}}$ is credible.

An easily available checking method is to start the algorithm at random initial values a few times and compare the obtained weight matrix estimates $\widehat{\mathbf{W}}$, if they are about

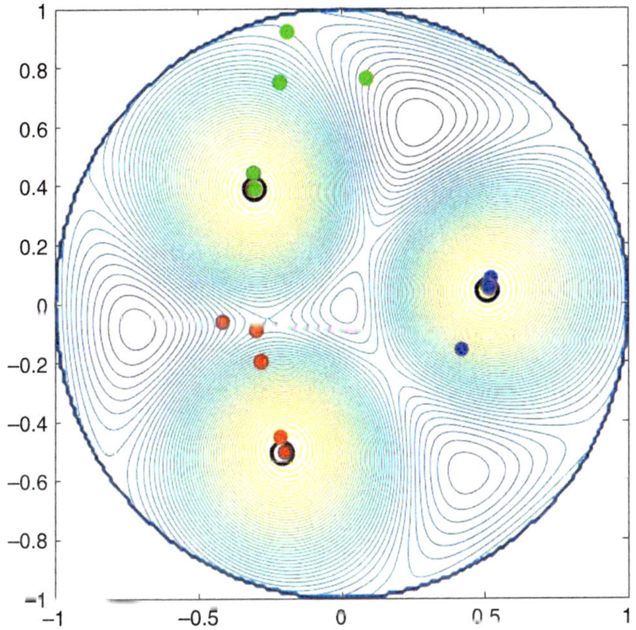

Figure 8.4

Contour plot of $J(\mathbf{w})$ with FastICA iteration. $J(\mathbf{w})$ has been formed with the same parameters as in figure 8.3. The true weight vectors $\mathbf{w}1$, $\mathbf{w}_2$, and $\mathbf{w}_3$ are marked with black circles. For finding $\mathbf{w}1$, $\mathbf{w}_2$, and $\mathbf{w}_3$, FastICA has been applied to the corresponding data matrix $\mathbf{Y}=\mathbf{AS}+\boldsymbol{\varepsilon}$ with 10 percent of additive noise. FastICA has converged in five iteration steps to the true weight vectors. The estimates in iteration steps for $\mathbf{w}_1$ have been marked by red balls, for $\mathbf{w}_2$ by green balls, and for $\mathbf{w}_3$ by blue balls.

equal, up to order and scaling. The approximate equality after several reruns often suggests convergence to the correct solution.

Usually, when failing to converge, FastICA finds several weight vectors correctly, and those are repeatedly found in several randomly started reruns. By comparing the results, one can identify them, say as weight-vector estimates $\widehat{\mathbf{w}}_1, \ldots, \widehat{\mathbf{w}}_j$. This yields time-course and corresponding mixing-matrix column estimates

$$\widehat{\mathbf{S}}(i,:)=\widehat{\mathbf{w}}_i{}^{\mathrm{T}}\mathbf{X} \text{ and } \widehat{\mathbf{A}}(:,i)=N^{-1}\mathbf{Y}_{\text{comp}}\widehat{\mathbf{S}}(i,:)^{\mathrm{T}},\ i=1,\ldots,j,$$

which often contain interesting information about the corresponding hidden sources.

8.4.1 Summary of the FastICA Algorithm

The separation of the $m \times N$ data matrix $\mathbf{Y} = \mathbf{AS} + \boldsymbol{\varepsilon}$ is carried out with FastICA as follows:

- Get the time-centered data matrix $\mathbf{Y}_c$ by

$$\mathbf{Y}_c(j,k) = \mathbf{Y}(j,k) - \frac{1}{N}\sum_{i=1}^{N} \mathbf{Y}(j,i),$$

for $j = 1, \ldots, m$, $k = 1, \ldots, N$.

- Estimate the number n of independent components (hidden sources) and get the compressed data matrix

$$\mathbf{Y}_{\text{comp}} = \mathbf{U}(:, 1:n)\,\mathbf{diag}\,(d_1, \ldots, d_n)\,\mathbf{V}(:, 1:n)^{\mathrm{T}},$$

where $\mathbf{UDV}^{\mathrm{T}} = \mathbf{Y}_c$ is the SVD of $\mathbf{Y}_c$.

- Whiten $\mathbf{Y}_{\text{comp}}$ to get the whitened data matrix

$$\mathbf{X} = \sqrt{N}\,\mathbf{V}(:, 1:n)^{\mathrm{T}},$$

where the orthonormal matrix $\mathbf{V}$ is as in the previous step.

- Choose the contrast function $G(t)$. If $G(t) = e^{-\frac{1}{2}t^2}$, then

$$G'(t) = -t\,e^{-\frac{1}{2}t^2} \text{ and } G''(t) = (t^2 - 1)e^{-\frac{1}{2}t^2}.$$

If $G(t) = \log\big(\cosh(t)\big)$, then

$$G'(t) = \tanh(t) \text{ and } G''(t) = \frac{1}{\cosh^2(t)} = 1 - \tanh^2(t).$$

- Choose an initial $\mathbf{W}_0$, for instance,

$$\mathbf{W}_0 = \mathbf{F}\,(\mathbf{F}^{\mathrm{T}}\mathbf{F})^{-\frac{1}{2}},$$

where $\mathbf{F} = [\mathbf{v}_1, \ldots, \mathbf{v}_n]$ and $\mathbf{v}_j \in \mathbb{R}^n$ are random vectors. Start the iteration.

- If $\mathbf{W}$ is the current estimate, then get the updated estimate $\mathbf{W}_{\text{new}}$ by updating each $\mathbf{u}_j = \mathbf{W}(:, j)$, $j = 1, \ldots, n$, by

$$\mathbf{u}_j^{\text{new}} = \nabla J_j - \alpha_j \mathbf{u}_j,$$

where

$$\nabla J_j = \frac{1}{N}\sum_{i=1}^{N} G'\Big(\mathbf{u}_j{}^{\mathrm{T}}\mathbf{X}(:, j)\Big)\mathbf{X}(:, j),$$

$$\alpha_j = \frac{1}{N}\sum_{i=1}^{N} G''\Big(\mathbf{u}_j{}^{\mathrm{T}}\mathbf{X}(:, j)\Big).$$

- Set $\widetilde{\mathbf{W}} = [\mathbf{u}_1^{\text{new}}, \ldots, \mathbf{u}_n^{\text{new}}]$, and get

$$\mathbf{W}_{\text{new}} = \widetilde{\mathbf{W}}\left(\widetilde{\mathbf{W}}^{\text{T}}\widetilde{\mathbf{W}}\right)^{-\frac{1}{2}}.$$

- Continue the iteration until the stopping condition is reached, and set $\widehat{\mathbf{W}}$ to be the last $\mathbf{W}_{\text{new}}$. As a stopping rule, for instance, use that at the end of the previous subsection.
- As estimates for $\mathbf{A}$ and $\mathbf{S}$, up to order and scaling, get

$$\widehat{\mathbf{S}} = \widehat{\mathbf{W}}^{\text{T}}\mathbf{X} \text{ and } \widehat{\mathbf{A}} = \frac{1}{N}\mathbf{Y}_{\text{comp}}\widehat{\mathbf{S}}^{\text{T}}.$$

8.4.2 ICA with Nonstationary Multi-Trial Data

ICA algorithms are, in principle, designed for stationary measurement data. In EEG and MEG, nonstationary data, however, are quite common, and with such data, ICA often fails.

If EEG or MEG is recorded in response to a specific stimulus or event, the measurement data are called *evoked (or event-related)* data. The stimulus can be a simple auditory tone or visual pattern, a sensory excitation, or a TMS (transcranial magnetic stimulation [147]) pulse. The stimulus is usually repeated several times, leading to *multi-trial data*. A *trial* refers to a measurement time series (i.e., measurements at several successive time points) after one single stimulus. The multi-trial data are usually collected during one measurement session. If the zero point of the time scale is the instant of the stimulus, the resulting measurement data are *time-locked (or stimulus-locked)*. The evoked responses typically yield *nonstationary* data. In this subsection, we discuss how ICA can be used with nonstationary multi-trial data.

Let the evoked EEG or MEG responses be measured with N_{c} channels (or sensors) at N_{t} time points and in N_{tr} trials. The resulting multi-trial data are collected in a *3-D array* $\mathbf{X}_{3\text{D}}$, where an array element $\mathbf{X}_{3\text{D}}(i, j, k)$ is the reading in the ith channel at the jth time point and in the kth trial. For 3-D array notation, see section 4.1.

We assume that the measurements are due to N_{h} *hidden (latent) components*, and the data obey the following linear model:

$$\mathbf{X}_{3\text{D}}(:, :, k) = \mathbf{A}\,\mathbf{S}_{3\text{D}}(:, :, k), \tag{8.45}$$

for $k = 1, \ldots, N_{\text{tr}}$, where $\mathbf{A}$ is a *mixing matrix*, not depending on time or trials, and $\mathbf{S}_{3\text{D}}$ is a *3-D time-course (or waveform) array* such that $\mathbf{A}(:, m)$ is the *topography* of the mth component and the row-vector $\mathbf{S}_{3\text{D}}(m, :, k)$ is the *time course* of the mth component in the kth trial (see figure 8.5).

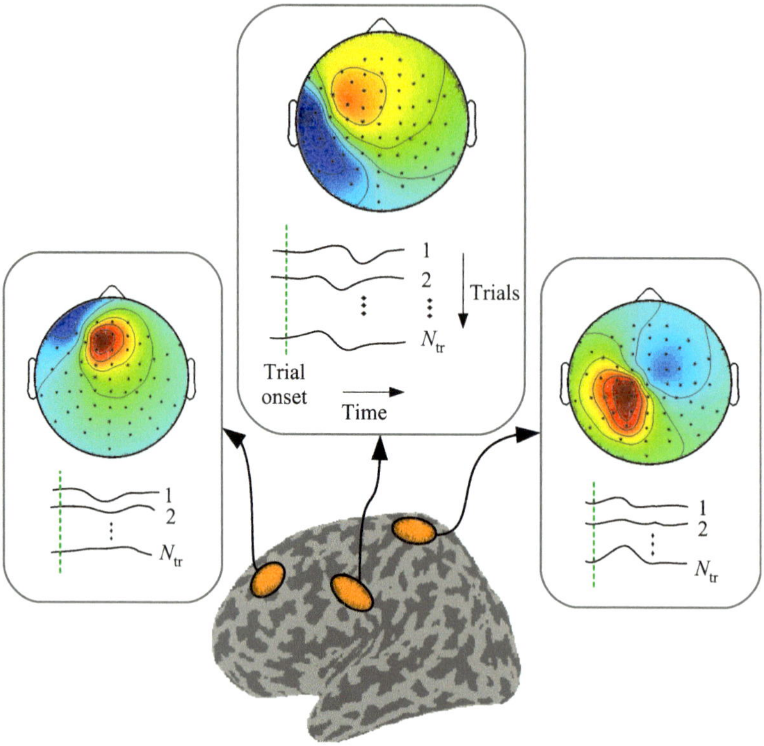

Figure 8.5
EEG multi-trial data concepts. Independently acting areas (corresponding to the hidden components) are shown on the cortex. They also may be distributed and overlapping. Each activity area corresponds to an EEG topography (mixing matrix column) and an activity time course (time course matrix row). All topographies are fixed, while time courses can change. The contour plots of topographies and the graphs of various time courses are presented in the subfigures. The vertical dashed lines represent the moment of the stimulus.

We next present three ways to preprocess the data $\mathbf{X}_{3D}$ into 2-D matrices: *concatenating, averaging, and mean-subtracting*. The two first ones are traditional ways of preprocessing multi-trial data, while the third one is more recent [204] and best suited to be used with ICA.

In the first way of preprocessing, we concatenate the trial data matrices $\mathbf{X}_{3D}(:,:,k)$ into an $N_c \times (N_t N_{tr})$ 2-D matrix

$$\mathbf{X}_{conc} = [\mathbf{X}_{3D}(:,:,1), \ldots, \mathbf{X}_{3D}(:,:,N_{tr})]. \tag{8.46}$$

The second way is to average the matrices $\mathbf{X}_{3D}(:,:,k)$ over trials $k = 1, \ldots, N_{tr}$, which yields the averaged $N_c \times N_t$ matrix

$$\mathbf{X}_{\text{ave}} = \frac{1}{N_{\text{tr}}} \sum_{k=1}^{N_{\text{tr}}} \mathbf{X}_{3\text{D}}(:,:,k), \tag{8.47}$$

that is, $\mathbf{X}_{\text{ave}}(i,j) = N_{\text{tr}}^{-1} \sum_{k=1}^{N_{\text{tr}}} \mathbf{X}_{3\text{D}}(i,j,k)$. Averaging multi-trial data is often used to suppress noise in the data, but it has the drawback of removing information contained in the variability over trials.

The third way is to subtract the average (or mean) $\mathbf{X}_{\text{ave}}$ from each $\mathbf{X}_{3\text{D}}(:,:,k)$ and to concatenate the differences into an $N_c \times (N_t N_{\text{tr}})$ 2-D mean-subtracted data matrix $\mathbf{X}_{\text{ms}}$ as follows:

$$\mathbf{X}_{\text{ms}} = [\mathbf{X}_{3\text{Dms}}(:,:,1), \ldots, \mathbf{X}_{3\text{Dms}}(:,:,N_{\text{tr}})], \quad \text{where} \tag{8.48}$$

$$\mathbf{X}_{3\text{Dms}}(:,:,k) = \mathbf{X}_{3\text{D}}(:,:,k) - \mathbf{X}_{\text{ave}}(:,:), \quad k = 1, \ldots, N_{\text{tr}}. \tag{8.49}$$

Accordingly, in the mean-subtracting, the average (mean) of the data over trials is removed while the information in the variability over trials is preserved in the preprocessed data.

In modifying $\mathbf{X}_{3\text{D}}$ in one of the above three ways, due to equation (8.45), $\mathbf{S}_{3\text{D}}$ also will be modified in the same way, and we get

$$\mathbf{X}_{kind} = \mathbf{A}\,\mathbf{S}_{kind} \ \text{ for } kind = \text{conc, ave, ms}, \tag{8.50}$$

where

$$\mathbf{S}_{\text{conc}} = [\mathbf{S}_{3\text{D}}(:,:,1), \ldots, \mathbf{S}_{3\text{D}}(:,:,N_{\text{tr}})], \tag{8.51}$$

$$\mathbf{S}_{\text{ave}} = \frac{1}{N_{\text{tr}}} \sum_{k=1}^{N_{\text{tr}}} \mathbf{S}_{3\text{D}}(:,:,k), \tag{8.52}$$

$$\mathbf{S}_{\text{ms}} = [\mathbf{S}_{3\text{Dms}}(:,:,1), \ldots, \mathbf{S}_{3\text{Dms}}(:,:,N_{\text{tr}})], \quad \text{where} \tag{8.53}$$

$$\mathbf{S}_{3\text{Dms}}(:,:,k) = \mathbf{S}_{3\text{D}}(:,:,k) - \mathbf{S}_{\text{ave}}(:,:), \quad k = 1, \ldots, N_{\text{tr}}. \tag{8.54}$$

For convenience, we assume that $\mathbf{X}_{kind}$ in equation (8.50) is row-centered, that is, the mean of every row of $\mathbf{X}_{kind}$ has zero mean, which implies that $\mathbf{S}_{kind}$ is also row-centered.

We want to apply ICA to the preprocessed data matrices. The evoked data are nonstationary and that property is (usually) inherited by the preprocessed data matrices so that we may not assume that they fulfill the ICA assumption. Therefore, instead of the ICA assumption, we assume that the original multi-trial data $\mathbf{X}_{3\text{D}}$ satisfies a weaker condition, which we call the *momentary independence* property, and define it in statistical terms as follows.

Consider the data $\mathbf{X}_{3D}(:, j, :)$ at a fixed time instant j, $1 \leq j \leq N_t$. We may assume that $\mathbf{X}_{3D}(:, j, :)$ is due to a hidden random N_h-vector $\mathbf{s}_j$ so that vectors $\mathbf{S}_{3D}(:, j, k)$, $k = 1, \ldots, N_{tr}$ are random samples of $\mathbf{s}_j$. We say that the data $\mathbf{X}_{3D}$ have the momentary independence property if the components $s_{j,1}, \ldots, s_{j,N_h}$ of $\mathbf{s}_j$ are statistically independent.

In practice, the momentary independence property of $\mathbf{X}_{3D}$ means that the underlying active neuronal sources, corresponding to the hidden components, are acting independently at every separate moment of time. It is somewhat counterintuitive that the momentary independence property of $\mathbf{X}_{3D}$ does not imply that the ICA assumption is valid to any of the modified data matrices $\mathbf{X}_{conc}$, $\mathbf{X}_{ave}$, and $\mathbf{X}_{ms}$; see Metsomaa et al. [204].

For ICA, the importance of the momentary independence is that if $\mathbf{X}_{3D}$ has this property, then FastICA, when applied to the mean-subtracted data $\mathbf{X}_{ms} = \mathbf{A}\,\mathbf{S}_{ms}$, can separate $\mathbf{X}_{ms}$ and find estimates $\widehat{\mathbf{A}}$ and $\widehat{\mathbf{S}}_{ms}$ for $\mathbf{A}$ and $\mathbf{S}_{ms}$ (up to the order and scaling of columns of $\mathbf{A}$ and rows of $\mathbf{S}_{ms}$). This happens, as shown in Metsomaa et al. [204], because for $\mathbf{X}_{ms}$, the momentary independence implies another property, called *null conditional mean* (NCM), which is a sufficient condition for FastICA for separation [204]. Numerical experiments suggest that NCM is also sufficient for several other commonly used ICA algorithms.

Because momentary independence does not imply NCM for $\mathbf{X}_{conc}$ and $\mathbf{X}_{ave}$ [204], their separation with ICA usually fails.

After having obtained estimates $\widehat{\mathbf{A}}$ and $\widehat{\mathbf{S}}_{ms}$ in separating the mean-subtracted data $\mathbf{X}_{ms} = \mathbf{A}\,\mathbf{S}_{ms}$, one can estimate $\mathbf{S}_{ave}$ by

$$\widehat{\mathbf{S}}_{ave} = \widehat{\mathbf{A}}^{\dagger}\,\mathbf{X}_{ave}, \tag{8.55}$$

or estimate the variance of the component $\mathbf{S}_{3D}(i, j, k)$ over $k = 1, \ldots, N_{tr}$, by

$$\widehat{\mathbf{S}}_{var}(i, j) = \frac{1}{N_{tr}} \sum_{k=1}^{N_{tr}} \widehat{\mathbf{S}}_{ms}(i, j, k)^2 \quad \text{for} \quad i = 1, \ldots, N_h,\ j = 1, \ldots, N_t, \tag{8.56}$$

where

$$\widehat{\mathbf{S}}_{ms}(:, :, k) = \widehat{\mathbf{A}}^{\dagger}\,\mathbf{X}_{3D}(:, :, k), \tag{8.57}$$

$k = 1, \ldots, N_{tr}$. The estimates $\widehat{\mathbf{S}}_{ave}$ and $\widehat{\mathbf{S}}_{var}(i, j)$ can be used in analyzing the time courses.

We complete this subsection by noting that the mean subtraction also effectively suppresses such strong artifacts in data that vary little over the trials, that is, they are phase-locked to the stimulus, such as muscular artifacts in TMS-evoked data. If those artifacts are not sufficiently suppressed, they may prevent the use of ICA.

8.5 Appendix: Weight Vectors Are among the Roots of the Lagrange Equation

Let $\mathbf{w}_j$ be one of the weight vectors $\mathbf{w}_1, \ldots, \mathbf{w}_n$. We want to show that

$$\nabla J(\mathbf{w}_j) + \beta \mathbf{w}_j = 0 \tag{8.58}$$

for some $\beta \in \mathbb{R}$. This equation is equivalent to

$$\nabla J(\mathbf{w}_j) = -\beta \mathbf{w}_j,$$

and that is equivalent to

$$\mathbf{v}^{\mathrm{T}} \nabla J(\mathbf{w}_j) = 0 \ \text{ for all } \ \mathbf{v} \in \mathbb{R}^n \text{ with } \mathbf{v}^{\mathrm{T}} \mathbf{w}_j = 0 \tag{8.59}$$

Because $\mathbf{w}_1, \ldots, \mathbf{w}_n$ form an orthonormal basis of $\mathbb{R}^n$, equation (8.59) is equivalent to

$$\mathbf{w}_k{}^{\mathrm{T}} \nabla J(\mathbf{w}_j) = 0 \ \text{ for all } k \neq j.$$

Using equations (8.21) and (8.25), we get

$$\begin{aligned}
\mathbf{w}_k{}^{\mathrm{T}} \nabla J(\mathbf{w}_j) &= \mathbf{w}_k{}^{\mathrm{T}} E\Big(G'(\mathbf{w}_j{}^{\mathrm{T}} \mathbf{x})\, \mathbf{x}\Big) \\
&= E\Big(G'(\mathbf{w}_j{}^{\mathrm{T}} \mathbf{x})\, \mathbf{w}_k{}^{\mathrm{T}} \mathbf{x}\Big) = E\Big(G'(s_j)\, s_k\Big) \\
&= E\Big(G'(s_j)\Big)\, E(s_k) = 0,
\end{aligned}$$

where we have used the fact that s_j are s_k are statistically independent, and so are $G'(s_j)$ and s_k, as well as $E(s_k) = 0$. The proof is completed.

8.6 Appendix: Hessian H(w) of Function J(w)

We want to show that for any weight vector $\mathbf{w}_k$, we have

$$\mathbf{H}(\mathbf{w}_k)\, \mathbf{h} = \alpha_k \mathbf{h} \ \text{ if } \mathbf{h}^{\mathrm{T}} \mathbf{w}_k = 0, \tag{8.60}$$

where $\alpha_k = E(G''(\mathbf{w}_k{}^{\mathrm{T}} \mathbf{x}))$.

By equations (8.30) and (8.21), and $\mathbf{h}^{\mathrm{T}} \mathbf{w}_k = 0$, we get

$$\begin{aligned}
\mathbf{H}(\mathbf{w}_k)\, \mathbf{h} &= E\Big(G''(\mathbf{w}_k{}^{\mathrm{T}} \mathbf{x})\, \mathbf{x}\, \mathbf{x}^{\mathrm{T}}\Big)\, \mathbf{h} = \sum_{i=1}^{n} \sum_{j=1}^{n} E\Big(G''(s_k)\, s_i\, s_j\Big)\, \mathbf{w}_i \mathbf{w}_j{}^{\mathrm{T}} \mathbf{h} \\
&= \sum_{i} \sum_{j \neq k} E\Big(G''(s_k)\, s_i s_j\Big)\, \mathbf{w}_i \mathbf{w}_j{}^{\mathrm{T}} \mathbf{h} \\
&= \sum_{i} \sum_{j \neq k} \delta_{i,j}\, E\Big(G''(s_k)\Big)\, E(s_i^2)\, \mathbf{w}_i \mathbf{w}_i{}^{\mathrm{T}} \mathbf{h} \\
&= \sum_{i \neq k} \alpha_k \mathbf{w}_i \mathbf{w}_i{}^{\mathrm{T}} \mathbf{h} = \alpha_k \mathbf{h},
\end{aligned}$$

where $\delta_{i,j}=1$ if $i=j$ and $\delta_{i,j}=0$ if $i\neq j$. Above, we have used the equation that if $j\neq k$, then

$$E\Big(G''(s_k)\, s_i\, s_j\Big)=\delta_{i,j}\, E\Big(G''(s_k)\Big)E(s_i^2)=\delta_{i,j}\,\alpha_k,$$

which is due to the statistical independence of $s_1,\ldots,s_n$ and the equations $E(s_i^2)=1$ and $E(s_i)=0$ for $i=1,\ldots,n$. We also used the fact that $\mathbf{w}_1,\ldots,\mathbf{w}_n$ form an orthogonal basis of $\mathbb{R}^n$. The proof of equation (8.60) is completed.

8.7 Appendix: Local Maxima and Minima of $J(\mathrm{w})$ for $\mathrm{w}=\mathrm{w}_k$

Consider $\mathbf{w},\ \mathbf{h}\in\mathbb{R}^n$ with $||\mathbf{w}||=1$ and $\mathbf{h}^{\mathrm{T}}\mathbf{w}=0$. We get

$$\begin{aligned}\frac{\mathbf{w}+\mathbf{h}}{||\mathbf{w}+\mathbf{h}||}&=\big(1-\frac{1}{2}||\mathbf{h}||^2+\mathcal{O}(||\mathbf{h}||^3)\big)(\mathbf{w}+\mathbf{h})\\&=\mathbf{w}+\mathbf{h}-\frac{1}{2}||\mathbf{h}||^2\mathbf{w}+\mathcal{O}(||\mathbf{h}||^3),\end{aligned}$$

where notation $\mathcal{O}(||\mathbf{h}||^3)$ means that $||\mathbf{h}||^{-3}\mathcal{O}(||\mathbf{h}||^3)$ remains bounded as $||\mathbf{h}||$ tends to 0. This yields the differential expansion

$$J\Big(\frac{\mathbf{w}+\mathbf{h}}{||\mathbf{w}+\mathbf{h}||}\Big)=J(\mathbf{w})+\nabla J(\mathbf{w})^{\mathrm{T}}\Big(\mathbf{h}-\frac{1}{2}||\mathbf{h}||^2\mathbf{w}\Big)+\frac{1}{2}\mathbf{h}^{\mathrm{T}}\mathbf{H}(\mathbf{w})\,\mathbf{h}+\mathcal{O}(||\mathbf{h}||^3). \tag{8.61}$$

Now, let $\mathbf{w}=\mathbf{w}_k$ be one of the weight vectors. Then, by Appendix 8.5, $\mathbf{w}_k$ is a root of equation $\nabla J(\mathbf{w})+\beta\,\mathbf{w}=0$, that is,

$$\nabla J(\mathbf{w}_k)=-\beta_k\mathbf{w}_k,$$

and so

$$\beta_k=-\mathbf{w}_k{}^{\mathrm{T}}\nabla J(\mathbf{w}_k)\ \text{ and }\ \mathbf{h}^{\mathrm{T}}\nabla J(\mathbf{w}_k)=0.$$

By Appendix 8.6,

$$\mathbf{H}(\mathbf{w}_k)\,\mathbf{h}=\alpha_k\mathbf{h}\ \text{ with }\ \alpha_k=E\Big(G''(\mathbf{w}_k{}^{\mathrm{T}}\mathbf{x})\Big).$$

Then, by equation (8.61), we get

$$\begin{aligned}J\Big(\frac{\mathbf{w}_k+\mathbf{h}}{||\mathbf{w}_k+\mathbf{h}||}\Big)&=J(\mathbf{w}_k)+\frac{1}{2}\mathbf{h}^{\mathrm{T}}\alpha_k\mathbf{h}+\frac{1}{2}\beta_k||\mathbf{h}||^2+\mathcal{O}(||\mathbf{h}||^3)\\&=J(\mathbf{w}_k)+\frac{1}{2}\mathbf{h}^{\mathrm{T}}(\alpha_k+\beta_k)\mathbf{h}+\mathcal{O}(||\mathbf{h}||^3).\end{aligned}$$

This implies that $\mathbf{w}_k$ is a local maximum point of $J(\mathbf{w})$ if $\alpha_k+\beta_k<0$, and it is a local minimum point if $\alpha_k+\beta_k>0$.

9 Blind Source Separation by Joint Diagonalization

The blind source separation (BSS) by *joint diagonalization* (JD) is a widely used separating method, which is different from ICA. In practice, an exact JD is possible only in a few special cases, and so the various numerical algorithms for JD only perform an *approximate joint diagonalization* (AJD). Popular BSS methods using AJD are, for instance, SOBI [28, 29, 137], where AJD is applied to autocorrelation matrices, and JADE [48, 137], where AJD is applied to fourth-order cumulants. Both methods are designed for a single stationary time-series measurement.

Here we present a BSS method, called *momentary-uncorrelated component analysis* (MUCA) [205], which makes use of AJD and is designed for nonstationary event-related multi-trial data, as in section 8.4.2.

MUCA decomposes EEG/MEG multi-trial data into *momentary-uncorrelated components*, that is, into components that are uncorrelated at each time instant after the stimulus. MUCA is based on the AJD of *data covariance matrices* at separate instants of the multi-trial data (covariance formed of the variance over trials). These time-dependent covariance matrices are called the *momentary covariance matrices* (MCMs).

The momentary-uncorrelated components can be recovered by MUCA from the multi-trial nonstationary data provided that the related MCMs have enough variability in time. The input into the MUCA algorithm are the MCMs, to which AJD is applied.

The MUCA algorithm also allows a two-step filtering procedure, which reduces the measurement and sampling noise, and often improves the performance of MUCA.

9.1 MUCA Algorithm

We consider multi-trial data measured at N_c channels (sensors) at N_t successive time points and in N_{tr} trials, as in section 8.4.2 for ICA with multi-trial data. The data are again collected in a three-dimensional array X, where the array element $X(i, j, k)$ is the signal in the ith channel at the jth instant in the kth trial. We assume that the measurement data obey a linear model where $\mathbf{X}$ is due to N_h *hidden components* and additive noise so that

$$\mathbf{X}(:,j,k) = \mathbf{A}\,\mathbf{S}(:,j,k) + \boldsymbol{\varepsilon}(:,j,k) \tag{9.1}$$

for $1 \le j \le N_t$, $1 \le k \le N_{tr}$, where $\mathbf{A}$ is an $N_c \times N_h$ *mixing matrix*, independent of time and trials; $\mathbf{S}$ is an $N_h \times N_t \times N_{tr}$ *time-course array*; and $\boldsymbol{\varepsilon}$ is the noise array. The columns $\mathbf{A}(:,l)$, $l = 1,\ldots,N_h$ are the *topographies* of the hidden components.

We assign the following statistical model to the measurement data $\mathbf{X}$. For a fixed time-instant (index) j, we denote $\mathbf{X}_j = \mathbf{X}(:,j,:)$ and $\mathbf{S}_j = \mathbf{S}(:,j,:)$, and so by equation (9.1),

$$\mathbf{X}_j = \mathbf{A}\,\mathbf{S}_j + \boldsymbol{\varepsilon}(:,j,k) \quad \text{for} \quad j = 1,\ldots,N_t. \tag{9.2}$$

We assume that the columns $\mathbf{S}_j(:,k)$, $k = 1,\ldots,N_{tr}$, consist of random samples of a *momentary random vectors* $\mathbf{s}_j$, and accordingly, the columns $\mathbf{X}_j(:,k)$, $k = 1,\ldots,N_{tr}$, consist of random samples of random vectors

$$\mathbf{x}_j = \mathbf{A}\,\mathbf{s}_j, \; j = 1,\ldots,N_t, \tag{9.3}$$

where we have ignored the noise.

Next, we make the basic assumption on which MUCA is based. We assume that the covariance matrices $\text{cov}(\mathbf{s}_j)$ of the hidden random vectors $\mathbf{s}_j$, $j = 1,\ldots,N_t$, are *diagonal*, that is, the components $s_{j,1},\ldots,s_{j,N_h}$ of $\mathbf{s}_j$ are uncorrelated. This is a weaker assumption than the momentary independence assumption in the previous section, because statistical independent components are always uncorrelated (but not reversely). Also, uncorrelatedness is easier to define and test than statistical independence.

Consider the (unnormalized) covariance matrices

$$\mathbf{C}_j = \text{Cov}(\mathbf{X}_j) = \frac{1}{N_{tr}} \sum_{k=1}^{N_{tr}} \left(\mathbf{X}_j(:,k) - \boldsymbol{\mu}_j\right)\left(\mathbf{X}_j(:,k) - \boldsymbol{\mu}_j\right)^{\mathrm{T}} \tag{9.4}$$

with the mean (over trials)

$$\boldsymbol{\mu}_j = \frac{1}{N_{tr}} \sum_{k=1}^{N_{tr}} \mathbf{X}_j(:,k), \tag{9.5}$$

and

$$\mathbf{D}_j = \text{Cov}(\mathbf{S}_j) = \sum_{k=1}^{N_{tr}} \left(\mathbf{S}_j(:,k) - \boldsymbol{\sigma}_j\right)\left(\mathbf{S}_j(:,k) - \boldsymbol{\sigma}_j\right)^{\mathrm{T}} \tag{9.6}$$

with the mean

$$\sigma_j = \frac{1}{N_{\mathrm{tr}}} \sum_{k=1}^{N_{\mathrm{tr}}} \mathbf{S}_j(:, k). \tag{9.7}$$

The normalized covariance matrix $\mathbf{D}_j$ approximates $\mathrm{cov}(\mathbf{s}_j)$, and so we may assume that matrices $\mathbf{D}_j$ also are (approximately) diagonal. Due to equation (9.1) (again noise is ignored),

$$\mathbf{C}_j = \mathrm{Cov}(\mathbf{X}_j) = \mathbf{A}\,\mathrm{Cov}(\mathbf{S}_j)\,\mathbf{A}^{\mathrm{T}} = \mathbf{A}\,\mathbf{D}_j\mathbf{A}^{\mathrm{T}} \tag{9.8}$$

for $j = 1, \ldots, N_{\mathrm{t}}$.

In the following, we let $\mathbf{A}$ be an $N_{\mathrm{h}} \times N_{\mathrm{h}}$ square matrix, that is, $N_{\mathrm{c}} = N_{\mathrm{h}}$; the case $N_{\mathrm{c}} > N_{\mathrm{h}}$ is treated later below. Then,

$$\mathbf{D}_j = \mathbf{A}^{-1}\mathbf{C}_j(\mathbf{A}^{\mathrm{T}})^{-1}, \; j = 1, \ldots, N_{\mathrm{t}}, \tag{9.9}$$

and so $\mathbf{A}^{-1}$ is the *joint diagonalizer* of matrices $\mathbf{C}_1, \ldots, \mathbf{C}_{N_{\mathrm{t}}}$. However, due to noise and other possible inaccuracies in equation (9.9), $\mathbf{A}^{-1}$ is usually only an approximate joint diagonalizer of $\mathbf{C}_1, \ldots, \mathbf{C}_{N_{\mathrm{t}}}$.

To find an estimate $\widehat{\mathbf{A}}$ for $\mathbf{A}$ on the basis of equation (9.8), we apply an appropriate AJD algorithm to matrices $\mathbf{C}_1, \ldots, \mathbf{C}_{N_{\mathrm{t}}}$ and let the algorithm yield $\widehat{\mathbf{A}}$ up to scaling and order of the columns of $\widehat{\mathbf{A}}$ (it cannot do it more specifically due to equation (9.2)). Thereafter, an estimate for $\mathbf{S}_j$ (up to scaling and order of its rows) is obtained by

$$\widehat{\mathbf{S}}_j = \widehat{\mathbf{A}}^{-1}\mathbf{X}_j, \quad j = 1, \ldots, N_{\mathrm{t}}, \tag{9.10}$$

and we have obtained the desired separations $\mathbf{X}_j \simeq \widehat{\mathbf{A}}\,\widehat{\mathbf{S}}_j$.

There are several AJD algorithms available like FFDiag, ACDC, and J-D, described in [290, 330, 334], respectively. These three algorithms assume that matrices $\mathbf{C}_1, \ldots, \mathbf{C}_{N_{\mathrm{t}}}$ are symmetric (not necessarily positive definite) square matrices that vary in time, that is, the data are not stationary. They are robust with respect to mild deviations from diagonality in matrices $\mathbf{D}_j$ and inaccuracies in equation (9.8). In Appendix 9.3, we present the FFDiag algorithm, which in numerical simulations has shown to be best suited for use in MUCA.

For multi-trial data, often the average time-course matrix

$$\mathbf{S}_{\mathrm{ave}} = \frac{1}{N_{\mathrm{tr}}} \sum_{k=1}^{N_{\mathrm{tr}}} \mathbf{S}(:, :, k) \tag{9.11}$$

is considered to present a trial-independent average activity of the hidden components. After having obtained the estimates $\widehat{\mathbf{S}}_j$, we get an estimate $\widehat{\mathbf{S}}_{\mathrm{ave}}$ for $\mathbf{S}_{\mathrm{ave}}$ by

$$\widehat{\mathbf{S}}_{\text{ave}}(:,j) = \frac{1}{N_{\text{tr}}} \sum_{k=1}^{N_{\text{tr}}} \widehat{\mathbf{S}}_j(:,k), \tag{9.12}$$

$j = 1, \ldots, N_{\text{t}}$.

We complete this section by showing how to treat the case where the $N_{\text{c}} \times N_{\text{h}}$ matrix $\mathbf{A}$ is not square, that is, $N_{\text{c}} > N_{\text{h}}$. Consider the concatenated matrix

$$\mathbf{X}_{\text{conc}} = [\mathbf{X}_1, \ldots, \mathbf{X}_{N_{\text{t}}}] = [\mathbf{A}\,\mathbf{S}_1, \ldots, \mathbf{A}\,\mathbf{S}_{N_{\text{t}}}] = \mathbf{A}\,[\mathbf{S}_1, \ldots, \mathbf{S}_{N_{\text{t}}}]. \tag{9.13}$$

Let

$$\mathbf{X}_{\text{conc}}\mathbf{X}_{\text{conc}}{}^{\text{T}} = \mathbf{U}\,\mathbf{D}\,\mathbf{V}^{\text{T}} \tag{9.14}$$

be the SVD of $\mathbf{X}_{\text{conc}}\mathbf{X}_{\text{conc}}{}^{\text{T}}$. Then,

$$\text{span}(\mathbf{A}) = \text{span}(\mathbf{X}_{\text{conc}}\mathbf{X}_{\text{conc}}{}^{\text{T}}) = \text{span}(\mathbf{U}(:, 1:N_{\text{h}})). \tag{9.15}$$

Because $\mathbf{U}(:, 1:N_{\text{h}})\,\mathbf{U}(:, 1:N_{\text{h}})^{\text{T}}$ is an orthogonal projection onto span($\mathbf{A}$), we have

$$\mathbf{U}(:, 1:N_{\text{h}})\,\mathbf{U}(:, 1:N_{\text{h}})^{\text{T}}\mathbf{A} = \mathbf{A}. \tag{9.16}$$

This implies that $\text{rank}(\mathbf{U}(:, 1:N_{\text{h}})^{\text{T}}\mathbf{A}) = N_{\text{h}}$, because $\text{rank}(\mathbf{U}(:, 1:N_{\text{h}})) = N_{\text{h}} = \text{rank}(\mathbf{A})$. We denote

$$\widetilde{\mathbf{X}}_j = \mathbf{U}(:, 1:N_{\text{h}})^{\text{T}}\mathbf{X}_j = \mathbf{U}(:, 1:N_{\text{h}})^{\text{T}}\mathbf{A}\,\mathbf{S}_j = \mathbf{B}\,\mathbf{S}_j, \quad \text{where} \tag{9.17}$$

$$\mathbf{B} = \mathbf{U}(:, 1:N_{\text{h}})^{\text{T}}\mathbf{A} \tag{9.18}$$

is an $N_{\text{h}} \times N_{\text{h}}$ square matrix, which is invertible because $\text{rank}(\mathbf{B}) = N_{\text{h}}$. We next apply AJD to the sequence

$$\widetilde{\mathbf{C}}_j = \widetilde{\mathbf{X}}_j\widetilde{\mathbf{X}}_j{}^{\text{T}} = \mathbf{B}\,\mathbf{S}_j\mathbf{S}_j{}^{\text{T}}\mathbf{B}^{\text{T}} = \mathbf{B}\,\mathbf{D}_j\mathbf{B}^{\text{T}}, \tag{9.19}$$

$j = 1, \ldots, N_{\text{t}}$. The chosen algorithm yields an estimate $\widehat{\mathbf{B}}$ for $\mathbf{B}$. The estimates $\widehat{\mathbf{A}}$ and $\widehat{\mathbf{S}}_j$ are then, by equations (9.16) to (9.18), obtained as

$$\widehat{\mathbf{A}} = \mathbf{U}(:, 1:N_{\text{h}})\,\widehat{\mathbf{B}} \tag{9.20}$$

$$\widehat{\mathbf{S}}_j = \widehat{\mathbf{B}}^{-1}\widetilde{\mathbf{X}}_j,\ j = 1, \ldots, N_{\text{t}}. \tag{9.21}$$

In practice, we usually do not know N_{h}, but we can estimate it with the SVD (9.14) ($\mathbf{X}_{\text{conc}}$ formed with noisy $\mathbf{X}_j$) as that index after which the sequence of singular values $d_1 \geq \cdots \geq d_{N_{\text{t}}}$, that is, the diagonal elements of $\mathbf{D}$, settles to the constant level of noise.

9.2 Two-Step Filtering for MUCA Algorithm

In MUCA, AJD is applied to momentary covariance matrices (MCMs) $\mathbf{C}_j$, which often contain a lot of noise and that may deteriorate the outcome of MUCA seriously. Therefore, it is often useful to apply the following two-step filtering to the MCMs before applying AJD.

Let us consider again the multi-trial data (9.1). If we denote that $\mathbf{X}_j = \mathbf{X}(:, j, :)$, $\mathbf{S}_j = \mathbf{S}(:, j, :)$, and $\boldsymbol{\varepsilon}_j = \boldsymbol{\varepsilon}(:, j, :)$, $j = 1, \ldots, N_t$, equation (9.1) gets the form

$$\mathbf{X}_j = \mathbf{A}\,\mathbf{S}_j + \boldsymbol{\varepsilon}_j, \; j = 1, \ldots, N_t. \tag{9.22}$$

Because $\boldsymbol{\varepsilon}_j$ is not correlated with $\mathbf{S}_j$, the MCMs are given by

$$\mathbf{C}_j = \mathbf{X}_j \mathbf{X}_j{}^{\mathrm{T}} = \mathbf{A}\,\mathrm{Cov}(\mathbf{S}_j)\,\mathbf{A}^{\mathrm{T}} + \mathrm{Cov}(\boldsymbol{\varepsilon}_j), \; j = 1, \ldots, N_t. \tag{9.23}$$

We assume that noise $\boldsymbol{\varepsilon}$ is stationary, or almost stationary, and so $\mathrm{Cov}(\boldsymbol{\varepsilon}_j)$ is constant or varies only slowly in time (index) j. Besides noise $\boldsymbol{\varepsilon}_j$, matrix $\mathbf{C}_j$ contains sampling error because of limited number N_{tr} of trials. These noise types are suppressed in two filtering steps.

Each filtering is carried out by taking moving averages (discrete convolutions) of matrices $\mathbf{C}_j$ with Gaussian weights, or kernels, $w_{k,j}$ given by

$$w_{k,j} = \frac{1}{\alpha_k} e^{-\frac{1}{2}\left(\frac{k-j}{d}\right)^2}, \quad \text{with} \quad \alpha_k = \sum_{j=1}^{N_t} e^{-\frac{1}{2}\left(\frac{k-j}{d}\right)^2}, \quad 1 \le k, j \le N_t, \tag{9.24}$$

where "standard deviation" $d > 0$ controls the width of the averaging (convoluting) kernel. The resulting filtered sequence $\mathbf{C}_1^{\mathrm{F}}, \ldots, \mathbf{C}_{N_t}^{\mathrm{F}}$ of the MCMs is then given by

$$\mathbf{C}_k^{\mathrm{F}} = \sum_{j=1}^{N_t} w_{k,j} \mathbf{C}_j, \quad k = 1, \ldots, N_t. \tag{9.25}$$

In the first filtering step, we filter $\mathbf{C}_1, \ldots, \mathbf{C}_{N_t}$ using weights $w_{k,j}$ with a small $d > 0$ and get the filtered sequence $\mathbf{C}_1^{(1)}, \ldots, \mathbf{C}_{N_t}^{(1)}$. The filtering suppresses the high-frequency sampling error in $\mathbf{C}_j$ but leaves the almost constant noise covariance matrices $\mathrm{Cov}(\boldsymbol{\varepsilon}_j)$ unchanged.

In the second step, we again filter the sequence $\mathbf{C}_1, \ldots, \mathbf{C}_{N_t}$ but now with a large $d > 0$ and get the sequence $\mathbf{C}_1^{(2)}, \ldots, \mathbf{C}_{N_t}^{(2)}$. Here, the large $d > 0$ is chosen so that the filtered matrices contain only the slowly varying parts of $\mathbf{C}_j$ mostly consisting of the noise covariance matrices $\mathrm{Cov}(\boldsymbol{\varepsilon}_j)$.

The final two-step filtered sequence is given by the differences

$$\tilde{\mathbf{C}}_k = \mathbf{C}_k^{(1)} - \mathbf{C}_k^{(2)}, \quad k = 1, \ldots, N_t, \tag{9.26}$$

which can also be written in the form

$$\widetilde{\mathbf{C}}_k = \mathbf{A}(\mathbf{D}_k^{(1)} - \mathbf{D}_k^{(2)})\,\mathbf{A}^{\mathrm{T}} + \mathrm{Cov}(\boldsymbol{\varepsilon}_k)^{(1)} - \mathrm{Cov}(\boldsymbol{\varepsilon}_k)^{(2)} \tag{9.27}$$

$$\simeq \mathbf{A}(\mathbf{D}_k^{(1)} - \mathbf{D}_k^{(2)})\,\mathbf{A}^{\mathrm{T}}, \quad k = 1, \ldots, N_{\mathrm{t}}, \tag{9.28}$$

where $\mathbf{D}_k^{(1)}$, $\mathbf{D}_k^{(2)}$, $\mathrm{Cov}(\boldsymbol{\varepsilon}_k)^{(1)}$, and $\mathrm{Cov}(\boldsymbol{\varepsilon}_k)^{(2)}$ are obtained in the filtering steps (1) and (2), respectively, and $\mathrm{Cov}(\boldsymbol{\varepsilon}_k)^{(1)} \simeq \mathrm{Cov}(\boldsymbol{\varepsilon}_k)^{(2)}$, because noise $\boldsymbol{\varepsilon}_j$ is almost stationary, and so the two filterings affect them very little or not at all.

Next, AJD is applied to the two-step filtered sequence (9.26). By equation (9.28), matrix $\mathbf{A}^{-1}$ is a joint diagonalizer of matrices $\widetilde{\mathbf{C}}_k$, and so AJD yields an estimate $\widehat{\mathbf{A}}$ for $\mathbf{A}$. Note that matrices $\widetilde{\mathbf{C}}_k$ are no longer positive definite, as $\mathbf{C}_k$ are, and therefore we must use an AJD algorithm that does not require positive definiteness (like the algorithms listed in the previous section).

With $\widehat{\mathbf{A}}$, we get estimates for the time-course matrices $\widehat{\mathbf{S}}_j$ by equation (9.10) and estimates $\widehat{\mathbf{S}}_{\mathrm{ave}}$ by equation (9.12).

If $\mathbf{A}$ is not a square matrix, as is usually the case, we must use the transformed matrices $\widetilde{\mathbf{X}}_j$ and $\widetilde{\mathbf{C}}_j$ given in equations (9.17) and (9.19); apply the two-step filtering to $\widetilde{\mathbf{C}}_j$, $j = 1, \ldots, N_{\mathrm{t}}$; apply AJD to the filtered sequence; and get estimates $\widehat{\mathbf{A}}$ and $\widehat{\mathbf{S}}_j$ by equations (9.20) and (9.21).

Finally, we note that, in practice, optimal values of parameter d in the two steps of filtering are best found experimentally, although values $d = 3, \frac{1}{4}N_{\mathrm{t}}$ for the first and second steps, respectively, will often do.

9.3 Appendix: FFdiag Algorithm

Let $\mathbf{A}$ be a regular $m \times m$ matrix, $\mathbf{D}_1, \ldots, \mathbf{D}_N$ symmetric, approximately diagonal $m \times m$ matrices and $\mathbf{C}_k = \mathbf{A}\,\mathbf{D}_k\mathbf{A}^{\mathrm{T}}$, $k = 1, \ldots, N$, symmetric matrices.

We want to find an approximation $\widehat{\mathbf{A}}$ for $\mathbf{A}$, up to scaling and order of the columns of $\mathbf{A}$, by jointly diagonalizing $\mathbf{C}_1, \ldots, \mathbf{C}_N$ with the FFdiag algorithm [334]. The algorithm finds a joint diagonalizer $\mathbf{B} \in \mathbb{R}^{m \times m}$ so that $\mathbf{B}\,\mathbf{C}_k\mathbf{B}^{\mathrm{T}}$, $k = 1, \ldots, N$, are approximately diagonal, and thereafter, we set $\widehat{\mathbf{A}} = \mathbf{B}^{-1}$.

The idea of the FFdiag algorithm is to iteratively construct a sequence of approximate diagonalizers $\mathbf{B}_j$, which converges to the desired diagonalizer $\mathbf{B}$, so that $\mathbf{B}$ is a regular matrix and minimizes the sum

$$\sum_{k=1}^{N} \Big(\sum_{i,j=1,\, i \neq j}^{m} \mathbf{F}_k(i,j)^2 \Big) \quad \text{with} \quad \mathbf{F}_k = \mathbf{B}\,\mathbf{C}_k\mathbf{B}^{\mathrm{T}}, \tag{9.29}$$

that is, the sum of all off-diagonal elements of all $\mathbf{B}\,\mathbf{C}_k\mathbf{B}^{\mathrm{T}}$. The iterative FFdiag algorithm is constructed as follows.

Starting at iteration step $n=1$:

- Choose the number M of iterations.
- Choose the value of parameter Θ, $0<\Theta<1$, that controls the performance and convergence rate of the algorithm; smaller Θ yields a slightly better accuracy of $\mathbf{B}$ but requires higher M. Usually, $M=20$ and $\Theta=0.7$ are a good choice.
- Set $\mathbf{B}_1=\mathbf{I}$, or use some educated guess for $\mathbf{B}_1$.

The nth iteration step, $2\leq n\leq M$:

- Set $\mathbf{F}_k=\mathbf{B}_{n-1}\mathbf{C}_k\mathbf{B}_{n-1}{}^{\mathrm{T}}$, $k=1,\dots,N$.
- Form $m\times m$ matrices $\mathbf{Y}$ and $\mathbf{Z}$ by setting

$$\mathbf{Y}(i,j)=\sum_{k=1}^{N}\mathbf{F}_k(i,j)\,\mathbf{F}_k(j,j)\,, \tag{9.30}$$

$$\mathbf{Z}(i,j)=\sum_{k=1}^{N}\mathbf{F}_k(i,i)\,\mathbf{F}_k(j,j)\,, \tag{9.31}$$

$$\text{for}\quad i,j=1,\dots,m. \tag{9.32}$$

- Next, form an $m\times m$ matrix $\mathbf{W}$ so that $\mathbf{W}(i,i)=0$, $i=1,\dots,m$, and for $i=2,\dots,m$, $j=1,\dots,i-1$,

$$\mathbf{W}(i,j)=\frac{1}{d_{i,j}}\Big(\mathbf{Z}(i,j)\,\mathbf{Y}(j,i)-\mathbf{Z}(i,i)\,\mathbf{Y}(i,j)\Big), \tag{9.33}$$

$$\mathbf{W}(j,i)=\frac{1}{d_{i,j}}\Big(\mathbf{Z}(i,j)\,\mathbf{Y}(i,j)-\mathbf{Z}(j,j)\,\mathbf{Y}(j,i)\Big),\quad\text{where} \tag{9.34}$$

$$d_{i,j}=\mathbf{Z}(i,i)\,\mathbf{Z}(j,j)-\mathbf{Z}(i,j)^2. \tag{9.35}$$

- With

$$s=\left(\sum_{i=1}^{N}\sum_{j=1}^{N}\mathbf{W}(i,j)^2\right)^{1/2}$$

set

$$\mathbf{B}_n=\begin{cases}(\mathbf{I}+\frac{\Theta}{s}\,\mathbf{W})\,\mathbf{B}_{n-1} & \text{if } s>\Theta,\\ (\mathbf{I}+\mathbf{W})\,\mathbf{B}_{n-1} & \text{otherwise.}\end{cases} \tag{9.36}$$

The nth iteration step is completed.

- Finally, set

$$\mathbf{B} = \mathbf{B}_M \quad \text{and} \quad \widehat{\mathbf{A}} = \mathbf{B}^{-1}. \tag{9.37}$$

- Note that the number M of iterations can also be chosen adaptively so that iteration is stopped when the sum

$$R_n = \sum_{k=1}^{N} \left(\sum_{i,j=1,\, i \neq j}^{m} \mathbf{F}_k(i,j)^2 \right), \quad \text{with} \quad \mathbf{F}_k = \mathbf{B}_n \mathbf{C}_k \mathbf{B}_n{}^{\mathrm{T}}, \tag{9.38}$$

does not decrease anymore.

Bibliography

[1] Adachi, Y., Miyamoto, M., Kawai, J., Kawabata, M., Higuchi, M., Oyama, D., Uehara, G., Ogata, H., Kado, H., Haruta, Y., Tesan, G., and Crain, S. Development of a whole-head child MEG system. In *17th International Conference on Biomagnetism: Advances in Biomagnetism–Biomag2010* (2010), S. Supek and A. Sušac, Eds., Springer, pp. 35–38.

[2] Adrian, E. D., and Matthews, B. H. C. The Berger rhythm: Potential changes from the occipital lobes in man. *Brain 57*, 4 (1934), 355–385.

[3] Aguera, P.-E., Jerbi, K., Caclin, A., and Bertrand, O. ELAN: A software package for analysis and visualization of MEG, EEG, and LFP signals. *Computational Intelligence and Neuroscience 2011*, Article ID 158970 (2011).

[4] Ahlfors, S., Ahonen, A., Ehnholm, G., Hämäläinen, M., Ilmoniemi, R., Kajola, M., Kiviranta, M., Knuutila, J., Lounasmaa, O. V., Simola, J., Tesche, C., and Vilkman, V. A 24-SQUID gradiometer for magnetoencephalography. *Physica B: Condensed Matter 165 & 166* (1990), 97–98.

[5] Ahlfors, S. P., and Mody, M. Overview of MEG. *Organizational Research Methods* (2016), 1094428116676344.

[6] Ahonen, A. I., Hämäläinen, M. S., Ilmoniemi, R. J., Kajola, M. J., Knuutila, J. E. T., Simola, J. T., and Vilkman, V. A. Sampling theory for neuromagnetic detector arrays. *IEEE Transactions on Biomedical Engineering 40*, 9 (1993), 859–869.

[7] Ahonen, A. I., Hämäläinen, M. S., Kajola, M. J., Knuutila, J. E. T., Lounasmaa, O. V., Simola, J. T., Tesche, C. D., and Vilkman, V. A. Multichannel SQUID systems for brain research. *IEEE Transactions on Magnetics 27*, 2 (1991), 2786–2792.

[8] Ahonen, A. I., Hämäläinen, M. S., Kajola, M. J., Knuutila, J. E. T., Laine, P. P., Lounasmaa, O. V., Parkkonen, L. T., Simola, J. T., and Tesche, C. D. 122-channel SQUID instrument for investigating the magnetic signals from the human brain. *Physica Scripta 1993*, T49A (1993), 198–205.

[9] Aine, C. J., Sanfratello, L., Ranken, D., Best, E., MacArthur, J. A., Wallace, T., Gilliam, K., Donahue, C. H., Montano, R., Bryant, J. E., Scott, A., and Stephen, J. M. MEG-SIM: A web portal for testing MEG analysis methods using realistic simulated and empirical data. *Neuroinformatics 10*, 2 (2012), 141–158.

[10] Aine, C. J. Highlights of 40 years of SQUID-based brain research and clinical applications. In *17th International Conference on Biomagnetism: Advances in Biomagnetism–Biomag2010* (2010), S. Supek and A. Sušac, Eds., Springer, pp. 9–34.

[11] Aittoniemi, K., Hari, R., Katila, T., Kuusela, M. L., and Varpula, T. Magnetic fields of human brain evoked by long auditory stimuli. In *Proceedings of the Third National Meeting in Biophysics and Medical Engineering in Finland, Lappeenranta* (1979), pp. 1–3.

[12] Aittoniemi, K., Karp, P., Katila, T., Kuusela, M., and Varpula, T. On balancing superconducting gradiometric magnetometers. *Journal de Physique Colloques 39*, C6 (1978), C6–1223–C6–1225.

[13] Akalin Acar, Z., and Makeig, S. Effects of forward model errors on EEG source localization. *Brain Topography 26*, 3 (2013), 378–396.

[14] Alem, O., Mhaskar, R., Jiménez-Martínez, R., Sheng, D., LeBlanc, J., Trahms, L., Sander, T., Kitching, J., and Knappe, S. Magnetic field imaging with microfabricated optically-pumped magnetometers. *Optics Express 25*, 7 (2017), 7849–7858.

[15] Andersen, L. M., Oostenveld, R., Pfeiffer, C., Ruffieux, S., Jousmäki, V., Hämäläinen, M., Schneiderman, J. F., and Lundqvist, D. Similarities and differences between on-scalp and conventional in-helmet magnetoencephalography recordings. *PloS One 12*, 7 (2017), e0178602.

[16] Andreev, A. I. The thermal conductivity of the intermediate state in superconductors. *Zhurnal Eksperimental'noj i Teoreticheskoj Fiziki 46*, 5 (1964), 1823–1828.

[17] Atluri, S., Frehlich, M., Mei, Y., Garcia Dominguez, L., Rogasch, N. C., Wong, W., Daskalakis, Z. J., and Farzan, F. TMSEEG: A MATLAB-based graphical user interface for processing electrophysiological signals during transcranial magnetic stimulation. *Frontiers in Neural Circuits 10* (2016), 78.

[18] Azizollahi, H., Aarabi, A., and Wallois, F. Effects of uncertainty in head tissue conductivity and complexity on EEG forward modeling in neonates. *Human Brain Mapping 37*, 10 (2016), 3604–3622.

[19] Bagic, A., Funke, M. E., and Ebersole, J. American clinical MEG society (ACMEGS) position statement: The value of magnetoencephalography (MEG)/magnetic source imaging (MSI) in noninvasive presurgical evaluation of patients with medically intractable localization-related epilepsy. *Journal of Clinical Neurophysiology 26*, 4 (2009), 290–293.

[20] Baillet, S. Magnetoencephalography for brain electrophysiology and imaging. *Nature Neuroscience 20*, 3 (2017), 327–339.

[21] Baillet, S., Friston, K., and Oostenveld, R. Academic software applications for electromagnetic brain mapping using MEG and EEG. *Computational Intelligence and Neuroscience 2011*, Article ID 97205 (2011).

[22] Baillet, S., Mosher, J. C., and Leahy, R. M. Electromagnetic brain mapping. *IEEE Signal Processing Magazine 18*, 6 (2001), 14–30.

[23] Bao, F. S., Liu, X., and Zhang, C. PyEEG: An open source Python module for EEG/MEG feature extraction. *Computational Intelligence and Neuroscience 2011*, Article ID 406391 (2011).

[24] Bardeen, J., Cooper, L. N., and Schrieffer, J. R. Theory of superconductivity. *Physical Review 108*, 5 (1957), 1175–1204.

[25] Barth, D. S., Sutherling, W., Engel, J., and Beatty, J. Neuromagnetic localization of epileptiform spike activity in the human brain. *Science 218*, 4575 (1982), 891–894.

[26] Baule, G., and McFee, R. Detection of the magnetic field of the heart. *American Heart Journal 66*, 1 (1963), 95–96.

[27] Bedard, C., Gomes, J.-M., Bal, T., and Destexhe, A. A framework to reconcile frequency scaling measurements, from intracellular recordings, local-field potentials, up to EEG and MEG signals. *Journal of Integrative Neuroscience 16*, 1 (2017), 3–18.

[28] Belouchrani, A., Abed-Meraim, K., Cardoso, J.-F., and Moulines, E. A blind source separation technique using second-order statistics. *IEEE Transactions on Signal Processing 45*, 2 (1997), 434–444.

[29] Belouchrani, A., and Amin, M. G. Blind source separation based on time-frequency signal representations. *IEEE Transactions on Signal Processing 46*, 11 (1998), 2888–2897.

[30] Berger, H. Über das Elektrenkephalogramm des Menschen. *Archiv für Psychiatrie und Nervenkrankheiten 87*, 1 (December 1929), 527–570.

[31] Bertrand, O., Perrin, F., and Pernier, J. A theoretical justification of the average reference in topographic evoked potential studies. *Electroencephalography and Clinical Neurophysiology/Evoked Potentials Section 62*, 6 (1985), 462–464.

[32] Bison, G., Castagna, N., Hofer, A., Knowles, P., Schenker, J.-L., Kasprzak, M., Saudan, H., and Weis, A. A room temperature 19-channel magnetic field mapping device for cardiac signals. *Applied Physics Letters 95*, 17 (2009), 173701.

[33] Bork, J., Hahlbohm, H. D., Klein, R., and Schnabel, A. The 8-layered magnetically shielded room of the PTB: Design and construction. In *Biomag2000, Proceedings of the 12th International Conference on Biomagnetism* (2001), J. Nenonen, R. J. Ilmoniemi, and T. Katila, Eds., Helsinki University of Technology, pp. 970–973.

[34] Borna, A., Carter, T. R., Goldberg, J. D., Colombo, A. P., Jau, Y.-Y., Berry, C., McKay, J., Stephen, J., Weisend, M., and Schwindt, P. D. D. A 20-channel magnetoencephalography system based on optically pumped magnetometers. *Physics in Medicine & Biology 62*, 23 (2017), 8909–8923.

[35] Boto, E., Holmes, N., Leggett, J., Roberts, G., Shah, V., Meyer, S. S., Muñoz, L. D., Mullinger, K. J., Tierney, T. M., Bestmann, S., Barnes, G. R., Bowtell, R., and Brookes, M. J. Moving magnetoencephalography towards real-world applications with a wearable system. *Nature 555* (2018), 657–661.

[36] Boto, E., Meyer, S. S., Shah, V., Alem, O., Knappe, S., Kruger, P., Fromhold, T. M., Lim, M., Glover, P. M., Morris, P. G., Bowtell, R., Barnes, G. R., and Brookes, M. J. A new generation of magnetoencephalography: Room temperature measurements using optically-pumped magnetometers. *NeuroImage 149* (2017), 404–414.

[37] Brazier, M. A. B. *A History of the Electrical Activity of the Brain: The First Half-Century*. Pitman, 1961.

[38] Brazier, M. A. B. *A History of Neurophysiology in the 17th and 18th Centuries: From Concept to Experiment*. Raven Press, New York, 1984.

[39] Brazier, M. A. B. *A History of Neurophysiology in the 19th Century*. Raven Press, New York, 1988.

[40] Brenner, D., Lipton, J., Kaufman, L., and Williamson, S. J. Somatically evoked magnetic fields of the human brain. *Science 199*, 4324 (1978), 81–83.

[41] Brenner, D., Williamson, S. J., and Kaufman, L. Visually evoked magnetic fields of the human brain. *Science 190*, 4213 (1975), 480–482.

[42] Brookes, M. J., Stevenson, C. M., Barnes, G. R., Hillebrand, A., Simpson, M. I. G., Francis, S. T., and Morris, P. G. Beamformer reconstruction of correlated sources using a modified source model. *NeuroImage 34*, 4 (2007), 1454–1465.

[43] Brunet, D., Murray, M. M., and Michel, C. M. Spatiotemporal analysis of multichannel EEG: CARTOOL. *Computational Intelligence and Neuroscience 2011*, Article ID 813870 (2011).

[44] Budker, D., and Romalis, M. Optical magnetometry. *Nature Physics 3*, 4 (2007), 227–234.

[45] Buzsáki, G. *Rhythms of the Brain*. Oxford University Press, 2006.

[46] Buzsáki, G., Anastassiou, C. A., and Koch, C. The origin of extracellular fields and currents—EEG, ECoG, LFP and spikes. *Nature Reviews Neuroscience 13*, 6 (2012), 407–420.

[47] Campi, C., Pascarella, A., Sorrentino, A., and Piana, M. Highly automated dipole estimation (HADES). *Computational Intelligence and Neuroscience 2011*, Article ID 982185 (2011).

[48] Cardoso, J.-F., and Souloumiac, A. Blind beamforming for non-Gaussian signals. *IEE Proceedings F (Radar and Signal Processing) 140*, 6 (1993), 362–370.

[49] Caruso, L., Wunderle, T., Lewis, C. M., Valadeiro, J., Trauchessec, V., Rosillo, J. T., Amaral, J. P., Ni, J., Jendritza, P., Fermon, C., Freitas, P. P., Fries, P., and Pannetier-Lecoeur, M. *In vivo* magnetic recording of neuronal activity. *Neuron 95*, 6 (2017), 1283–1291.

[50] Caton, R. The electric currents of the brain. *British Medical Journal 2*, 765 (August 28, 1875), 278.

[51] Chapman, R. M., Ilmoniemi, R. J., Barbanera, S., and Romani, G. L. Selective localization of alpha brain activity with neuromagnetic measurements. *Clinical Neurophysiology 58*, 6 (1984), 569–572.

[52] Chaumon, M., Bishop, D. V. M., and Busch, N. A. A practical guide to the selection of independent components of the electroencephalogram for artifact correction. *Journal of Neuroscience Methods 250* (2015), 47–63.

[53] Christodoulou, E. G., Sakkalis, V., Tsiaras, V., and Tollis, I. G. BrainNetVis: An open-access tool to effectively quantify and visualize brain networks. *Computational Intelligence and Neuroscience 2011*, Article ID 747290 (2011).

[54] Clarke, J., and Braginski, A. I. *The SQUID Handbook: Applications of SQUIDs and SQUID Systems.* John Wiley & Sons, 2006.

[55] Cohen, D. Magnetoencephalography: Evidence of magnetic fields produced by alpha-rhythm currents. *Science 161*, 3843 (1968), 784–786.

[56] Cohen, D. Large-volume conventional magnetic shields. *Revue de Physique Appliquee 5*, 1 (1970), 53–58.

[57] Cohen, D. Magnetoencephalography: Detection of the brain's electrical activity with a superconducting magnetometer. *Science 175*, 4022 (1972), 664–666.

[58] Cohen, D. Measurements of the magnetic fields produced by the human heart, brain, and lungs. *IEEE Transactions on Magnetics 11*, 2 (1975), 694–700.

[59] Cohen, D., Edelsack, E. A., and Zimmerman, J. E. Magnetocardiograms taken inside a shielded room with a superconducting point-contact magnetometer. *Applied Physics Letters 16*, 7 (1970), 278–280.

[60] Cohen, D., and Halgren, E. Magnetoencephalography (neuromagnetism). In *Encyclopedia of Neuroscience*, L. R. Squire, Ed. Elsevier (2003), 1–7.

[61] Cohen, M. X. Where does EEG come from and what does it mean? *Trends in Neurosciences 40*, 4 (2017), 208–218.

[62] Collura, T. F. History and evolution of electroencephalographic instruments and techniques. *Journal of Clinical Neurophysiology 10*, 4 (1993), 476–504.

[63] Colon, A. J., Ossenblok, P., Nieuwenhuis, L., Stam, K. J., and Boon, P. Use of routine MEG in the primary diagnostic process of epilepsy. *Journal of Clinical Neurophysiology 26*, 5 (2009),326–332.

[64] Comon, P., and Jutten, C. *Handbook of Blind Source Separation: Independent Component Analysis and Applications*. Academic Press, Amsterdam, 2010.

[65] Cuffin, B. N., and Cohen, D. Magnetic fields of a dipole in special volume conductor shapes. *IEEE Transactions on Biomedical Engineering 24*, 4 (1977), 372–381.

[66] Curio, G., Mackert, B.-M., Burghoff, M., Koetitz, R., Abraham-Fuchs, K., and Härer, W. Localization of evoked neuromagnetic 600 Hz activity in the cerebral somatosensory system. *Electroencephalography and Clinical Neurophysiology 91*, 6 (1994), 483–487.

[67] Dabek, J., Kalogianni, K., Rotgans, E., van der Helm, F. C. T., Kwakkel, G., van Wegen, E. E. H., Daffertshofer, A., and de Munck, J. C. Determination of head conductivity frequency response *in vivo* with optimized EIT-EEG. *NeuroImage 127* (2016), 484–495.

[68] Dalal, S. S., Zumer, J. M., Guggisberg, A. G., Trumpis, M., Wong, D. D. E., Sekihara, K., and Nagarajan, S. S. MEG/EEG source reconstruction, statistical evaluation, and visualization with NUTMEG. *Computational Intelligence and Neuroscience 2011*, Article ID 758973 (2011).

[69] Dale, A. M., Liu, A. K., Fischl, B. R., Buckner, R. L., Belliveau, J. W., Lewine, J. D., and Halgren, E. Dynamic statistical parametric mapping: Combining fMRI and MEG for high-resolution imaging of cortical activity. *Neuron 26*, 1 (2000), 55–67.

[70] Darvas, F., Pantazis, D., Kucukaltun-Yildirim, E., and Leahy, R. M. Mapping human brain function with MEG and EEG: Methods and validation. *NeuroImage 23* (2004), S289–S299.

[71] Davis, H., Davis, P. A., Loomis, A. L., Harvey, E. N., and Hobart, G. Electrical reactions of the human brain to auditory stimulation during sleep. *Journal of Neurophysiology 2*, 6 (1939), 500–514.

[72] Davis, P. A. Effects of acoustic stimuli on the waking human brain. *Journal of Neurophysiology 2*, 6 (1939), 494–499.

[73] De Tiège, X., Lundqvist, D., Beniczky, S., Seri, S., and Paetau, R. Current clinical magnetoencephalography practice across Europe: Are we closer to use MEG as an established clinical tool? *Seizure 50* (2017), 53–59.

[74] Deaver Jr., B. S., and Fairbank, W. M. Experimental evidence for quantized flux in superconducting cylinders. *Physical Review Letters 7*, 2 (1961), 43–46.

[75] Degen, C. L., Reinhard, F., and Cappellaro, P. Quantum sensing. *Reviews of Modern Physics 89*, 3 (2017), 035002.

[76] Degenhart, A. D., Kelly, J. W., Ashmore, R. C., Collinger, J. L., Tyler-Kabara, E. C., Weber, D. J., and Wang, W. Craniux: A LabVIEW-based modular software framework for brain-machine interface research. *Computational Intelligence and Neuroscience 2011*, Article ID 363565 (2011).

[77] Del Gratta, C., Pizzella, V., Tecchio, F., and Romani, G. L. Magnetoencephalography - a noninvasive brain imaging method with 1 ms time resolution. *Reports on Progress in Physics 64*, 12 (2001), 1759–1814.

[78] Del Gratta, C., Pizzella, V., Torquati, K., and Romani, G. L. New trends in magnetoencephalography. *Electroencephalography and Clinical Neurophysiology. Supplement 50* (1999), 59–73.

[79] Della Penna, S., Pizzella, V., and Romani, G. L. Impact of SQUIDs on functional imaging in neuroscience. *Superconductor Science and Technology 27*, 4 (2014), 044004.

[80] Delorme, A., and Makeig, S. EEGLAB: An open source toolbox for analysis of single-trial EEG dynamics including independent component analysis. *Journal of Neuroscience Methods 134*, 1 (2004), 9–21.

[81] Delorme, A., Mullen, T., Kothe, C., Acar, Z. A., Bigdely-Shamlo, N., Vankov, A., and Makeig, S. EEGLAB, SIFT, NFT, BCILAB, and ERICA: New tools for advanced EEG processing. *Computational Intelligence and Neuroscience 2011*, Article ID 130714 (2011).

[82] Desmedt, J. E., Chalklin, V., and Tomberg, C. Emulation of somatosensory evoked potential (SEP) components with the 3-shell head model and the problem of 'ghost potential fields' when using an average reference in brain mapping. *Electroencephalography and Clinical Neurophysiology/Evoked Potentials Section 77*, 4 (1990), 243–258.

[83] Diwakar, M., Huang, M.-X., Srinivasan, R., Harrington, D. L., Robb, A., Angeles, A., Muzzatti, L., Pakdaman, R., Song, T., Theilmann, R. J., and Lee, R. R. Dual-core beamformer for obtaining highly correlated neuronal networks in MEG. *NeuroImage 54*, 1 (2011), 253–263.

[84] Diwakar, M., Tal, O., Liu, T. T., Harrington, D. L., Srinivasan, R., Muzzatti, L., Song, T., Theilmann, R. J., Lee, R. R., and Huang, M.-X. Accurate reconstruction of temporal correlation for neuronal sources using the enhanced dual-core MEG beamformer. *NeuroImage 56*, 4 (2011), 1918–1928.

[85] Doll, R., and Näbauer, M. Experimental proof of magnetic flux quantization in a superconducting ring. *Physical Review Letters 7*, 2 (1961), 51–52.

[86] Draganova, R., Eswaran, H., Murphy, P., Huotilainen, M., Lowery, C., and Preissl, H. Sound frequency change detection in fetuses and newborns, a magnetoencephalographic study. *NeuroImage 28*, 2 (2005), 354–361.

[87] Draganova, R., Eswaran, H., Murphy, P., Lowery, C., and Preissl, H. Serial magnetoencephalographic study of fetal and newborn auditory discriminative evoked responses. *Early Human Development 83*, 3 (2007), 199–207.

[88] Ehnholm, G. J., Ilmoniemi, R. J., and Wiik, T. O. A seven channel SQUID magnetometer for brain research. *Physica B + C 107*, 1–3 (1981), 29–30.

[89] Eichele, T., Rachakonda, S., Brakedal, B., Eikeland, R., and Calhoun, V. D. EEGIFT: Group independent component analysis for event-related EEG data. *Computational Intelligence and Neuroscience 2011*, Article ID 129365 (2011).

[90] Evans, J. R., and Abarbanel, A. *Introduction to Quantitative EEG and Neurofeedback*. Elsevier, 1999.

[91] Farrell, D. E., Tripp, J. H., Norgren, R., and Teyler, T. J. A study of the auditory evoked magnetic field of the human brain. *Clinical Neurophysiology 49*, 1 (1980), 31–37.

[92] Fedele, T., Scheer, H. J., Burghoff, M., Curio, G., and Körber, R. Ultra-low-noise EEG/MEG systems enable bimodal non-invasive detection of spike-like human somatosensory evoked responses at 1 kHz. *Physiological Measurement 36*, 2 (2015), 357–368.

[93] Ferrier, D. On the localisation of the functions of the brain. *British Medical Journal 2*, 729 (1874), 766–767.

[94] Friston, K. J., Ashburner, J. T., Kiebel, S. J., Nichols, T. E., and Penny, W. D. *Statistical Parametric Mapping: The Analysis of Functional Brain Images*. Elsevier, 2007.

[95] Fritsch, G., and Hitzig, E. Ueber die elektrische Erregbarkeit des Grosshirns. *Archiv fuer Anatomie, Physiologie und wissenschaftliche Medicin 37* (1870), 300–332.

[96] Galvani, L. De viribus electricitatis in motu musculari: Commentarius. *Bononiae: Ex Typographia Istituto Scientiarum* (1791).

[97] Gavaret, M., Badier, J.-M., Bartolomei, F., Bénar, C.-G., and Chauvel, P. MEG and EEG sensitivity in a case of medial occipital epilepsy. *Brain Topography 27*, 1 (2014), 192–196.

[98] Gençer, N. G., and Acar, C. E. Sensitivity of EEG and MEG measurements to tissue conductivity. *Physics in Medicine & Biology 49*, 5 (2004), 701–717.

[99] Geselowitz, D. On the magnetic field generated outside an inhomogeneous volume conductor by internal current sources. *IEEE Transactions on Magnetics 6*, 2 (1970), 346–347.

[100] Geselowitz, D. B. On bioelectric potentials in an inhomogeneous volume conductor. *Biophysical Journal 7*, 1 (1967), 1–11.

[101] Giazotto, F., Peltonen, J. T., Meschke, M., and Pekola, J. P. Superconducting quantum interference proximity transistor. *Nature Physics 6*, 4 (2010), 254–259.

[102] Gramfort, A., Kowalski, M., and Hämäläinen, M. Mixed-norm estimates for the M/EEG inverse problem using accelerated gradient methods. *Physics in Medicine & Biology 57*, 7 (2012), 1937–1961.

[103] Gramfort, A., Luessi, M., Larson, E., Engemann, D. A., Strohmeier, D., Brodbeck, C., Parkkonen, L., and Hämäläinen, M. S. MNE software for processing MEG and EEG data. *NeuroImage 86* (2014), 446–460.

[104] Gramfort, A., Papadopoulo, T., Olivi, E., and Clerc, M. Forward field computation with OpenMEEG. *Computational Intelligence and Neuroscience 2011*, Article ID 923703 (2011).

[105] Gramfort, A., Strohmeier, D., Haueisen, J., Hämäläinen, M. S., and Kowalski, M. Time-frequency mixed-norm estimates: Sparse M/EEG imaging with non-stationary source activations. *NeuroImage 70* (2013), 410–422.

[106] Gross, J., Baillet, S., Barnes, G. R., Henson, R. N., Hillebrand, A., Jensen, O., Jerbi, K., Litvak, V., Maess, B., Oostenveld, R., Parkkonen, L., Taylor, J. R., van Wassenhove, V., Wibral, M., and Schoffelen, J.-M. Good practice for conducting and reporting MEG research. *NeuroImage 65* (2013), 349–363.

[107] Grynszpan, F., and Geselowitz, D. B. Model studies of the magnetocardiogram. *Biophysical Journal 13*, 9 (1973), 911–925.

[108] Güllmar, D., Haueisen, J., and Reichenbach, J. R. Influence of anisotropic electrical conductivity in white matter tissue on the EEG/MEG forward and inverse solution. A high-resolution whole head simulation study. *NeuroImage 51*, 1 (2010), 145–163.

[109] Hagemann, D., Naumann, E., and Thayer, J. F. The quest for the EEG reference revisited: A glance from brain asymmetry research. *Psychophysiology 38*, 5 (2001), 847–857.

[110] Hämäläinen, M., and Hari, R. Magnetoencephalographic (MEG) characterization of dynamic brain activation. In *Brain Mapping: The Methods*, A. Toga and J. Mazziotta, Eds. Academic Press, 2002, pp. 227–255.

[111] Hämäläinen, M., Hari, R., Ilmoniemi, R. J., Knuutila, J., and Lounasmaa, O. V. Magnetoencephalography—theory, instrumentation, and applications to noninvasive studies of the working human brain. *Reviews of Modern Physics 65*, 2 (1993), 413–497.

[112] Hämäläinen, M. S., and Ilmoniemi, R. J. Interpreting measured magnetic fields of the brain: Estimates of current distributions. Tech. Rep TKK-F-A559, Helsinki University of Technology, Espoo, 1984.

[113] Hämäläinen, M. S., and Ilmoniemi, R. J. Interpreting magnetic fields of the brain: Minimum norm estimates. *Medical & Biological Engineering & Computing 32*, 1 (1994), 35–42.

[114] Hämäläinen, M. S., Lin, F.-H., and Mosher, J. C. Anatomically and functionally constrained minimum-norm estimates. In *MEG: An Introduction to Methods*, P. Hansen, M. Kringelbach, and R. Salmelin, Eds. Oxford University Press, New York, 2010, pp. 186–215.

[115] Hämäläinen, M. S., and Sarvas, J. Realistic conductivity geometry model of the human head for interpretation of neuromagnetic data. *IEEE Transactions on Biomedical Engineering 36*, 2 (1989), 165–171.

[116] Handy, T. C. *Brain Signal Analysis: Advances in Neuroelectric and Neuromagnetic Methods*. MIT Press, 2009.

[117] Hansen, P., Kringelbach, M., and Salmelin, R. *MEG: An Introduction to Methods*. Oxford University Press, 2010.

[118] Hari, R., Aittoniemi, K., Järvinen, M.-L., Katila, T., and Varpula, T. Auditory evoked transient and sustained magnetic fields of the human brain localization of neural generators. *Experimental Brain Research 40*, 2 (1980), 237–240.

[119] Hari, R., Baillet, S., Barnes, G., Burgess, R., Forss, N., Gross, J., Hämäläinen, M., Jensen, O., Kakigi, R., Mauguière, F., Nakasato, N., Puce, A., Romani, G. L., Schnitzler, A., and Taulu, S. IFCN-endorsed practical guidelines for clinical magnetoencephalography (MEG). *Clinical Neurophysiology 129*, 8 (2018), 1720–1747.

[120] Hari, R., and Ilmoniemi, R. J. Cerebral magnetic fields. *CRC Critical Reviews in Biomedical Engineering 14*, 2 (1986), 93–126.

[121] Hari, R., and Puce, A. *MEG–EEG Primer*. Oxford University Press, New York, 2017.

[122] Hari, R., and Salmelin, R. Magnetoencephalography: From SQUIDs to neuroscience: Neuroimage 20th anniversary special edition. *NeuroImage 61*, 2 (2012), 386–396.

[123] Haueisen, J., Ramon, C., Eiselt, M., Brauer, H., and Nowak, H. Influence of tissue resistivities on neuromagnetic fields and electric potentials studied with a finite element model of the head. *IEEE Transactions on Biomedical Engineering 44*, 8 (1997), 727–735.

[124] Hauk, O. Keep it simple: A case for using classical minimum norm estimation in the analysis of EEG and MEG data. *NeuroImage 21*, 4 (2004), 1612–1621.

[125] He, B., Dai, Y., Astolfi, L., Babiloni, F., Yuan, H., and Yang, L. eConnectome: A MATLAB toolbox for mapping and imaging of brain functional connectivity. *Journal of Neuroscience Methods 195*, 2 (2011), 261–269.

[126] Heinonen, P., Tuomola, M., Lokkala, J., and Malmivuo, J. Properties of a thick-walled conducting enclosure in low-frequency magnetic shielding. *Journal of Physics E: Scientific Instruments 13*, 5 (1980), 569–570.

[127] Heller, L., and van Hulsteyn, D. B. Brain stimulation using electromagnetic sources: Theoretical aspects. *Biophysical Journal 63*, 1 (1992), 129–138.

[128] Hernández Pavón, J. C., Sarvas, J., Ilmoniemi, R. J., and Stenroos, M. Beamformer with temporally correlated sources and iterative RAP search. *Submitted.*

[129] Hillebrand, A., Singh, K. D., Holliday, I. E., Furlong, P. L., and Barnes, G. R. A new approach to neuroimaging with magnetoencephalography. *Human Brain Mapping 25*, 2 (2005), 199–211.

[130] Hodgkin, A. L., and Huxley, A. F. A quantitative description of membrane current and its application to conduction and excitation in nerve. *The Journal of Physiology 117*, 4 (1952), 500–544.

[131] Hu, S., Lai, Y., Valdes-Sosa, P. A., Bringas-Vega, M. L., and Yao, D. How do reference montage and electrodes setup affect the measured scalp EEG potentials? *Journal of Neural Engineering 15*, 2 (2018), 026013.

[132] Hui, H. B., Pantazis, D., Bressler, S. L., and Leahy, R. M. Identifying true cortical interactions in MEG using the nulling beamformer. *NeuroImage 49*, 4 (2010), 3161–3174.

[133] Huotilainen, M., Ilmoniemi, R. J., Tiitinen, H., Lavikainen, J., Alho, K., Kajola, M., and Näätänen, R. The projection method in removing eye-blink artefacts from multichannel MEG measurements at the site of pain. In *Biomagnetism: Fundamental Research and Clinical Applications*, C. Baumgartner, L. Deecke, G. Stroink, and S. J. Williamson, Eds. IOS Press, 1995. pp. 363–367.

[134] Huotilainen, M., Kujala, A., Hotakainen, M., Parkkonen, L., Taulu, S., Simola, J., Nenonen, J., Karjalainen, M., and Näätänen, R. Short-term memory functions of the human fetus recorded with magnetoencephalography. *NeuroReport 16*, 1 (2005), 81–84.

[135] Huotilainen, M., Kujala, A., Hotakainen, M., Shestakova, A., Kushnerenko, E., Parkkonen, L., Fellman, V., and Näätänen, R. Auditory magnetic responses of healthy newborns. *NeuroReport 14*, 14 (2003), 1871–1875.

[136] Hyvärinen, A. Fast and robust fixed-point algorithms for independent component analysis. *IEEE Transactions on Neural Networks 10*, 3 (1999), 626–634.

[137] Hyvärinen, A., Karhunen, J., and Oja, E. *Independent Component Analysis*, vol. 46 of *Adaptive and Learning Systems for Signal Processing, Communications, and Control*. John Wiley & Sons, New York, 2001.

[138] Iivanainen, J., Stenroos, M., and Parkkonen, L. Measuring MEG closer to the brain: Performance of on-scalp sensor arrays. *NeuroImage 147* (2017), 542–553.

[139] Ilmoniemi, R. Neuromagnetism—Theory, Techniques, and Measurements. PhD Thesis, Helsinki University of Technology, Espoo, 1985.

[140] Ilmoniemi, R. Models of source currents in the brain. *Brain Topography 5*, 4 (1993), 331–336.

[141] Ilmoniemi, R. Method and apparatus for separating the different components of evoked response and spontaneous activity brain signals as well as of signals measured from the heart, August 1997. US Patent 5,655,534.

[142] Ilmoniemi, R. J., Hämäläinen, M. S., and Knuutila, J. The forward and inverse problems in the spherical model. In *Biomagnetism: Applications & Theory* (New York, 1985), H. Weinberg, G. Stroink, and T. Katila, Eds., Pergamon Press, pp. 278–282.

[143] Ilmoniemi, R., Hari, R., and Reinikainen, K. A four-channel SQUID magnetometer for brain research. *Electroencephalography and Clinical Neurophysiology 58*, 5 (1984), 467–473.

[144] Ilmoniemi, R. J. Holes in the skull do not affect the neuromagnetic field. In *4th International Symposium of the International Society for Brain Electromagnetic Topography (ISBET)*. P. Valdés Sosa, Ed., Havana, 1993, p. 28.

[145] Ilmoniemi, R. J. Radial anisotropy added to a spherically symmetric conductor does not affect the external magnetic field due to internal sources. *Europhysics Letters 30*, 5 (1995), 313–316.

[146] Ilmoniemi, R. J. The triangle phantom in magnetoencephalography. *The Journal of Japan Biomagnetism and Bioelectromagnetics Society 22* (2009), 44–45.

[147] Ilmoniemi, R. J., and Kičić, D. Methodology for combined TMS and EEG. *Brain Topography 22*, 4 (2010), 233–248.

[148] Ilmoniemi, R. J., and Lounasmaa, O. V. Magnetoencephalography at the Helsinki University of Technology. *Physica Scripta 1989*, T25 (1989), 243–246.

[149] Im, C.-H., Jun, S. C., and Sekihara, K. Recent advances in biomagnetism and its applications. *Biomedical Engineering Letters 7* (2017), 183–184.

[150] Imada, T., Zhang, Y., Cheour, M., Taulu, S., Ahonen, A., and Kuhl, P. K. Infant speech perception activates Broca's area: A developmental magnetoencephalography study. *NeuroReport 17*, 10 (2006), 957–962.

[151] Ioannides, A. A. Magnetoencephalography as a research tool in neuroscience: State of the art. *The Neuroscientist 12*, 6 (2006), 524–544.

[152] Jabdaraghi, R. N. *Magnetometry by a proximity Josephson junction interferometer*. PhD thesis, Aalto University, Espoo, Finland, 2018.

[153] Jackson, J. D. *Electrodynamics*. John Wiley & Sons, New York, 1975.

[154] Jiménez-Martínez, R., and Knappe, S. Microfabricated optically-pumped magnetometers. In *High Sensitivity Magnetometers*. G. Asaf, M. S. Haji-Sheikh, and S. C. Mukhopadhyay, Eds., Springer, 2017, pp. 523–551.

[155] John, E. R., Karmel, B., Corning, W., Easton, P., Brown, D., Ahn, H., John, M., Harmony, T., Prichep, L., Toro, A., Gerson, I., Bartlett, F., Thatcher, R., Kaye, H., Valdes, P., and Schwartz, E. Neurometrics. *Science 196*, 4297 (1977), 1393–1410.

[156] Johnson, B. W., Crain, S., Thornton, R., Tesan, G., and Reid, M. Measurement of brain function in pre-school children using a custom sized whole-head MEG sensor array. *Clinical Neurophysiology 121*, 3 (2010), 340–349.

[157] Johnson, C., Schwindt, P. D., and Weisend, M. Magnetoencephalography with a two-color pump-probe, fiber-coupled atomic magnetometer. *Applied Physics Letters 97*, 24 (2010), 243703.

[158] Josephson, B. D. Possible new effects in superconductive tunnelling. *Physics Letters 1*, 7 (1962), 251–253.

[159] Josephson, B. D. Theoretical discovery of the Josephson effect. In *Josephson Junctions: History, Devices, and Applications*, E. L. Wolf, G. B. Arnold, M. A. Gurvitch, and J. F. Zasadzinski, Eds. Pan Stanford, 2017, pp. 1–15.

[160] Kajola, M., Ahlfors, S., Ehnholm, G. J., Hällström, J., Hämäläinen, M. S., Ilmoniemi, R. J., Kiviranta, M., Knuutila, J., Lounasmaa, O. V., Tesche, C. D., and Vilkman, V. A 24-channel magnetometer for brain research. In *Advances in Biomagnetism*, S. J. Williamson, M. Hoke, G. Stroink, and M. Kotani, Eds. Springer, 1989, pp. 673–676.

[161] Karadas, M., Wojciechowski, A. M., Huck, A., Dalby, N. O., Andersen, U. L., and Thielscher, A. Feasibility and resolution limits of opto-magnetic imaging of neural network activity in brain slices using color centers in diamond. *Scientific Reports 8*, 1 (2018), 4503.

[162] Kariniemi, V., Ahopelto, J., Karp, P. J., and Katila, T. E. The fetal magnetocardiogram. *Journal of Perinatal Medicine 2* (1974), 214–216.

[163] Katila, T., Maniewski, R., Poutanen, T., Varpula, T., and Karp, P. J. Magnetic fields produced by the human eye. *Journal of Applied Physics 52*, 3 (1981), 2565–2571.

[164] Keil, A., Debener, S., Gratton, G., Junghöfer, M., Kappenman, E. S., Luck, S. J., Luu, P., Miller, G. A., and Yee, C. M. Committee report: Publication guidelines and recommendations for studies using electroencephalography and magnetoencephalography. *Psychophysiology 51*, 1 (2014), 1–21.

[165] Kelhä, V., Pukki, J., Peltonen, R., Penttinen, A., Ilmoniemi, R., and Heino, J. Design, construction, and performance of a large-volume magnetic shield. *IEEE Transactions on Magnetics 18*, 1 (1982), 260–270.

[166] Kemppainen, P. K., and Ilmoniemi, R. J. Channel capacity of multichannel magnetometers. In *Advances in Biomagnetism*, S. Williamson, M. Hoke, G. Stroink, and M. Kotani, Eds. Springer, 1989, pp. 635–638.

[167] Kharkar, S., and Knowlton, R. Magnetoencephalography in the presurgical evaluation of epilepsy. *Epilepsy & Behavior 46* (2015), 19–26.

[168] Kiebel, S. J., Garrido, M. I., Moran, R., Chen, C.-C., and Friston, K. J. Dynamic causal modeling for EEG and MEG. *Human Brain Mapping 30*, 6 (2009), 1866–1876.

[169] Kleiner, R., Koelle, D., Ludwig, F., and Clarke, J. Superconducting quantum interference devices: State of the art and applications. *Proceedings of the IEEE 92*, 10 (2004), 1534–1548.

[170] Klem, G. H., Lüders, H. O., Jasper, H. H., and Elger, C. The ten–twenty electrode system of the International Federation. *Electroencephalography and Clinical Neurophysiology 52*, 3 (1999), 3–6.

[171] Knuutila, J., Ahlfors, S., Ahonen, A., Hällström, J., Kajola, M., Lounasmaa, O. V., Vilkman, V., and Tesche, C. Large-area low-noise seven-channel dc SQUID magnetometer for brain research. *Review of Scientific Instruments 58*, 11 (1987), 2145–2156.

[172] Kobayashi, K., Yoshinaga, H., Ohtsuka, Y., and Gotman, J. Dipole modeling of epileptic spikes can be accurate or misleading. *Epilepsia 46*, 3 (2005), 397–408.

[173] Koenig, T., Kottlow, M., Stein, M., and Melie-García, L. Ragu: A free tool for the analysis of EEG and MEG event-related scalp field data using global randomization statistics. *Computational Intelligence and Neuroscience 2011*, Article ID 938925 (2011).

[174] Kominis, I. K., Kornack, T. W., Allred, J. C., and Romalis, M. V. A subfemtotesla multichannel atomic magnetometer. *Nature 422*, 6932 (2003), 596–599.

[175] Komssi, S., Huttunen, J., Aronen, H. J., and Ilmoniemi, R. J. EEG minimum-norm estimation compared with MEG dipole fitting in the localization of somatosensory sources at S1. *Clinical Neurophysiology 115*, 3 (2004), 534–542.

[176] Körber, R., Storm, J.-H., Seton, H., Mäkelä, J. P., Paetau, R., Parkkonen, L., Pfeiffer, C., Riaz, B., Schneiderman, J. F., Dong, H., Hwang, S., You, L., Inglis, B., Clarke, J., Espy, M. A., Ilmoniemi, R. J., Magnelind, P. E., Matlashov, A. N., Nieminen, J. O., Volegov, P. L., Zevenhoven, K. C. J., Höfner, N., Burghoff, M., Enpuku, K., Yang, S. Y., Chieh, J.-J., Knuutila, J., Laine, P., and Nenonen, J. SQUIDs in biomagnetism: A roadmap towards improved healthcare. *Superconductor Science and Technology 29*, 11 (2016), 113001.

[177] Kornhuber, H. H., and Deecke, L. Hirnpotentialänderungen beim Menschen vor und nach Willkürbewegungen dargestellt mit Magnetbandspeicherung und Rückwärtsanalyse. *Pflüger's Archiv für die gesamte Physiologie des Menschen und der Tiere 281*, 1 (1964), 52.

[178] Kornhuber, H. H., and Deecke, L. Hirnpotentialänderungen bei Willkürbewegungen und passiven Bewegungen des Menschen: Bereitschaftspotential und reafferente Potentiale. *Pflüger's Archiv für die gesamte Physiologie des Menschen und der Tiere 284*, 1 (1965), 1–17.

[179] Kress, R. *Linear Integral Equations*, vol. 82 of *Applied Mathematical Sciences*. Springer, New York, 1989.

[180] Kujala, A., Huotilainen, M., Hotakainen, M., Lennes, M., Parkkonen, L., Fellman, V., and Näätänen, R. Speech-sound discrimination in neonates as measured with MEG. *NeuroReport 15*, 13 (2004), 2089–2092.

[181] Lacoss, R. T. Adaptive combining of wideband array data for optimal reception. *IEEE Transactions on Geoscience Electronics 6*, 2 (1968), 78–86.

[182] Lau, S., Flemming, L., and Haueisen, J. Magnetoencephalography signals are influenced by skull defects. *Clinical Neurophysiology 125*, 8 (2014), 1653–1662.

[183] Leclercq, Y., Schrouff, J., Noirhomme, Q., Maquet, P., and Phillips, C. fMRI artefact rejection and sleep scoring toolbox. *Computational Intelligence and Neuroscience 2011*, Article ID 598206 (2011).

[184] Lee, Y. H., Yu, K. K., Kwon, H., Kim, J. M., Kim, K., Park, Y. K., Yang, H. C., Chen, K. L., Yang, S. Y., and Horng, H. E. A whole-head magnetoencephalography system with compact axial gradiometer structure. *Superconductor Science and Technology 22*, 4 (2009), 045023.

[185] Lew, S., Sliva, D. D., Choe, M.-s., Grant, P. E., Okada, Y., Wolters, C. H., and Hämäläinen, M. S. Effects of sutures and fontanels on MEG and EEG source analysis in a realistic infant head model. *NeuroImage 76* (2013), 282–293.

[186] Lewine, J. D., and Orrison, W. W. Magnetic source imaging: Basic principles and applications in neuroradiology. *Academic Radiology 2*, 5 (1995), 436–440.

[187] Lin, F.-H., Belliveau, J. W., Dale, A. M., and Hämäläinen, M. S. Distributed current estimates using cortical orientation constraints. *Human Brain Mapping 27*, 1 (2006), 1–13.

[188] Lin, F.-H., Vesanen, P. T., Hsu, Y.-C., Nieminen, J. O., Zevenhoven, K. C. J., Dabek, J., Parkkonen, L. T., Simola, J., Ahonen, A. I., and Ilmoniemi, R. J. Suppressing multi-channel ultra-low-field MRI measurement noise using data consistency and image sparsity. *PloS One 8*, 4 (2013), e61652.

[189] Lin, F.-H., Witzel, T., Ahlfors, S. P., Stufflebeam, S. M., Belliveau, J. W., and Hämäläinen, M. S. Assessing and improving the spatial accuracy in MEG source localization by depth-weighted minimum-norm estimates. *NeuroImage 31*, 1 (2006), 160–171.

[190] Litvak, V., Mattout, J., Kiebel, S., Phillips, C., Henson, R., Kilner, J., Barnes, G., Oostenveld, R., Daunizeau, J., Flandin, G., Penny, W., and Friston, K. EEG and MEG data analysis in SPM8. *Computational Intelligence and Neuroscience 2011*, Article ID 852961 (2011).

[191] Lounasmaa, O. V. Medical applications of SQUIDs in neuro- and cardiomagnetism. *Physica Scripta T66* (1996), 70–79.

[192] Lowery, C. L., Eswaran, H., Murphy, P., and Preissl, H. Fetal magnetoencephalography. *Seminars in Fetal and Neonatal Medicine 11*, 6 (2006), 430–436.

[193] Lu, Z.-L., and Kaufman, L. *Magnetic Source Imaging of the Human Brain.* Taylor & Francis, 2003.

[194] Luck, S. J. *An Introduction to the Event-Related Potential Technique.* MIT Press, 2014.

[195] Luomahaara, J., Vesterinen, V., Grönberg, L., and Hassel, J. Kinetic inductance magnetometer. *Nature Communications 5* (2014), 4872.

[196] Mäkelä, N., Stenroos, M., Sarvas, J., and Ilmoniemi, R. J. Locating highly correlated sources from MEG with (recursive)(R) DS-MUSIC. *bioRxiv* (2017), 230672.

[197] Mäkelä, N., Stenroos, M., Sarvas, J., and Ilmoniemi, R. J. Truncated RAP-MUSIC (TRAP-MUSIC) for MEG and EEG source localization. *NeuroImage 167* (2018), 73–83.

[198] Mäki, H., and Ilmoniemi, R. J. Projecting out muscle artifacts from TMS-evoked EEG. *NeuroImage 54*, 4 (2011), 2706–2710.

[199] Mäkinen, A. J., Zevenhoven, K. C. J., and Ilmoniemi, R. J. Fully automatic calibration of ultra-low-field MRI for high spatial accuracy in hybrid MEG–MRI. *IEEE Transactions on Medical Imaging* (2019), in press.

[200] Malmivuo, J., Lekkala, J., Kontro, P., Suomaa, L., and Vihinen, H. Improvement of the properties of an eddy current magnetic shield with active compensation. *Journal of Physics E: Scientific Instruments 20*, 2 (1987), 151–164.

[201] Malmivuo, J., and Plonsey, R. *Bioelectromagnetism: Principles and Applications of Bioelectric and Biomagnetic Fields*. Oxford University Press, 1995.

[202] Mantini, D., Penna, S. D., Marzetti, L., De Pasquale, F., Pizzella, V., Corbetta, M., and Romani, G. L. A signal-processing pipeline for magnetoencephalography resting-state networks. *Brain Connectivity 1*, 1 (2011), 49–59.

[203] McDermott, R., Lee, S., Ten Haken, B., Trabesinger, A. H., Pines, A., and Clarke, J. Microtesla MRI with a superconducting quantum interference device. *Proceedings of the National Academy of Sciences of the United States of America 101*, 21 (2004), 7857–7861.

[204] Metsomaa, J., Sarvas, J., and Ilmoniemi, R. J. Multi-trial evoked EEG and independent component analysis. *Journal of Neuroscience Methods 228* (2014), 15–26.

[205] Metsomaa, J., Sarvas, J., and Ilmoniemi, R. J. Blind source separation of event-related EEG/MEG. *IEEE Transactions on Biomedical Engineering 64*, 9 (2017), 2054–2064.

[206] Michel, C. M. *Electrical Neuroimaging*. Cambridge University Press, 2009.

[207] Millett, D. Hans Berger: From psychic energy to the EEG. *Perspectives in Biology and Medicine 44*, 4 (2001), 522–542.

[208] Moiseev, A., Gaspar, J. M., Schneider, J. A., and Herdman, A. T. Application of multi-source minimum variance beamformers for reconstruction of correlated neural activity. *NeuroImage 58*, 2 (2011), 481–496.

[209] Mosher, J. C., and Leahy, R. M. Recursive MUSIC: A framework for EEG and MEG source localization. *IEEE Transactions on Biomedical Engineering 45*, 11 (1998), 1342–1354.

[210] Mosher, J. C., and Leahy, R. M. Source localization using recursively applied and projected (RAP) MUSIC. *IEEE Transactions on Signal Processing 47*, 2 (1999), 332–340.

[211] Mosher, J. C., Lewis, P. S., and Leahy, R. M. Multiple dipole modeling and localization from spatio-temporal MEG data. *IEEE Transactions on Biomedical Engineering 39*, 6 (1992), 541–557.

[212] Murakami, S., and Okada, Y. Contributions of principal neocortical neurons to magnetoencephalography and electroencephalography signals. *The Journal of Physiology 575*, 3 (2006), 925–936.

[213] Mutanen, T. P., Kukkonen, M., Nieminen, J. O., Stenroos, M., Sarvas, J., and Ilmoniemi, R. J. Recovering TMS-evoked EEG responses masked by muscle artifacts. *NeuroImage 139* (2016), 157–166.

[214] Mutanen, T. P., Metsomaa, J., Liljander, S., and Ilmoniemi, R. J. Automatic and robust noise suppression in EEG and MEG: The SOUND algorithm. *NeuroImage 166* (2018), 135–151.

[215] Näätänen, R. *Attention and Brain Function*. Psychology Press, 1992.

[216] Näätänen, R., Gaillard, A. W., and Mäntysalo, S. Early selective-attention effect on evoked potential reinterpreted. *Acta Psychologica 42*, 4 (1978), 313–329.

[217] Näätänen, R., Ilmoniemi, R. J., and Alho, K. Magnetoencephalography in studies of human cognitive brain function. *Trends in Neurosciences 17*, 9 (1994), 389–395.

[218] Näätänen, R., Lehtokoski, A., Lennes, M., Cheour, M., Huotilainen, M., Iivonen, A., Vainio, M., Alku, P., Ilmoniemi, R. J., Luuk, A., Allik, J., Sinkkonen, J., and Alho, K. Language-specific phoneme representations revealed by electric and magnetic brain responses. *Nature 385*, 6615 (1997), 432–434.

[219] Naik, G. R., and Kumar, D. K. An overview of independent component analysis and its applications. *Informatica 35*, 1 (2011).

[220] Nakasato, N., and Yoshimoto, T. Somatosensory, auditory, and visual evoked magnetic fields in patients with brain diseases. *Journal of Clinical Neurophysiology 17*, 2 (2000), 201–211.

[221] Nam, C. S., Nijholt, A., and Lotte, F. *Brain–Computer Interfaces Handbook: Technological and Theoretical Advances*. CRC Press, 2018.

[222] Nevalainen, P., Lauronen, L., and Pihko, E. Development of human somatosensory cortical functions – what have we learned from magnetoencephalography: A review. *Frontiers in Human Neuroscience 8* (2014), 158.

[223] Niedermeyer, E., and da Silva, L. F. H. *Electroencephalography: Basic Principles, Clinical Applications, and Related Fields*. Lippincott Williams & Wilkins, 2005.

[224] Nieminen, J. O., Koponen, L. M., and Ilmoniemi, R. J. Experimental characterization of the electric field distribution induced by TMS devices. *Brain Stimulation: Basic, Translational, and Clinical Research in Neuromodulation 8*, 3 (2015), 582–589.

[225] Nieminen, J. O., Zevenhoven, K. C. J., Vesanen, P. T., Hsu, Y.-C., and Ilmoniemi, R. J. Current-density imaging using ultra-low-field MRI with adiabatic pulses. *Magnetic Resonance Imaging 32*, 1 (2014), 54–59.

[226] Nolte, G., and Curio, G. The effect of artifact rejection by signal-space projection on source localization accuracy in MEG measurements. *IEEE Transactions on Biomedical Engineering 46*, 4 (1999), 400–408.

[227] Nolte, G., and Hämäläinen, M. S. Partial signal space projection for artefact removal in MEG measurements: A theoretical analysis. *Physics in Medicine & Biology 46*, 11 (2001), 2873–2887.

[228] Nunez, P. L., and Srinivasan, R. *Electric Fields of the Brain: The Neurophysics of EEG*. Oxford University Press, 2006.

[229] Öisjöen, F., Schneiderman, J. F., Figueras, G. A., Chukharkin, M. L., Kalabukhov, A., Hedström, A., Elam, M., and Winkler, D. High-T_c superconducting quantum interference device recordings of spontaneous brain activity: Towards high-T_c magnetoencephalography. *Applied Physics Letters 100*, 13 (2012), 132601.

[230] Okada, Y., Hämäläinen, M., Pratt, K., Mascarenas, A., Miller, P., Han, M., Robles, J., Cavallini, A., Power, B., Sieng, K., Sun, L., Lew, S., Doshi, C., Ahtam, B., Dinh, C., Esch, L., Grant, E., Nummenmaa, A., and Paulson, D. BabyMEG: A whole-head pediatric magnetoencephalography system for human brain development research. *Review of Scientific Instruments 87*, 9 (2016), 094301.

[231] Okada, Y., Lauritzen, M., and Nicholson, C. MEG source models and physiology. *Physics in Medicine & Biology 32*, 1 (1987), 43–51.

[232] Okada, Y., Pratt, K., Atwood, C., Mascarenas, A., Reineman, R., Nurminen, J., and Paulson, D. BabySQUID: A mobile, high-resolution multichannel magnetoencephalography system for neonatal brain assessment. *Review of Scientific Instruments 77*, 2 (2006), 024301.

[233] Okada, Y. C., Wu, J., and Kyuhou, S. Genesis of MEG signals in a mammalian CNS structure. *Electroencephalography and Clinical Neurophysiology 103*, 4 (1997), 474–485.

[234] Onnes, H. K. Further experiments with liquid helium. G. On the electrical resistance of pure metals etc. VI. On the sudden change in the rate at which the resistance of mercury disappears. In *Koninklijke Nederlandsche Akademie van Wetenschappen Proceedings* (1911), vol. 14, pp. 818–821.

[235] Oostenveld, R., Fries, P., Maris, E., and Schoffelen, J.-M. FieldTrip: Open source software for advanced analysis of MEG, EEG, and invasive electrophysiological data. *Computational Intelligence and Neuroscience 2011*, Article ID 156869 (2011).

[236] Oostenveld, R., and Praamstra, P. The five percent electrode system for high-resolution EEG and ERP measurements. *Clinical Neurophysiology 112*, 4 (2001), 713–719.

[237] Oyama, D., Adachi, Y., Hatsusaka, N., Kasahara, T., Yumoto, M., Hashimoto, I., and Uehara, G. Evaluation of an isosceles-triangle-coil phantom for magnetoencephalography. *IEEE Transactions on Magnetics 47*, 10 (2011), 3853–3856.

[238] Oyama, D., Adachi, Y., and Uehara, G. Evaluation method of magnetic sensors using the calibrated phantom for magnetoencephalography. *Journal of the Magnetics Society of Japan 41*, 4 (2017), 70–74.

[239] Oyama, D., Adachi, Y., Yumoto, M., Hashimoto, I., and Uehara, G. Dry phantom for magnetoencephalography—Configuration, calibration, and contribution. *Journal of Neuroscience Methods 251* (2015), 24–36.

[240] Pang, E. W. *Magnetoencephalography*. InTech, 2011.

[241] Pannetier, M., Fermon, C., Le Goff, G., Simola, J., and Kerr, E. Femtotesla magnetic field measurement with magnetoresistive sensors. *Science 304*, 5677 (2004), 1648–1650.

[242] Pannetier-Lecoeur, M., Parkkonen, L., Sergeeva-Chollet, N., Polovy, H., Fermon, C., and Fowley, C. Magnetocardiography with sensors based on giant magnetoresistance. *Applied Physics Letters 98*, 15 (2011), 153705.

[243] Papadelis, C., Grant, P. E., Okada, Y., and Preissl, H. *Magnetoencephalography: An Emerging Neuroimaging Tool for Studying Normal and Abnormal Human Brain Development*. Frontiers Media SA, 2015.

[244] Pascual-Marqui, R. D. Standardized low-resolution brain electromagnetic tomography (sLORETA): Technical details. *Methods & Findings in Experimental & Clinical Pharmacology 24*, Suppl D (2002), 5–12.

[245] Pascual-Marqui, R. D., Michel, C. M., and Lehmann, D. Segmentation of brain electrical activity into microstates: Model estimation and validation. *IEEE Transactions on Biomedical Engineering 42*, 7 (1995), 658–665.

[246] Pernet, C. R., Chauveau, N., Gaspar, C., and Rousselet, G. A. LIMO EEG: A toolbox for hierarchical LInear MOdeling of ElectroEncephaloGraphic data. *Computational Intelligence and Neuroscience 2011*, Article ID 831409 (2011).

[247] Petrashov, V. T., Antonov, V. N., Delsing, P., and Claeson, T. Phase controlled conductance of mesoscopic structures with superconducting "mirrors". *Physical Review Letters 74*, 26 (1995), 5268–5271.

[248] Peyk, P., De Cesarei, A., and Junghöfer, M. ElectroMagnetoEncephalography software: Overview and integration with other EEG/MEG toolboxes. *Computational Intelligence and Neuroscience 2011*, Article ID 861705 (2011).

[249] Pfurtscheller, G., and Cooper, R. Frequency dependence of the transmission of the EEG from cortex to scalp. *Electroencephalography and Clinical Neurophysiology 38*, 1 (1975), 93–96.

[250] Piccolino, M. Animal electricity and the birth of electrophysiology: The legacy of Luigi Galvani. *Brain Research Bulletin 46*, 5 (1998), 381–407.

[251] Pizzella, V., Marzetti, L., Della Penna, S., de Pasquale, F., Zappasodi, F., and Romani, G. L. Magnetoencephalography in the study of brain dynamics. *Functional Neurology 29*, 4 (2014), 241–253.

[252] Porcaro, C., Zappasodi, F., Barbati, G., Salustri, C., Pizzella, V., Rossini, P. M., and Tecchio, F. Fetal auditory responses to external sounds and mother's heart beat: Detection improved by independent component analysis. *Brain Research 1101*, 1 (2006), 51–58.

[253] Puce, A., and Hämäläinen, M. S. A review of issues related to data acquisition and analysis in EEG/MEG studies. *Brain Sciences 7*, 6 (2017), 58.

[254] Ramírez, R. R. Source localization. *Scholarpedia 3*, 11 (2008), 1733.

[255] Ramírez, R. R., Kopell, B. H., Butson, C. R., Hiner, B. C., and Baillet, S. Spectral signal space projection algorithm for frequency domain MEG and EEG denoising, whitening, and source imaging. *NeuroImage 56*, 1 (2011), 78–92.

[256] Regan, D. *Human Brain Electrophysiology: Evoked Potentials and Evoked Magnetic Fields in Science and Medicine*. Elsevier, New York, 1989.

[257] Reite, M., Edrich, J., Zimmerman, J. T., and Zimmerman, J. E. Human magnetic auditory evoked fields. *Electroencephalography and Clinical Neurophysiology 45*, 1 (1978), 114–117.

[258] Riaz, B., Pfeiffer, C., and Schneiderman, J. F. Evaluation of realistic layouts for next generation on-scalp MEG: Spatial information density maps. *Scientific Reports 7*, 1 (2017), 6974.

[259] Roberts, T. P., Paulson, D. N., Hirschkoff, G., Pratt, K., Mascarenas, A., Miller, P., Han, M., Caffrey, J., Kincade, C., Power, W., Murray, R., Chow, V., Fisk, C., Ku, M., Chudnovskaya, D., Dell, J., Golembski, R., Lam, P., Blaskey, L., Kuschner, E., Bloy, L., Gaetz, W., and Edgar, J. C. Artemis 123: Development of a whole-head infant and young child MEG system. *Frontiers In Human Neuroscience 8* (2014), 99.

[260] Robinson, S. E., and Vrba, J. Functional neuroimaging by synthetic aperture magnetometry (SAM). In *Recent Advances in Biomagnetism*, T. Yoshimoto, M. Kotani, S. Kuriki, H. Karibe, and N. Nakasato, Eds. (1999), pp. 302–305.

[261] Romani, G. L., Williamson, S. J., and Kaufman, L. Tonotopic organization of the human auditory cortex. *Science 216*, 4552 (1982), 1339–1340.

[262] Rose, D. F., Smith, P. D., and Sato, S. Magnetoencephalography and epilepsy research. *Science 238*, 4825 (1987), 329–335.

[263] Rykunov, S. D., Oplachko, E. S., Ustinin, M. N., and Llinás, R. Methods for magnetic encephalography data analysis in MathBrain cloud service. *Mathematical Biology and Bioinformatics 12*, 1 (2017), 176–185.

[264] Saarinen, M., Karp, P. J., Katila, T. E., and Siltanen, P. The magnetocardiogram in cardiac disorders. *Cardiovascular Research 8*, 6 (1974), 820–834.

[265] Salmelin, R., and Hari, R. Characterization of spontaneous MEG rhythms in healthy adults. *Electroencephalography and Clinical Neurophysiology 91*, 4 (1994), 237–248.

[266] Salmelin, R., Hari, R., Lounasmaa, O. V., and Sams, M. Dynamics of brain activation during picture naming. *Nature 368*, 6470 (1994), 463–465.

[267] Sander, T. H., Preusser, J., Mhaskar, R., Kitching, J., Trahms, L., and Knappe, S. Magnetoencephalography with a chip-scale atomic magnetometer. *Biomedical Optics Express 3*, 5 (2012), 981–990.

[268] Sanei, S., and Chambers, J. A. *EEG Signal Processing*. John Wiley & Sons, 2013.

[269] Sarvas, J. Basic mathematical and electromagnetic concepts of the biomagnetic inverse problem. *Physics In Medicine & Biology 32*, 1 (1987), 11–22.

[270] Sato, S. *Magnetoencephalography*, vol. 54. Lippincott Williams & Wilkins, 1990.

[271] Sato, S., Balish, M., and Muratore, R. Principles of magnetoencephalography. *Journal of Clinical Neurophysiology 8*, 2 (1991), 144–156.

[272] Sato, S., and Smith, P. D. Magnetoencephalography. *Journal of Clinical Neurophysiology 2*, 2 (1985), 173–192.

[273] Scherg, M. Fundamentals of dipole source potential analysis. In *Auditory Evoked Magnetic Fields and Electric Potentials*. F. Grandori, M. Hoke, and G. L. Romani, Eds., vol. 6 of *Advances in Audiology* (1990), 40–69.

[274] Scherg, M., and Von Cramon, D. Two bilateral sources of the late AEP as identified by a spatio-temporal dipole model. *Electroencephalography and Clinical Neurophysiology/Evoked Potentials Section 62*, 1 (1985), 32–44.

[275] Schleger, F., Landerl, K., Muenssinger, J., Draganova, R., Reinl, M., Kiefer-Schmidt, I., Weiss, M., Wacker-Gußmann, A., Huotilainen, M., and Preissl, H. Magnetoencephalographic signatures of numerosity discrimination in fetuses and neonates. *Developmental Neuropsychology 39*, 4 (2014), 316–329.

[276] Schmidt, R. Multiple emitter location and signal parameter estimation. *IEEE Transactions on Antennas and Propagation 34*, 3 (1986), 276–280.

[277] Schneiderman, J. F. Information content with low-vs. high-T_c SQUID arrays in MEG recordings: The case for high-T_c SQUID-based MEG. *Journal of Neuroscience Methods 222* (2014), 42–46.

[278] Schomer, D. L., and Da Silva, F. L. *Niedermeyer's Electroencephalography: Basic Principles, Clinical Applications, and Related Fields*. Lippincott Williams & Wilkins, 2018.

[279] Schwartz, D., Lemoine, D., Poiseau, E., and Barillot, C. Registration of MEG/EEG data with 3D MRI: Methodology and precision issues. *Brain Topography 9*, 2 (1996), 101–116.

[280] Seeck, M., Koessler, L., Bast, T., Leijten, F., Michel, C., Baumgartner, C., He, B., and Beniczky, S. The standardized EEG electrode array of the IFCN. *Clinical Neurophysiology 128*, 10 (2017), 2070–2077.

[281] Sekihara, K., Kawabata, Y., Ushio, S., Sumiya, S., Kawabata, S., Adachi, Y., and Nagarajan, S. S. Dual signal subspace projection (DSSP): A novel algorithm for removing large interference in biomagnetic measurements. *Journal of Neural Engineering 13*, 3 (2016), 036007.

[282] Sekihara, K., and Nagarajan, S. S. *Adaptive Spatial Filters for Electromagnetic Brain Imaging*. Springer Science & Business Media, Berlin, 2008.

[283] Sekihara, K., and Nagarajan, S. S. *Electromagnetic Brain Imaging: A Bayesian Perspective*. Springer, 2015.

[284] Shahbazi, F., Ewald, A., and Nolte, G. Self-consistent MUSIC: An approach to the localization of true brain interactions from EEG/MEG data. *NeuroImage 112* (2015), 299–309.

[285] Sheng, J., Wan, S., Sun, Y., Dou, R., Guo, Y., Wei, K., He, K., Qin, J., and Gao, J.-H. Magnetoencephalography with a Cs-based high-sensitivity compact atomic magnetometer. *Review of Scientific Instruments 88*, 9 (2017), 094304.

[286] Sheridan, C. J., Matuz, T., Draganova, R., Eswaran, H., and Preissl, H. Fetal magnetoencephalography—achievements and challenges in the study of prenatal and early postnatal brain responses: A review. *Infant and Child Development 19*, 1 (2010), 80–93.

[287] Silver, A. H., and Zimmerman, J. E. Quantum states and transitions in weakly connected superconducting rings. *Physical Review 157*, 2 (1967), 317–341.

[288] Simpson, G. V., Pflieger, M. E., Foxe, J. J., Ahlfors, S. P., Vaughan Jr., H. G., Hrabe, J., Ilmoniemi, R. J., and Lantos, G. Dynamic neuroimaging of brain function. *Journal of Clinical Neurophysiology 12*, 5 (1995), 432–449.

[289] Sorenson, H. W. *Parameter Estimation: Principles and Problems*, vol. 9 of *Control and Systems Theory*. M. Dekker, New York and Boston, 1980.

[290] Souloumiac, A. Nonorthogonal joint diagonalization by combining givens and hyperbolic rotations. *IEEE Transactions on Signal Processing 57*, 6 (2009), 2222–2231.

[291] Srinivasan, R., Tucker, D. M., and Murias, M. Estimating the spatial Nyquist of the human EEG. *Behavior Research Methods, Instruments, & Computers 30*, 1 (1998), 8–19.

[292] Stefan, H., and Trinka, E. Magnetoencephalography (MEG): Past, current and future perspectives for improved differentiation and treatment of epilepsies. *Seizure 44* (2017),121–124.

[293] Stenbacka, L., Vanni, S., Uutela, K., and Hari, R. Comparison of minimum current estimate and dipole modeling in the analysis of simulated activity in the human visual cortices. *NeuroImage 16*, 4 (2002), 936–943.

[294] Stenroos, M., Hunold, A., and Haueisen, J. Comparison of three-shell and simplified volume conductor models in magnetoencephalography. *NeuroImage 94* (2014), 337–348.

[295] Stenroos, M., and Sarvas, J. Bioelectromagnetic forward problem: Isolated source approach revis(it)ed. *Physics in Medicine & Biology 57*, 11 (2012), 3517–3535.

[296] Sternickel, K., and Braginski, A. I. Biomagnetism using SQUIDs: Status and perspectives. *Superconductor Science and Technology 19*, 3 (2006), S160–S171.

[297] Stinstra, J. G., and Peters, M. J. The volume conductor may act as a temporal filter on the ECG and EEG. *Medical and Biological Engineering & Computing 36*, 6 (1998), 711–716.

[298] Storm, J.-H., Hömmen, P., Drung, D., and Körber, R. An ultra-sensitive and wideband magnetometer based on a superconducting quantum interference device. *Applied Physics Letters 110*, 7 (2017), 072603.

[299] Stufflebeam, S. M., Tanaka, N., and Ahlfors, S. P. Clinical applications of magnetoencephalography. *Human Brain Mapping 30*, 6 (2009), 1813–1823.

[300] Sudre, G., Parkkonen, L., Bock, E., Baillet, S., Wang, W., and Weber, D. J. rtMEG: A real-time software interface for magnetoencephalography. *Computational Intelligence and Neuroscience 2011*, Article ID 327953 (2011).

[301] Supek, S., and Aine, C. J. *Magnetoencephalography: From Signals to Dynamic Cortical Networks.* Springer, 2014.

[302] Tadel, F., Baillet, S., Mosher, J. C., Pantazis, D., and Leahy, R. M. Brainstorm: A user-friendly application for MEG/EEG analysis. *Computational Intelligence and Neuroscience 2011*, Article ID 879716 (2011).

[303] Tarkiainen, A., Liljeström, M., Seppä, M., and Salmelin, R. The 3D topography of MEG source localization accuracy: Effects of conductor model and noise. *Clinical Neurophysiology 114*, 10 (2003), 1977–1992.

[304] Taulu, S., and Simola, J. Spatiotemporal signal space separation method for rejecting nearby interference in MEG measurements. *Physics in Medicine & Biology 51*, 7 (2006), 1759–1768.

[305] Taulu, S., Simola, J., and Kajola, M. Applications of the signal space separation method. *IEEE Transactions on Signal Processing 53*, 9 (2005), 3359–3372.

[306] Tesche, C. D., Uusitalo, M. A, Ilmoniemi, R. J., Huotilainen, M., Kajola, M., and Salonen, O. Signal-space projections of MEG data characterize both distributed and well-localized neuronal sources. *Electroencephalography and Clinical Neurophysiology 95*, 3 (1995), 189–200.

[307] Tesche, C. D., Uusitalo, M. A., Ilmoniemi, R. J., and Kajola, M. J. Characterizing the local oscillatory content of spontaneous cortical activity during mental imagery. *Cognitive Brain Research 2*, 4 (1995), 243–249.

[308] Teyler, T. J., Cuffin, B. N., and Cohen, D. The visual evoked magnetoencephalogram. *Life Sciences 17*, 5 (1975), 683–691.

[309] Tian, X., Poeppel, D., and Huber, D. E. TopoToolbox: Using sensor topography to calculate psychologically meaningful measures from event-related EEG/MEG. *Computational Intelligence and Neuroscience 2011*, Article ID 674605 (2011).

[310] Tripp, J. H. Physical concepts and mathematical models. In *Biomagnetism*. Springer, 1983, pp. 101–139.

[311] Ustinin, M. N., Sytchev, V. V., Walton, K. D., and Llinás, R. New methodology for the analysis and representation of human brain function: MEGMRIAn. *Mathematical Biology and Bioinformatics 9*, 2 (2014), 464–481.

[312] Uusitalo, M. A., and Ilmoniemi, R. J. Signal-space projection method for separating MEG or EEG into components. *Medical & Biological Engineering & Computing 35*, 2 (1997),135–140.

[313] Uutela, K., Hämäläinen, M., and Somersalo, E. Visualization of magnetoencephalographic data using minimum current estimates. *NeuroImage 10*, 2 (1999), 173–180.

[314] Van Veen, B. D., and Buckley, K. M. Beamforming: A versatile approach to spatial filtering. *IEEE ASSP Magazine 5*, 2 (1988), 4–24.

[315] Van Veen, B. D., Van Drongelen, W., Yuchtman, M., and Suzuki, A. Localization of brain electrical activity via linearly constrained minimum variance spatial filtering. *IEEE Transactions on Biomedical Engineering 44*, 9 (1997), 867–880.

[316] Vesanen, P. T., Nieminen, J. O., Zevenhoven, K. C., Dabek, J., Parkkonen, L. T., Zhdanov, A. V., Luomahaara, J., Hassel, J., Penttilä, J., Simola, J., Ahonen, A., Mäkelä, J. P., and Ilmoniemi, R. J. Hybrid ultra-low-field MRI and magnetoencephalography system based on a commercial whole-head neuromagnetometer. *Magnetic Resonance in Medicine 69*, 6 (2013),1795–1804.

[317] Vesanen, P. T., Nieminen, J. O., Zevenhoven, K. C., Hsu, Y.-C., and Ilmoniemi, R. J. Current-density imaging using ultra-low-field MRI with zero-field encoding. *Magnetic Resonance Imaging 32*, 6 (2014), 766–770.

[318] Vidaurre, C., Sander, T. H., and Schlögl, A. BioSig: The free and open source software library for biomedical signal processing. *Computational Intelligence and Neuroscience 2011*, Article ID 935364 (2011).

[319] Vrba, J., Fife, A. A., Burbank, M. B., Weinberg, H., and Brickett, P. A. Spatial discrimination in SQUID gradiometers and 3rd order gradiometer performance. *Canadian Journal of Physics 60*, 7 (1982), 1060–1073.

[320] Vrba, J., and Robinson, S. E. SQUID sensor array configurations for magnetoencephalography applications. *Superconductor Science and Technology 15*, 9 (2002), R51.

[321] Walter, W. G., Cooper, R., Aldridge, V. J., McCallum, W. C., and Winter, A. L. Contingent negative variation: An electric sign of sensori-motor association and expectancy in the human brain. *Nature 203*, 4943 (1964), 380–384.

[322] Wang, C., Sun, L., Lichtenwalter, B., Zerkle, B., and Okada, Y. Compact, ultra-low vibration, closed-cycle helium recycler for uninterrupted operation of MEG with SQUID magnetometers. *Cryogenics 76* (2016), 16–22.

[323] Wikswo Jr., J. P., Gevins, A., and Williamson, S. J. The future of the EEG and MEG. *Electroencephalography and Clinical Neurophysiology 87*, 1 (1993), 1–9.

[324] Wipf, D., and Nagarajan, S. A unified Bayesian framework for MEG/EEG source imaging. *NeuroImage 44*, 3 (2009), 947–966.

[325] Wolpaw, J. R., and Wood, C. C. Scalp distribution of human auditory evoked potentials. I. Evaluation of reference electrode sites. *Electroencephalography and Clinical Neurophysiology 54*, 1 (1982), 15–24.

[326] Wu, W., Keller, C. J., Rogasch, N. C., Longwell, P., Shpigel, E., Rolle, C. E., and Etkin, A. ARTIST: A fully automated artifact rejection algorithm for single-pulse TMS-EEG data. *Human Brain Mapping* (2018).

[327] Wyllie, R., Kauer, M., Wakai, R. T., and Walker, T. G. Optical magnetometer array for fetal magnetocardiography. *Optics Letters 37*, 12 (2012), 2247–2249.

[328] Xia, H., Ben-Amar Baranga, A., Hoffman, D., and Romalis, M. V. Magnetoencephalography with an atomic magnetometer. *Applied Physics Letters 89*, 21 (2006), 211104.

[329] Yao, D. Electric potential produced by a dipole in a homogeneous conducting sphere. *IEEE Transactions on Biomedical Engineering 47*, 7 (2000), 964–966.

[330] Yeredor, A. Non-orthogonal joint diagonalization in the least-squares sense with application in blind source separation. *IEEE Transactions on Signal Processing 50*, 7 (2002), 1545–1553.

[331] Zevenhoven, K. C. J., Busch, S., Hatridge, M., Öisjöen, F., Ilmoniemi, R. J., and Clarke, J. Conductive shield for ultra-low-field magnetic resonance imaging: Theory and measurements of eddy currents. *Journal of Applied Physics 115*, 10 (2014), 103902.

[332] Zhang, X., Lei, X., Wu, T., and Jiang, T. A review of EEG and MEG for brainnetome research. *Cognitive Neurodynamics 8*, 2 (2014), 87–98.

[333] Zhang, Z. A fast method to compute surface potentials generated by dipoles within multilayer anisotropic spheres. *Physics in Medicine & Biology 40*, 3 (1995), 335–349.

[334] Ziehe, A., Laskov, P., Nolte, G., and Müller, K.-R. A fast algorithm for joint diagonalization with non-orthogonal transformations and its application to blind source separation. *Journal of Machine Learning Research 5* (2004), 777–800.

[335] Zotev, V. S., Matlashov, A. N., Volegov, P. L., Savukov, I. M., Espy, M. A., Mosher, J. C., Gomez, J. J., and Kraus Jr., R. H. Microtesla MRI of the human brain combined with MEG. *Journal of Magnetic Resonance 194*, 1 (2008), 115–120.

Index